# Maryland
# 1860 Agricultural Census
## Volume 1

Linda L. Green

WILLOW BEND BOOKS
2007

**WILLOW BEND BOOKS**
*AN IMPRINT OF HERITAGE BOOKS, INC.*

**Books, CDs, and more—Worldwide**

For our listing of thousands of titles see our website
at
www.HeritageBooks.com

Published 2007 by
HERITAGE BOOKS, INC.
Publishing Division
65 East Main Street
Westminster, Maryland 21157-5026

International Standard Book Number: 978-0-7884-3828-X

# Introduction

This census names only the head of the household. Often times when an individual was missed on the regular U. S. Census, they would appear on this agricultural census. So you might try checking this census for your missing relatives. Unfortunately, many of the Agricultural Census records have not survived. But, they do yield unique information about how people lived. There are 48 columns of information. I chose to transcribe only six of the columns. The six are: Name of the Owner, Improved Acreage, Unimproved Acreage, Cash Value of the Farm, Value of Farm Implements and Machinery, and Value of Livestock. Below is a list of other types of information available on this census.

Linda L. Green
13950 Ruler Court
Woodbridge, VA 22193

# Other Data Columns

Column/Title

6. Horses
7. Asses and Mules
8. Milch Cows
9. Working Oxen
10. Other Cattle
11. Sheep
12. Swine
14. Wheat, bushels of
15. Rye, bushels of
16. Indian Corn, bushels of
17. Oats, bushels of
18. Rice, lbs of
19. Tobacco, lbs of
20. Ginned cotton, bales of 400 lbs each
21. Wood, lbs of
22. Peas and beans, bushels of
23. Irish potatoes, bushels of
24. Sweet potatoes, bushels of
25. Barley, bushels of
26. Buckwheat, bushels of
27. Value of Orchard products in dollars
28. Wine, gallons of
29. Value of Products of Market Gardens
30. Butter, lbs of
31. Cheese, lbs of
32. Hay, tons of
33. Clover seed, bushels of
34. Other grass seeds, bushels of
35. Hops, lbs of
36. Dew Rotten Hemp, tons of
37. Water Rotted Hemp, tons of
38. Other Prepared Hemp
39. Flax, lbs of
40. Flaxseed, bushels of
41. Silk cocoons, lbs of
42. Maple sugar, lbs of
43. Cane Sugar, hunds of 1,000 lbs
44. Molasses, gallons of
45. Beeswax, lbs of
46. Honey, lbs of
47. Value of Home Made Manufactures
48. Value of Animals Slaughtered

√

# Table of Contents

Alleghany County Maryland
1860 Agricultural Census

The University of North Carolina Library under a grant from the National Science Foundation microfilmed agricultural Census records. Records were filmed at the University of North Carolina from original records at the Maryland State Library.

Columns 1, 2, 3, 4, 5, and 13 represent the following information on the census:
1. Name of Owner, Agent or Manager of Farm
2. Acres of Improved Land
3. Acres of Unimproved Land
4. Cash Value of the Farm
5. Value of Farming Implements and Machinery
13. Value of Livestock

Pages in this county are out of sequence.

Richard Selby, 10, 40, 100, 30, 130
John Hartman, 100, 94, 1500, 150, 506
Joseph Codrington, 100, 130, 2000, 175, 390
Jacob Turney, 20, 180, 800, 60, 254
Levey Switzer, 50, 50, 1000, 50, 132
John Hancock, 85, 30, 900, 40, 264
Jacob Thomas, 50, 50, 900, 80, 170
Michael, Slossnagle, 100, 100, 1500, 247, 580
Andrew Brown, 15, 185, 1000, 80, 237
Abraham Brown, 100, 50, 2000, 40, 95
Charles Slossnagle, 45, 40, 600, 125, 269
Michael Crouse, 25, 85, 600, 50, 320
Franklin Friend, 13, 200, 700, 5, 90
Godfrey Fox, 50, 16, 700, 65, 232
Jacob Shetzer, 50, 16, 700, 100, 330
Jonathan Friend, 50, 13, 315, 150, 337
Mathew Fazee, 50, 50, 800, 100, 357
Isaiah Friend, 5, 195, 200, 100, 112

Abraham Steele, 50, 138, 588, 115, 220
Elijah Friend, 10, 15, 300, 50, 103
John Green, 20, 67, 500, 60, 188
Thomas Casteele, 5, 75, 160, -, 33
Emanuel Custer, 75, 125, 1800, 150, 150
William Juloes, 15, 85, 500, -, 213
Joseph DeWitt, 40, 400, 1500, 303, 229
John DeWitt, 30, 30, 800, 100, 275
Giden Furgeson, -, -, -, 60, 96
Robert Furgeson, 40, 85, 375, -, 120
Jacob Flagel, 3, 97, 200, 56, 85
Lucian Danhan, -, -, -, -, 300
Archibald Delbitt of J, 40, 10, 1000, 54, 121
Barnabus Delbitt, 8, 29, 300, 50, 177
William Delbitt, 50, 153, 800, 40, 257
John Brasly, 10, 12, 800, 200, 85

1

Ann Mattingly, 70, 306, 2500, 150, 495

Jacob Mryer, 175, 25, 3000, -, 210

Elizabeth McGittigen, 100, 40, 2000, 162, 422

James Fresh, -, -, -, 50, 72

John Stine, 50, -, 300, 50, 95

George Upkold, -, -, -, 50, 30

Tobias Godee, 110, 260, 2000, 300, 462

Moses Schrock, 100, 100, 3700, 250, 584

Wm. Glotfelty, 200, 789, 3000, 300, 1052

Daniel Hostettler, 175, 225, 3500, 300, 1176

Nimrod Glotfelty, 6, 243, 600, 100, 250

John Waver (Warner), 25, 175, 1000, 150, 196

Alex Fresh, 50, 150, 800, 150, 292

Henry Shaffer, 25, 175, 800, -, 21

Joseph Lfiker, 50, 50, 650, 120, 306

Dennis O'Brian, 30, 170, 1000, 50, 85

David Bowman, 30, 35, 800, 150, 105

Leonard Fraty, 40, 83, 800, 30, 125

Andrew Detrick, 50, 50, 800, 100, 181

John George, 100, 150, 1500, 200, 231

Christian Zenger, 15, 235, 400, 50, 92

Michael Englehardt, 50, 200, 450, 150, 198

George Stark, 100, 100, 2000, 100, 585

Adam Stark, 50, 50, 800, 150, 370

Augwaltz Kensey, 50, 75, 700, 75, 240

Edward Minkroff, 10, 65, 500, -, 110

Daniel Glatez, 35, 80, 600, 100, 135

Conrad Spenloine, 90, 110, 1500, 150, 422

Janos Moses, 160, 40, 3600, 150, 622

Jeremiah Durst, 130, 70, 3800, 150, 361

Jeremiah Beachley, 80, 20, 2000, 125, 384

Turnan West, 200, 100, 7000, 150, 591

Richard West, -, -, -, 150, 259

Alex. Farvak (Farvale), 70, 40, 1000, 100, 178

Jesse Lovingood, 50, -, 500, 150, 275

Samuel Secter, 70, 40, 1000, 100, 438

Michael Boyer, 50, -, 1000, 50, 172

Walter Willburn, 80, 120, 1000, 100, 90

John Shek, 80, 120, 2000, 100, 84

Benjamin Keller, 65, 15, 2000, 100, 179

Jacob Pysal, 100, 450, 2000, 110, 295

Samuel Blackley, 80, 80 2500, 100, 217

Levin Jenkins, 100, 100, 1000, 110, 247

Thomas Casteele, 100, -, 2000, 125, 482

Richard Frazer, 80, 50, 1000, 100, 367

Moses Glass, 40, 160, 1200, 115, 169

John Liston, 50, 56, 800, 75, 967

Jacob Cust, 100, 62, 1200, 150, 221

Gabrial Forsythe, 25, 25, 200, 28, 134

Jacob B. Friend, 25, 60, 400, 40, 184

John S. Friend, 45, 20, 1200, 150, 480

Andrew Friend, 50, 50, 1500, 150, 27

Joshua Friend, 75, 75, 1200, 70, 102

John Hook, 250, 154, 12000, 200, 1391

David Kent, 8, 52, 1200, 150, 409

Franklin Suter, 50, 150, 1000, 150, 291

Joseph Reclaim, 100, 15, 1000, 125, 295

James Ross, 300, 150, 6000, 125, 765

Cornelius Garry, 100, 115, 1000, 150, 582

Stephen Ryley, 50, 62, 1000, 115, 178

Rebecca Frantz, 50, 50, 700, -, 1222

John Savage, 100, 100, 1200, 200, 305

Samuel Savage, 100, 80, 800, 100, 227

Abraham Friend, 30, 48, 800, 75, 211

Isaac Womble, 25, 25, 500, 175 56

Tobias Huff, 20, 20, 1200, -, 101

David Vanshield, 50, 50, 1000, 150, 496

Andrew Friend, 25, 25, 104, 75, 173

Timothy Miller, 100, 100, 2000, 50, 512

George Culf, 25, 75, 580, 25, 192

Henry Culf, 25, 75, 500, 25, 187

Michael Miller, 190, 210, 1500, 200, 370

Henry Spaker, 40, 20, 600, -, 97

Melchser Weller, 200, 800, 8000, 212, 482

Henry J. Miller, 75, 133, 2500, 125, 298

Azariah Hembertson, 20, 100, 1000, -, 110

Alex Poland, 200, 100, 6000, 100, 576

Wm. Warson, 80, 320, 2000, 25, 476

Benjamin Green, 100, 100, 1200, 75, 397

Elizabeth Boyer, 100, 75, 1500, 50, 253

Levi Frihe, 200, 50, 2250, 225, 688

Francis Griffith, 60, 40, 1000, 75, 326

Peter Speilman, 250, 150, 8000, 200, 474

Isaac Frazer, 150, 350, 3700, 175, 529

William Richner, -, -, -, 100, 249

Solomon Kemp, 75, 75, 2500, 100, 3

Richd. White, 125, 125, 1000, 100, 431

Jonathan Frazer, 50, 100, 1000, 200, 344

Edwd. Speilman, -, -, -, 75, 282

Chris Frike, 125, 110, 4000, 200, 1000

John Frike, 150, 150, 6000, 400, 1493

Daniel Bitner, 50, 50, 500, 25, 293

William Frike, 60, 90, 500, 50, 301

Thomas Frantz, 10, -, 300, -, 82

George Glaser, 75, 90, 1500, 50, 282

John Loudermilk, 8, 4, 300, -, 217

Ralph Thayer, 70, 46, 3000, 30, 295

Jacob Kizler, 80, 200, 2900, 50, 321

Samuel Nealy, 20, 80, 2000, -, 93

John Frantz, 60, 140, 2000, 100, 148

Noah Hemberson, 50, 130, 1200, 75, 326

George Mathews, 200, 125, 10000, 15, 873

Sylvester Ryland, 200, 200, 7000, 200, 486

Elisha Leighty, 15, 23, 150, -, 88

Thomas Savage, 20, 30, 400, -, 299

Abraham Weitz, 30, 100,700, 100, 336

James Garraphan (Garraghan), 78, 175, 2000, 30, 248

Catherine Garraghan, 20, 80, 100, 20, 345

John Friend, 40, 60, 800, 20, 288

Henry Riley, 30, 47, 500, 25, 108

Joseph Thomas, 30, 120, 600, 25, 95

Amos Thomas, 50, 50, 700, 25, 171

Adam Schroyer, 50, 100, 450, 25, 94

R. B. Jamison, 100, 90, 300, 100, 361

James Fortune, 50, -, 300, 100, 405

Abraham Hoff, 60, 55, 1200, 20, 291
Joseph Feams, 40, 50, 1200, 100, 386
Mary Thomas, 24, -, 100, -, 83
David Evans, 70, 130, 2000, 100, 362
John Binke, 35, 50, 500, -, 30
Levi Wolf, 15, 15, 50, 25, 154
John Vansick, 100, 150, 2000, 50, 457
Catherine Vandike, 100, 75, 1750, -, 156
Cornelius Friend, 100, 200, 2000, 100, 490
Nathaniel Coates, 225, 115, 3000, 25, 390
Jeremiah Enlow, 30, 20, 400, 20, 286
Harrison Friend, 50, 29, 800, 15, 132
Charles Friend, 30, 270, 1800, 5, 50
Wm. McCabe, 20, 60, 400, 10, 150
Henry McCabe, 60, 40, 1000, 50, 364
James Forsythe, 50, 60, 800, 50, 160
Archd. Casteele, 80, 90, 5000, 75, 195
William Browning, 80, 120, 400, 125, 655
Allen Browning, 77, 200, 1500, 20, 181
Joseph DeWitt, 15, 35, 400,15, 127
George Wimer, 100, 150, 3000, 50, 293
Julian Denham, 125, 275, 5000, 125, 615
Edward Hage, 75, 140, 3400, 150, 494
Joseph Gorder, 15, 135, 2000, 100, 347
Jacob Flagel, 15, 85, 200, 100, 258
Jane Browning, 30, 120, 700, 30, 497
John P. Hone, 25, 75, 500, 30, 221
George Rinkle(Hinkle, 30, 100, 500, 25, 162
Samuel Spricht, 50, 74, 300, 100, 619

Hiram P. Fasker, 175, 1325, 6000, 150, 1200
Joseph Slomghbaugh, 100, 403, 2000, 50, 548
Margaret Soar, 150, 50, 1600, -, 25
Thornton Gilpin, -, -, -, 50, 425
Els. Hall, 100, 700, 1000, 350, 1600
William Soar, 15, 85, 1000, 100, 333
Margaret Bolster, 15, 35, 1000, 5, 195
John Haun (Hann), 45, 25, 1000, 25, 272
Robert Porter, 150, 50, 3000, 150, 502
John Spilker, 75, 75, 2000, 100, 644
Abraham Spilker, 30, 70, 1400, 25, 445
Alfred C. Brosk (Brook), 50, -, 400, 100, 192
Henry Richel, 100, 100, 3000, 225, 700
Jacob Spilker, 50, 50, 1000, 100, 169
John E. Otts, 75, 50, 1000, 50, 468
William White, 5, 150, -, 1000, 100, 389
Nathaniel Harvey, 50, 140, 1000, 50, 330
Joseph Shaffers, 8, 43, 500, -, 96
Henry Shaffers, 40, 110, 1000, 25, 273
Ruben Moore, 20, 30, 300, 50, 220
Reason Freeman, 50, 180, 600, 50, 228
John Edgar, 12, 348, 400, 10, 70
Louis Buskirk, 175, 100, 8250, 100, 400
Wm. J. Ward, 350, 100, 15000, 250, 700
Patrick Rooney, 50, 50, 1000, 25, 285
JoAnn Ryley, 10, 10, 500, 5, 60
Feline McNally, 8, -, 500, -, 700
Geo. Jeffries, 200, 200, 12000, 60, 390
John Porter, 60, -, 1200, 30, 130
Chas. Wolf, 5, 185, 1000, 50, 125

Solomon Harred, 50, 60, 2500, 50, 255

E. K. Huntley, 230, 100, 6600, -, -

John Crone (Crose), 40, -, 800, -, 161

Wm. Long, 250, 150, 12000, 60, 2640

Clay McCulloh, -, -, 10000, 50, 600

Thos. Johnson, 70, 3460, 14120, 200, 740

Richard Beall, 10, -, 200, 10, -

Frances Hamner, 12, 37, 250, 20, 20

David Juskup, 70, -, 2100, -, 400

Hankins & Gore, 20, -, 605, -, 110

Henry Juskup, 70, 70, 1400, -, -

E. H. Tancry, 100, -, 3000, -, 100

Henry Knapp, 20, 60, 800, -, 103

Lewis Long, 200, 200, 6000, 75, 304

Wm. Ross, 18, -, 300, 10, 150

Geo. Parker, 75, -, 1500, 18, 320

J. Wesly Porter, 300, 150, 13500, 100, 700

Laurence Logsdon, 40, -, 400, 50, 750

Harry L. Porter, 100, 30, 3000, 40, 400

Patrick Mullrony, 80, -, 800, 40, -

Moses Poland, 100, -, 2000, 50, 825

Elizabeth Lease, 100, -, 1000, 20, 385

Jane Reardon, 8, -, 80, -, 40

Bridget Quinn, 10, -, 100, -, 75

Michael Tegley (Fegley), 15, -, 150, 30, 200

John Conners, 15, -, 150, 20, 250

Joseph Miller, 40, -, 400, 30, 200

Henry Bankan, 200, 200, 6000, 75, 745

Wm. _. Barman, 60, 100, 2500, 60, 400

Davis Barnard, 60, 51, 2220, 50, 310

Danl. Conrad, 100, -, 1000, 100, 320

Wm. Staples, 100, -, 1000, 300, 430

George Colmes, 100, -, 800, 100, 285

Meshac Frost, 276, -, 2760, 160, 880

Isaac Wincbenner, 75, 125, 2000, 75, 320

Jas. McKerr, 30, -, 300, 10, 212

Saml. Buskirk, 75, -, 750, 50, 265

Jas. Omalley(O'Malley),3, -,60, -, 40

Phillip Papp, 60, 90, 1500, 40, 246

Geo. Winteroff, 200, 159, 11000, 100, 710

Jasse Tomlinson, 100, -, 1000, 40, 255

Cintain Graham, 400, 85, 14550, 150, 1252

Christopher Ore, 300, 150, 5000, 75, 649

Salene Hamberson, 125, 200, 5000, 80, 604

Jacob Winter, 70, 296, 4000, 50, 304

Jacob Winter Jr., 30, -, 500, 25, 450

John Winter, 200, 100, 5000, 100, 1227

Noah Matthews, 100, 150, 5000, 30, 233

John Lancaster, 100, -, 1000, 50, 460

Andrew Bluebaugh, 70, 110, 1500, 50, 315

John Llwellyn, 100, 136, 2360, 50, 843

Wm. Van Buskirk, 50, -, 500, 10, 304

Saml. Miller Sr., 73, -, 730, 10, 160

Andrew Miller, 70, 67, 1360, 30, 485

James Dye, 60, 10, 1100, 10, 270

William Miller, 50, -, 500, 5, 194

Saml. Miller Jr., 80, -, 800, 5, 85

H__ Burnstine, 30, 45, 300, 15, 210

Geo. Br___, 150, 140, 1450, 25, 271

Jacob Buman (Barman), 10, -, 50, 5, 145

John Koontz, 200, 650, 8500, 80, 1220

Wm. Parker, 100, 150, 2500, 50, 418

Wm. Workman, 100, 86, 2000, 25, 340

Levi Lancaster, 60, 15, 750, 5, 154

Henry Cutler, 34, -, 350, 10, 121

Elenor Coleman, 50, -, 500, -, 150

Joseph Martin, 35, -, 750, 8, 105
Jos. Dornigel, 70, -, 700, 15, 310
Benj. Coleman, 60, -, 600, 54, 329
Jno. M. Porter, 100, 84, 2000, 50, 705
Moses Porter, 10, -, 100, 5, 95
Jane Once, 50, 50, 1000, 30, 195
John Crous, 50, 60, 1100, 40, 250
Christ Hariekoo, 90, 30, 1000, 25, 216
Jacob Folk, 50, -, 450, 5,70
Jesse Arnold, 100, 30, 1950, 30, 395
John Porter, 200, 44, 5000, 40, 765
Josiah Porter, 100, 25, 2500, 25, 430
Joseph Tippett, 50, -, 1000, 5, 70
Henry Riphon, 60, 100, 1600, 50, 340
Harry Trimble, 200, 400, 6000, 50, 446
Joseph Trimble, 200, 65, 2650, 50, 675
Wm. Robeson, 84, 26, 1000, 50, 472
Eliza Meitzell, 70, 20, 800, 20, 244
Danl. Plinched, 120, 103, 2230 80, 1140
Windle Firte (Fiste), 150, 100, 3000, 75, 431
Michael Porter, 100, 50, 1200, 100, 625
John Kimberley, 40, -, 400, 30, 330
Justus Gitenskin, 30, 50, 1100, 15, 155
Conrad Race, 100, 150, 2500, 50, 420
Elisha Workman, 100, 300, 4000, 15, 392
Daniel Brod, 35, 62, 2000, 40, 260
Martin Clark, 40, 35, 1000, 20, 435
Saml. Wilhelm, 100, 62, 1944, 25, 642
Leonidas Logdson, 75, 56, 2000, 75, 464
George Hardin, 120, 10, 5000, 200, 910
Wm. Logsden, 180, 50, 2500, 50, 688

Simon Arnold, 300, 300, 16000, 100, 620
Mary Dean, 150, 48, 6000, 100, 646
Michael Hogan (Hagan), 55, 7, 2000, 50, 258
William Engle, 80, 20, 4000, 50, 260
John Engle, 50, 50, 1000, 50, 245
George Winter, 75, 89, 2500, 100, 551
Blankhard E. Bluebaugh, 100, 117, 2500, 30, 664
Henry F. Weld, 75, 75, 10000, 85, 393
Levi Baughman, 60, 148, -, 25, 100
Mary Raine, 80, 45, 1300, 30, 230
Henry Snowden, 11, -, 100, -, 25
Anthony McDonald, 125, 55, 500, 50, 208
Martin River, 100, 50, 2000, 50, 380
James Calahan, 75, 25, 1000, 50, 290
Ferdinand Olakes, 80, 50, 1300, 50, 240
Jno. G. Wilt (Witt), 24, 37, 1000, 50, 210
Henry Mattingley, 150, 84, 6000, 100, 474
Ly___, P. Mattingley, 125, 25, 6000, 100, 775
Francis Mattingley, 150, 64, 6000, 100, 788
Phillip Miller, 30, -, 600, 5, 100
John Shilling, 40, 305, 1000, 25, 170
Henry Wagner, 60, 140, 1200, 25, 310
John Crowe, 30, 270, 1000, 5, 130
Jacob Crowe, 30, 310, 2000, 25, 295
Keleta Coleman, 25, 75, 400, 30, 220
Sam. McKinsay, 18, 18, 250, 21, 74
Jno. R. Durst, 25, 50, 450, 25, 290
Jesse W. Chaney, 80, 920, 1500, 25, 335
Chas. Brodwater, 75, 125, 600, 20, 50
William Brodwater, 85, 75, 640, 30, 482

John M. Brodwater, 125, 95, 1000, 20, 200

Saml. Brownaman, 150, 50, 2000, 175, 592

John Wright, 150, 50, 3000, 25, 160

Jacob Otto, 125, 75, 2500, 100, 606

Andrew Blocher, 100, 90, 1000, 100, 211

Harry Durst, 100, 417, 1034, 125, 511

Samuel Durst, 50, 169, 450, 70, 295

Michael Durst, 50, 200, 500, 50, 212

Morgan Roberson, 100, -, 520, 60, 285

Saml. F. McKinley, 100, 25, 1000, 25, 146

Jenny McKinley, 70, 71, 1500, 20, 156

Leo McKinley, 62, 240, 1500, 20, 182

Isadore McKinley, 45, 115, 500, 20, 166

Isaac Crowe, 20, 170, 500, 10, 138

Wm. N. Broadwater, 40, 100, 600, 15, 312

Henry Broadwater, 75, 85, 640, 25, 361

Amos Broadwater, 125, 75, 800, 200, 689

Israel Gartely, 75, 75, 450, 15, 175

John Gartley, 75, 75, 450, 15, 175

Henry Gartley, 60, 60, 240, -, 151

Henry Gartley, 30, 55, 300, 25, 130

Samuel Gartley, 25, 50, 225, -, 127

Jno. Muce (Meece), 16, 19, 350, -, 141

Benj. Wilhelm, 20, 157, 400, -, 111

Jacob Fresh, 25, 289, 333, -, 142

Ellenora Merrill, 40, 8, 240, 10, 153

W. A. Broadwater, 60, -, 600, 10, 211

Theophilus Wilt (Witt), 75, 75, 750, 25, 423

John Witt, 75, -, 375, -, 75

Wm. Clipstine, 50, 100, 450, -, 118

Danl. Rhodes, 5, -, 30, -, 25

Wm. Coleman, 30, -, 150, 5, 68

Geo. T. Witt, 25, -, 75, -, 86

David Witt, 50, -, 250, 25, 210

George W. Witt, 70, -, 350, 25, 175

Hamont Brown, 300, -, 6000, 100, 475

Louisa Blocher, 50, 50, 1000, 50, 190

John Criner, 20, -, 200, -, 152

Wm. Rickner, 40, -, 300, 50, 292

John Bacon, 37, -, 300, 25, 136

Wm. Frost, 30, -, 300, 20, 460

Henry Beall, 45, 305, 7000, 75, 448

Bazil Gartley, 700, 524, 12000, 50, 244

John Newman, 100, 150, 2000, 75, 515

Joseph Masirt (Mariot), 100, 110, 3150, 100, 527

James Caton, 45, -, 450, 25, 315

Abraham Masirt (Mariot), 100, 1150, 3750, 80, 492

Samuel Miller, 100, 230, 4000, 100, 499

Saml. Beachy, 150, 150, 4000, 25, 301

Wm. Stanton, 300, 600, 14500, 100, 1380

Elmer Riddey, 200, 270, 3000, 100, 715

Philip Durst, 18, -, 500, 50, 157

Levi Kensinger, 90, 210, 2000, 50, 253

Isariah Borchus, 150, 400, 5000, 50, 373

Elias Beachy, 90, 242, 5000, 75, 331

Elizabeth Miller, 30, 70, 500, 5, 102

Jacob Blocher, 180, 89, 8000, 100, 882

Mathias Rush, 18, 132, 200, 5, 134

Elijah Brown, 130, 22, 3000, 50, 155

Al__ Shultz, 75, 225, 3000, 75, 254

Peter Kensinger, 100, 122, 2000, 40, 48

Archibald Stimer, 140, 88, 4000, 100, 606

Joel Beachy, 200, 84, 4000, 100, 714
Moses Kemp, 60, 55, 2000, 50, 224
Adam Shultz, 150, 150, 12000, 100, 785
Andrew Arndt, 90, 224, 2000, 50, 295
John Binder, 90, 110, 3000, 10, 324
David Crampton, 150, 175, 1500, 100, 662
Jacob Lisagay, 140, 80, 3000, 50, 475
John Callan, 200, 353, 3000, 50, 559
Saml. Custer, 100, 87, 2000, 5, 416
Saml. Engle, 150, 55, 5000, 150, 1044
Jacob Gartley, 100, 660, 3700, 50, 328
Henry Hall, 60, 136, 784, 10, 122
Jacob Ott, 130, 100, 2300, 50, 273
Solomon Balser, 80, 20, 2000, 50, 293
Christian Beachy, 75, 125, 3000, 50, 457
David Slabaugh, 25, 35, 240, 25, 167
Peter Miller, 125, 100, 5000, 50, 490
John Garner, 80, 150, 3000, 50, 1094
Reuben Taylor, 15, 130, 350, 30, 215
Gabriel Bourer, 40, 510, 1500, -, 466
Edmond Gracy, 160, 255, 4500, 150, 472
John Durst, 50, 112, 3211, 50, 169
Mahala Durst, 60, 40, 1700, -, 156
Chas. Broadwater, 125, 225, 3000, 100, 783
Ashford Broadwater, 40, 60, 800, 25, 256
Wm. Broadwater, 120, 230, 3000, 100, 693
Hiram Duckworth, 60, 140, 1200, 50, 496
John Warnick, 125, 250, 3000, 100, 858
Saml Yoder, 80, 139, 3000, 110, 495
Maria Shamburg, 50, 30, 500, -, 135
Valentine & Huey, 25, 34, 250, 50, 226

John Schmidt, 75, 325, 2000, 50, 376
Wm. D. Harem, 80, 50, 2000, 25, 445
Chauncey Lang, 33, 157, 1200, 10,175
Henry Varner, 80, 120, 800, 50, 228
Daniel Peck, 15, 135, 500, -, 90
Andrew Achtel, 30, -, 120, 10, 86
Sylvanus Butler, 20, 88, 600, 5, 83
Truman Fairville, 280, 70, 3000, 75, 490
Samuel Grass, 120, 60, 2500, 150, 150
Samuel Lochel, 31, 100, 800, 10, 161
John Swartz, 40, 21, 800, 10, 100
Laurence Knight, 48, 98, 500, 100, 225
Benj. Edwards, 70, 130, 1500, 50, 473
Henry H. Miller, 160, 131, 4000, 50, 472
Patrick McAndrew, 20, 30, 250, -, 60
George Stephen, 26, 52, 527, 5, 108
To___ Biddinger, 12, 13, 100, 5, 98
Jacob Echenberger, 65, 41, 1500, -, 31
Eli Engle, 83, 1, 1500, 30, 303
Sarah Bitner, 60, 5, 1000, 10, 277
Peter Starke, 9, -, 150, -, 156
John Custar, 75, 105, 3000, 30, 522
David Custar, 137, -, 1200, 50, 337
Jacob Durst, 100, 150, 1500, 25, 290
Wm. Castle, 30, 70, 500, 30, 165
James McGaffin, 50, 114, 1200, 10, 225
Jacob Hoover, 40, 85, 800, 10, 92
Alex Myler, 45, 107, 1000, 20, 145
Henry Difenbaugh, 16, 10, 250, 50, 230
Walter Engle, 45, 155, 1500, 50, 189
Abraham Miller, 100, 60, 3000, 15, 175
Henry Starke, 30, 60, 500, 5, 153
Noah Biddinger, 60, 140, 1500, 5, 170

Benedict Beachy, 90, 110, 1800, 25, 570

John Lord, 40, 110, 800, 10, 158

Elisha Bittinger, 20, 30, 300, -, 69

Peter Hautauk (Hantank), 30, 120, 600, 15, 151

Thomas Stanton, 200, 200, 4000, 150, 789

Adam Harman, 150, 80, 3000, 50, 444

Sampson Sibert, 100, 100, 1000, 30, 324

Barney Wertleman, 60, 170, 600, 30, 297

Holmes Wiley, 65, 485, 700, 25, 154

George Blatter, 100, 100, 1000, 20, 269

Thos. Wilburn, 20, 30, 150, 5, 70

Jno. Barkholder, 20, 170, 600, -, 233

John Tice, 25, 175, 1000, -, 158

Danl. Brimenan, 50, 150, 1000, 20, 209

John Peck, 35, 60, 800, 15, 210

Christ. Beachy, 150, 77, 2000, 150, 968

Saml. J. Beachy, 50, 580, 1500, 25, 200

Mary Bailey, 40, 600, 400, -, 112

Charles Genff, 19, -, 1500, 100, 200

Alexander King, 500, 500, 20000, 500, 1100

Saml. M. Simms, 40, 75, 7000, 500, 528

Saml. P. Smith, 200, 80, 12000, 150, 900

John Steel, 30, 10, 1200, 50, 125

R. D. Johnson, 70, 15, 2000, 50, 400

H. D. Black, 40, -, 10000, 150, 300

Adam Holbrook, 10, -, 1200, 50, 65

Henry Schirer, 11, -, 2000, 50, 125

John Smith, 100, 7, 9000, 300, 300

Geo. Banker, 36, 5, 2000, 50, 250

Mary Vampa, 30, -, 1500, 25, 200

Dennis Dougherty, 275, 60, 7000, 100, 800

William Rest, 15, -, 700, 50, 100

Joseph Lang, 25, 5, 600, 50, 200

Jas. Anderson, 50, 10, 3000, 100, 800

Henry Weaver, 15, -, 700, 50, 150

Nelson C. Read, 160, 40, 6000, 100, 600

HenrySpokeman, 115, 100, 2500,75, 400

Charles Carner (Cavner), 80, 110, 200, 50, 350

J. W. H. Pollock, 500, 500, 25000, 600, 5765

J. W. H. Pollock, 137, 80, 2000, 100, 150

John Young, 60, 140, 1000, 100, 250

Margaret Adams, 40, 10, 1500, 50, 225

Frank Lang, 20, 20, 1250, 100, 150

Harman Brown, 25, 25, 1200, 80, 150

Zadock Brashiers, 40, 175, 2000, 50, 125

Patrick Stanton, 20, -, 500, 80, 200

James Ferrell, 40, 5, 1500, 50, 100

Nelson C. Read, 100, 130, 7000, 150, 200

John Wentler, 80, 20, 1000, 50, 400

William Hill, 33, 72, 4000, 40, 80

Henreitta Roby, 25, 70, 800, 40, 100

Henry Graves, 35, 30, 800, 50, 100

Walter Selley (Selby), 100, 180, 5000, 100, 600

Francis Zimerly, 60, 90, 1500, 100, 300

John Christa, 100, 65, 1500, 50, 400

Robert Christa, 80, 25, 2000, 100, 400

Andrew Rice, 50, 40, 1800, 50, 250

George P. Hinkle, 75, 135, 3000, 75, 380

John Folck, 200, 50, 20000, 200, 700

Richd. Hendrickson, 150, 90, 3500, 125, 550

Joseph Dilley, 160, 320, 8000, 200, 1000

9

Peter Ranik, 100, 100, 2500, 100, 700

Peter Miller, 80, 60, 2500, 100,700

George Myers, 40, 35, 1000, 80, 500

Elizabeth Taylor, 50, 15, 1200, 50, 150

Samuel Garcy, 40, 70, 1500, 50, 100

Samuel Garcy, 60, 40, 1800, 50, 80

James Taylor, 35, 65, 800, 25, 70

Laurence Spielman, 100, 100, 1800, 50, 200

Frank Apple, 30, 20, 800, 30, 40

John Birch, 30, 60, 700, 25, 80

Frank Gelhause, 20, 10, 800, 30, 150

Thomas Smith, 60, 30, 1000, 30, 100

Jacob Devote, 60, 70, 1800, 125, 300

J. Everstine, 100, 283, 3000, 50, 350

Charles Boetcher, 20, 7, 1000, 50, 150

Samuel Eckels, 60, 940, 20000, 150, 400

Henry Knapp, 20, 30, 1000, 25, 300

Wm. A. Bruce, 100, 327, 4000, 50, 400

Saml. P. Smith, 80, 120, 2000, 75, 400

Martin Roman, 70, 70, 1000, 50, 350

Elizabeth McKenzie, 60, 80, 1000, 50, 400

Jesse McKenzie, 60, 65, 1800, 80, 500

Moses McKenzie, 60, 50, 1000, 50, 400

Ann S. Lissell, 150, 250, 3000, 80, 600

John Jones, 40, 150, 800, 50, 550

Charles Ridgway, 80, 170, 2000, 50, 50

Wm. Neill, 250, 100, 7000, 400, 100

Michael Fosenmyer, 40, 73, 5000, 100, -

L.M. Bruce, 60, 30, 1800, 60, 300

Andrew Stagmeyer, 42, 4, 3000, -, 300

James Reynolds, 58, 30, 1000, 50, 150

Daniel We__am, 100, 25, 7000, 300, 700

David Shriver, 30, 20, 3000, 50, 250

George Creek Co., 60, 25, 6000, 80, 400

John Smelsing, 50, 150, 5000, 500, 350

George McKenzie, 50, 10, 2000, -, 325

Henry Millaman, 60, -, 6000, 50, 400

Carroll Miller, 40, 360, 8000, 50, 300

John Oliphant, 60, 140, 5000, 80, 500

Albert Rigri, 200, 200, 10000, 100, 602

George Rigri, 190, 200, 9000, 100, 900

Saml. D. Brady, 700, 550, 35000, 1000, 7105

Jno. Greenwade, 300, 200, 14000, 300, 600

Lois McMullin, 130, 100, 3000, 40, 500

Debra Swearageri, 250, 150, 15000, 100, 1800

Blake Wilson, 150, 150, 8000, 150, 1400

George W. Wilson, 250, 250, 10000, 300, 1870

John Welch, 70, 225, 10000, 200, 500

Rebecca Cresap, 250, 270, 1500, 300, 1300

Henry Hudson, 350, 1000, 28000, 300, 2700

Israel Miller, 100, 100, 6000, 100, 500

Aquilla McCarty, 300, 300, 25000, 1000, 2500

James McCarty, 250, 300, 25000, 800, 2500

M. Greenwade, 200, 150, 1000, 100, 400

Jno. McCowan, 60, 100, 1800, 80, 700

Jno. Greenwade, 65, 150, 3000, 75, 400

Patrick Dayton, 60, 140, 2500, 50, 575

Mary Dayton, 100, 200, 6000,75, 650

Mary Juskip, 100, 150, 3000, 75, 200

George Mackery, 60, 80, 1500, 50, 225

Daniel Clarke, 40, 10, 500, 30, 250

Thos. W. Dawson, 45, 215, 4000, 50, 300

Hamilton Poland, 60, 140, 3000, 80, 400

Philip Sigler, 100, 96, 3000, 100, 500

The American Co., 75, 100, 3000, 50, 200

G. W. Kildare Mgr., 175, 25, 5000, 50, 400

W. H. Miller Agt., 150, 150, 4000, 100, 700

J. Broadwater, 40, 200, 3000, 15, 200

Galantin Jacobs, 75, 121, 5000, 50, 500

Adam Green Mgr., 100, 300, 5000, 50, 800

J. A. Graham, 100, 300, 6000, 60, 900

John Lovett, 60, 240, 2000, 50, 430

James Poland, 75, 100, 2000, 50, 200

Jesse Graves, 30, 170, 2000, 50, 200

Dennis Graves, 100, 100, 3000, 60, 300

Robert Green, 40, 60, 1600, 50, 200

Terence Malone, 48, 60, 1800, 50, 350

John Davis, 95, 200, 4000, 80, 500

John Daniels, 95, 500, 5000, 80, 500

Price Twigg, 50, 140, 1500, 30, 150

John Squires, 25, 175, 800, 50, 300

Dana Twigg, 75, 100, 3500, 80, 400

Oliver Twigg, 100, 360, 4000, 60, 800

Thomas Twigg, 100, 50, 4000, 50, 500

J. T. Stallings, 80, 120, 800, 50, 300

Mat Stallings, 80, 120, 800, 50, 300

Charles Stallings, 75, 100, 1800, 40, 250

Nat. Robinett, 75, 132, 1200, 50, 260

J. H. Stallings, 50, 250, 2000, 80, 350

J. C. Hughes, 560, 125, 1200, 50, 1500

Thos. Hiddong, 30, 20, 80, 25, 200

Saml. D. Ferlon, 100, 300, 2000, 40, 550

John Ruby, 40, 82, 700, 50, 150

Charles Little, 80, 155, 255, 80, 600

E. Galisbenner, 50, 150, 2000, 60, 400

John Shaster, 100, 232, 1500, 50, 400

Jesse Malone, 50, 100, 1000, 40, 350

Jno. Patterson, 60, 200, 5000, 50, 250

Margaret Barnes, 100, 150, 5000, 80, 400

Henry Green, 50, 10, 1000, 50, 350

Ezra Koontz, 75, 100, 2000, 60, 500

Henry Speker, 60, 80, 1000, 45, 200

Jacob Koontz, 250, 600, 10000, 80, 800

Matthew Jacobs, 150, 50, 6000, 80, 800

Henry T. Shaw, 120, 138, 10000, 75, 700

Thos Pickel M. Co., 60, 40, 4000, 8, 300

John Barnes, 130, 80, 10000, 100, 1400

Jefferson Barnes, 50, 120, 2500, 80, 600

The Lenaway Co., 40, 200, 800, 50, 275

Peter Colman, 90, 136, 2000, 50, 600

John Miller, 125, 175, 6000, 80, 900

Geo. W Shaw Agt., 200, 120, 4000, 40, 860

Jer__ Miller Agt., 100, 130, 1400, 50, 450

Jos. Juskip Agt. 100, 106, 5000, 60, 500

Jesse Malone, 100, 100, 500, 40, 250

William Dean, 200, 700, 5500, 300, 800

Luke Alderton Agt., 80, 100, 1800, 50, 700

_. T. Stewart, 50, 100, 1000, 30, 150

David Bender Mgr., 40, 80, 400, 40, 200

Wm. Morgan, 50, 100, 1000, 100, 700

Henry Keifer, 20, 30, 600, 50, 100

Robert Twigg Mgr., 60, 50, 1000, 40, 500

David Crosly, 50, 250, 1000, 30, 150

Jasper Middleton, 40, 50, 500, 40, 250

Adam Bawyer, 80, 100, 800, 40, 300

Daniel Frager (Frazer) Mgr., 150, 300, 3000, 80, 870

Geo. Robinett, 100, 50, 2000, 60, 250

David Riel, 75, 265, 1500, 150, 200

Thos. Riel, 50, 120, 1000, 80, 600

John Wilson, 175, 150, 3000, 100, 785

William Conley, 18, 20, 750, 25, 150

James Crabtree, 200, 100, 1500, 50, 150

James Twigg, 200, 200, 1800, 50, 100

Isaac Twigg, 85, -, 4000, 100, 300

C. Kerlly, 350, 850, 22000, 400, 2600

Jacob Stump, 350, 150, 4000, 100, 600

Jacob Stump Jr., 70, 60, 1600, 80, 50

Thomas Ruby, 300, 300 6000, 100, 800

Jno. Ruby, 175, 125, 2500, 60, 600

Samuel T. Stablings, 60, 170, 600, 40, 150

E. Stattling, 65, 65, 700, 50, 200

Charles Twigg, 50, 150, 200, 60, 600

Griffin Twigg, 100, 300, 2000, 50, 400

Griffin Twigg Jr., 85, 470, 2600, 60, 300

J. A. Switzer, 80, 120, 1000, 50, 300

Jasper Huff, 125, 75, 2000, 50, 500

Charles Buey, 250, 550, 5000, 80, 600

Nathan Ruby, 40, 170, 1000, 60, 400

John Young, 80, 60, 1600, 50, 300

L. G. Shryock, 150, 125, 4000, 50, 350

Darniel McLaughlin, 100, 400, 4000, 60, 700

Lewis Crabtree, 150, 200, 3000, 80, 600

Elias Twigg, 50, 150, 1800, 60, 250

_. Diffenbaugh, 150, 350, 4000, 80, 600

_. Diffenbaugh, 100, 150, 1500, 60, 1000

Moses G. Robinett, 170, 150, 4000, 50, 915

Moses Rawlings, 350, 500, 8000, 250, 1800

Charles Statting, 80, 70, 500, 50, 500

Reby Twigg, 80, 140, 600, 30, 400

Daniel Crabtree, 100, 100, 1200, 50, 300

S. L. Wagoner, 200, 337, 3000, 80, 750

C. Wagoner, 90, 260, 3000, 100, 350

David Koontz, 40, -, 500, 50, 100

James Harris, 50, 50, 500, 30, 200

Conrod Bennett, 50, 100, 600, 30, 300

Henry Hamrick, 100, 68, 1500, 70, 300

L. M. Cresap, 600, 800, 12600, 200, 2100

James Mathews, 130, 370, 1500, 40, 600

James Crabtree, 90, 120, 1200, 40, 400

Harison Twigg, 25, 30, 1000, 40, 300

John Herman, 120, 20, 2500, 50, 650

Luther Givan, 160, 872, 7000, 100, 960

Christopher Bard, 100, 150, 2000, 80, 700

James J. Harris, 40, 15, 500, 30, 200

Jesse Davis Agt, 30, 60, 600, 30, 100

William Morrow, 40, 65, 1000, 50, 200

J. Diffenbaugh, 100, 200, 4000, 80, 780

Geo. Athey, 60, 340, 900, 50, 400

Geo. Athey Jr., 70, 103, 1200, 40, 200

Peter Diffenbaugh, 175, 775, 6000, 300, 427

Adam Bard, 100, 200, 2000, 100, 600

Riley Hartley, 30, 50, 500, 40, 200

Evan Hartley, 70, 200, 7500, 60, 700

D. Diffenbaugh, 125, 25, 800, 40, 350

E. Reams, 22, 30, 250, 30, 250

Peter Diffenbaugh, 100, 140, 1800, 40, 400

Amos Kerns, 40, 60, 450, 40, 200

Petr Alderton, 50, 70, 1000, 40, 300

E. Berkley, 100, 300, 1000, 60, 500

Elizabeth Robinson, 100, 200, 800, 30, 80

J. Robinson, 70, 180, 800, 40, 200

B. Cortes, 80, 100, 1000, 30, 150

Benjamin Bruce, 100, 200, 1000, 30, 200

Nathan McTee, 40, 59, 400, 40, 160

Jams Dean, 80, 100, 2000, 50, 600

J. _. Morrison, 125, 1400, 30000, 300, 1050

William Neff, 100, 130, 2000, 200, 400

Cat. McEnally, 80, 220, 3000, 50, 300

M. Fosebaker, 45, 65, 1200, 60, 400

J. Michael, 150, 150, 2000, 50, 350

K. Michaels, 50, 50, 250, 60, 400

W. Michaels, 80, 20, 600, 50, 400

Jesse Michael, 77, 120, 1200, 85, 333

George Michaels, 80, 70, 1600, 80, 650

W. Michaels, 100, 600, 3000, 800, 100

E. L. Michaels, 45, 70, 800, 50, 500

Phillip Michel, 40, 110, 1000, 50, 400

Jesse Michaels, 100, 96, 3000, 80, 600

Geo. McConnell, 60, 80, 2000, 50, 250

W. Michaels, 40, 80, 800, 50, 300

E. Knight, 50, -, 800, 80, 460

Mary Harnet, 100, 172, 2000, 50, 600

Franklin Company, 200, 87, 3000, 80, 900

L. K. Mullen, 160, 200, 4000, 100, 600

R. Ravenscroft, 300, 200, 5000, 80, 1200

C. Miller, 60, 120, 100, 50, 450

George Duckworth, 346, -, 3500, 80, 1200

Jacob Jacobs, 200, 300, 5000, 100, 900

Henry Turner, 50, -, 1200, 150, 600

B. B. Davison, 80, 50, 2000, 80, 600

William Polan, 60, 80, 1000, 50, 200

Samuel Miller, 50, 30, 1000, 60, 500

Wm. Harrowbaugh, 60, 80, 1200, 80, 300

Asa Jacobs, 60, 100, 1500, 75, 800

J. M. Kletsman, 25, -, 800, 4, 300

Henry Gauser, 35, 25, 1000, 60, 350

Aaron Potts, 120, 100, 2000, 80, 400

C. Weaver, 60, 67, 800, 60, 300

Asa Twigg, 50, 40, 1000, 80, 100

F. Twigg, 40, 80, 1200, 80, 400

Thomas Dicken, 50, 200, 1200, 80, 40

B. Davis, 50, 20, 2000, 80, 300

Patrick Irons, 80, 150, 1000, 80, -

Geo. L. Zimerley, 25, 10, 300, 60, 15

Isaac Rue, 100, 400, 5000, 100, 600

Jeni Knight, 40, 22, 600, 50, 300

Margaret Smith, 60, 80, 1200, 60, 300

Samuel Rice, 50, 50, 1400, 80, 400

William Shaw, 200, 700, 10000, 600, 3000

Wesley Dawson, 300, 50, 5000, 300, 900

Eli W. Dawson, 100, 100, 3000, 150, 700

J. F. Dawson, 80, 120, 200, 60, 400

George Winters, 60, 80, 1500, 80, 700

Jacob Winters, 40, 90, 1600, 80, 500

Joseph McKenzie, 60, 190, 1400, 60, 400

Henry Headinger, 40, 40, 2000, 100, 400

J. Wilson, 60, 190, 2000, 100, 300

J. H. Smith, 150, 100, 2800, 60, 1256

Jesse Hinkle, 200, 150, 6000, 150, 250

Alfred Rice, 25, 20, 1000, 60, 300

A. Grass, 150, 200, 3000, 150, 680

David Hinkle, 80, 100, 2000, 100, 200

A. Hinkle, 150, 150, 5000, 100, 500

S. Hinkle, 150, 75, 2000, 80, 700

Edward Buey, 125, 160, 2500, 60, 450

S. Rawdens (Bawden), 100, 25, 1500, 80, 800

E. Neal, 100, 25, 1500, 75, -

George Buey, 40, 10, 400, 80, 500

F. Bawden, 150, 150, 3000, 70, 300

J. E. Robinett, 60, 60, 600, 50, 200

Samuel Zimerly, 100, 100, 4000, 70, 400

W. Wilkerson, 60, 150, 1200, -, -

L. Perren, 150, -, 2000, 10, 950

Ann Hardsock, 80, 40, 1500, 80, 400

Henry Hartsock, 75, 50, 1200, 50, 400

Thomas Street, 80, 120, 2500, 80, 800

E. Willisson, 100, 125, 2000, 175, 320

H. Street, 80, 120, 2000, 60, 300

A. Willisson, 102, 60, 1200, 80, 400

Danl. Willisson, 50, 15, 1500, 60, 300

James Wilson, 200, 200, 10000, 400, 700

J. Piper, 110, 260, 8000, 180, 600

Amos Robinett, 200, 200, 8000, 250, 1100

Jesse Wilson, 108, 108, 4000, 400, 900

J. Branning, 75, 80, 1500, 50, 300

H. Biggs, 30, 100, 1000, 50, 125

Geor. Voeifer Sr., 36, 60, 400, 60, 100

L. Shireliff, 100, 70, 2000, 25, 350

N. Slider mgr. 50, 60, 1200, 60, 200

Wm. Shireliff, 25, 80, 1500, 50, 300

Daniel Ryon, 25, 135, 500, 60, 200

E. Roley agt., 30, 40, 500, 30, 300

Wm. Young agt., 50, 180, 1000, 60, 2560

W. M. Alender, 40, 130, 1500, 50, 300

Wm. Mathews, 650, 120, 1800, 60, 300

J. Gross, 70, 200, 3500, 80, 600

A. Keenan, 110, 200, 5510, 100, 800

A. M. Bevans, 130, 150, 3000, 80, 300

W. Brinker, 50, 200, 1810, 60, 400

R. Higgins, 125, 125, 3000, 90, 600

J. Callan, 50, 200, 1000, 60, 800

Thos. Callan, 140, 160, 1500, 150, 1100

P. Stottlemyre, 65, 300, 500, 60, 200

A. Sigler, 25, 25, 400, 30, 200

C. Norris, 60, 50, 1800, 40, 300

M. Doran, 100, 100, 1500, 30, 400

C. Norris Jr., 60, 100, 100, 30, 250

N. Trail, 100, 850, 2000, 80, 700

A. Trail, 80, 102, 600, 50, 400

M. Neall, 150, 650, 5000, 100, 1000

J. Watson, 100, 800, 2500, 170, 500

E. Bevans, 200, 300, 3000, 80, 900

Geo. W. Price, 60, 150, 1200, 100, 700

J. T. Harley, 15, 30, 800, 30, 125

Isaiah Myers, 60, 100, 1800, 60, 400

Amos Ash (mgr.), 125, 175, 2500, 80, 500

James Cramford, 150, 135, 2400, 125, 300

Geo. Robinett, 175, 75, 5500, 400, 1170

A. Wilson, 200, 82, 6000, 250, 840

Geo. Twigg, 125, 150, 3000, 80, 300

S. Wilson, 100, 200, 4000, 30, 370

E. Wilson, 80, 113, 2500, 100, 290

R. Wilson, 14, -, 800, 25, 150

A. Wilson, 100, 100, 2000, 200, 600

Isaac Wilson, 70, 100, 5000, 200, 450

Geo. Stein, 100, 20, 4000, 60, 500

Jas. H. Stein (mgr.), 80, 20, 400, 80, 400

Elias Wilson, 300, 300, 6000, 450, 870

Jno. H. Wilson (agt.), 150, 110, 3500, 50, 365

David McElfish, 150, 43, 2200, 80, 700

David Cheney, 60, 30, 2000, 80, 700

Thomas Twigg, 80, 200, 3000, 90, 500

Argyle Twigg, 30, 115, 1000, 80, 120

F. A. Jamison, 140, 24, 7000, 450, 85

William Robinett, 200, 200, 4000, 200, 900

William Hendrixson, 50, 150, 200, 80, 500

E. A. Qjmes, 40, 60, 600, 30, 200

Daniel Willisson, 70, 30, 2000, 100, 350

J. A. House, 300, 50, 5000, 100, 650

H. Willisson, 200, 400, 4000, 300, 350

M. Dean, 60, 40, 2000, 50, 260

Joseph Dean, 48, 75, 2000, 60, 450

Amos Stattings (mgr.), 40, 25, 1500, 30, 200

John Miller, 250, 100, 10000, 200, 600

John North, 150, 100, 2000, 150, 450

J. McElfish (mgr.), 80, 100, 2500, 80, 300

A. Keifer, 40, 100, 1600, 80, 300

James Wilson, 140, 160, 4400, 100, 600

J. McElfish, 100, 100, 200, 80, 150

Isaace Smith, 70, 130, 1800, 50, 400

David House, 45, 30, 800, 30, 200

H. A. Jamison, 225, 265, 15000, 600, 1900

Wm. H. Hamilton, 100, 224, 600, 80, 400

Lanty Twigg, 50, 100, 1000, 80, 300

M. L. Robinett, 30, 50, 800, 15, 400

G. Twigg, 50, 100, 100, 60, 500

Joel Middleton, 50, 52, 1000, 80, 400

Thos. McElfish, 30, 130, 1500, 100, 250

E. Dran (mgr.), 70, 100, 2000, 80, 300

M. H. Myers (mgr.), 100, 200, 3000, 60, 400

Jeremiah Wilson, 80, 100, 3000, 80, 700

Robert Twigg, 50, 100, 1500, 60, 500

Henry Buey, 60, 240, 1500, 80, 400

John Springsted, 40, 160, 800, 50, 300

Gillis Robinett, 50, 100, 1200, 45, 350

Daniel Koontz, 30, 78, 500, 50, 300

Wm. McCulloh, 60, 200, 2000, 60, 150

Wm. Hartsock (mgr.), 200, 430, 8000, 125, 950

Thos. Davis, 100, 190, 4500, 100, 700

Jno. Fletcher, 90, 185, 2200, 150, 750

Amos Willisson, 75, 152, 3000, 60, 500

Jacob Lashley, 75, 175, 1020, 80, 500

P. Kermard, 100, 200, 700, 80, 264

M. Roberts (mgr.), 80, 100, 1500, 75, 200

Henry Buey, 60, 50, 1200, 80, 250

W. & E. Montgomery, 102, 200, 4000, 120, 550

Isaac Wilson, 100, 3000, 3000, 80, 300

Robt. Lashley, 100, 200, 1800, 60, 600

E. Browning, 125, 125, 2000, 80, 100

L. Chaney, 55, 50, 1500, 75, 300

S. Chaney, 80, 210, 3000, 80, 700

Daniel Gorden, 40, 340, 1000, 60, 150

E. W. Ash, 100, 200, 2000, 80, 600

A. Bennett, 30, 100, 1000, 40, 100

William Ellis, 80, 50, 800, 60, 450

B. Johnson (mgr.), 30, 30, 100, 40, 100

Amrbose Chaney, 30, 80, 500, 40, 200

Geo. Roberts, 30, 136, 800, 40, 100

John Hires, 60, 118, 400, 40, 350

E. W. Ash, 75, 95, 1000, 50, 200

Wm. Slider, 35, 65, 800, 40, 200

Wm. McCoy, 40, 160, 800, 50, 300

Jacob Kifer, 50, 150, 500, 80, 400

William Frey, 100, 150, 6000, 125, 712

S. K. Frey, 92, 163, 2000, 125, 735

John Neff, 25, 25, 250, -, 164

Michael Neff, 25, 25, 600, 25, 103

John Sees, 50, 40, 450, 10, 137

Isaac Umbel, 30, 30, 700, -, 110

John A. Thomas, 95, 25, 500, -, 88

William Welch, -, -, -, -, 130

Elijah Thomas, 80, 40, 800, 10, 95

Jacob A Thomas, 70, 61, 1048, 30, -

G. J. Thomas, -, -, -, -, 280

Jackson Umbel, 40, 110, 1000, 50, 221

Daniel Chrise, 40, 60, 400, -, 12

Elias Thomas, 65, 95, 1500, 25, 118

Jacob Marst, 65, 72, 1000, 150, 415

Jacob Mories, 25, 125, 800, 15, 186

H. B. Demmitt, -, -, -, -, 69

Squire Frazer, 40, 69, 700, 15, 220

Humnoma Frazer, 100, 125, 1000, 20, 367

John W. Shafer, 15, 95, 450, 10, 108

Isaac J. Frazer, 65, 135, 1000, 20, 315

Elisha Frazer, 100, 100, 1200, 40, 360

W. R. Stuck, -, -, -, 15, 202

John J. Frazer, 140, 160, 1275, 100, 467

John Augustine, 60, 100, 3000, 50, 572

Jos. G. Frazer, 100, 50, 1200, 75, 358

Elijah Mason, 2, 48, 100, -, 26

Elijah Friend, 125, 1485, 4000, 100, 462

William Sines, 40, 60, 500, 30, 125

Henry Sines, 40, 60, 500, 25, 36

Henry Friend, 20, 80, 500, 30, 116

Rececca Johnson, 5, 5, 50, 20, 135

John Riley, 100, 40, 3000, 150, 490

Stephen W. Friend, 24, 110, 288, -, 152

Wm. A. Savage, 30, 70, 400, 12, 228

Cornelius Savage, 30, 70, 400, 8, 284

William Johnson, 15, 122, 300, 2, 133

Harrison H. Johnson, 5, 45, 100, -, -

Thomas Browning, 50, 43, 2000, 50, 239

James Browning, 75, 215, 2000, 200, 487

Adam Nethkin, 20, 80, 900, 25, 185

Jeremiah Browning, 45, 55, 900, 50, 257

Israel Frankhouse, 100, 100, 1600, 54, 451

David Somers, 18, 1287, 2000, 20, 136

Samuel Willhelan, 12, 108, 2000, 20, 150

Alpheus Machian, 40, 40, 1250, 75, 431

John E. Jilo, 40, 30, 1250, 100, 577

Lewis Wolf, 40, 160, 800, 30, 309

Lewis Lemmyer, 150, 50, 8000, 300, 933

Francis Browning, 2, 998, 7000, 110, 450

Ceolis Wilt, 35, 65, 600, 15, 90

W. R. Sollen, 200, 856, 5000, 200, 291

George Stazer (Stozer), 100, 200, 3500, 100, 280

Fred___ Lee, 75, 200, 1200, 75, 490

John T. Harrington, 300, 5700, 30000, 300, 675

John M. Davis, 200, 415, 6000, 125, 1510

Richd. M. Sprig, 225, 445, 6000, 200, 790

Peter Martin (Maston), 80, 80, 1000, 150, 460

Henry Loran, 40, 85, 500, 12, 38

Wm. W. Ashby, 90, 370, 1500, 50, 439

Jesse Ashbey, 150, 120, 2700, 50, 301

Thos. W. Ashbey, 80, 166, 1300, 40, 5329

Francis Maston, 40, 160, 1000, 5, 279

Jacob Gutzy, 230, 75, 2150, 175, 618

Archibald Chisem, 45, 155, 1200, 20, 210

Peter Gardner, 90, 110, 700, 40, 343

John C. Lower, 125, 35, 1280, 30, 419

George Sliger, 75, 75, 600, 15, 190

Abraham Spilker, 50, 100, 1500, 100, 505

Lewis Thompson, 100, 30, 1500, 75, 647

John Spiker, 50, 125, 1000, 100, 418

Aston Styer, 100, 200, 1000, 200, 562

John Horn, 60, 15, 900, 15, 228

Nelson Savage, 15, 30, 500, 10, 100

Nicholas Pugh (Laugh), 30, 100, 1200, 25, 218

Randolph Beckman, 50, 50, 500, 25, 502

Benjamin Drin, 25, 40, 600, 50, 291

William Buey, 25, 135, 200, 25, 314

Henry Tusker (Tucker, Tasker), 30, 70, 700, 10, 165

John L. Browning, 200, 170, 8000, 300, 919

Elijah P. Branst, 35, 265, 2500, 785, 546

Joseph Faith, -, -, -, -, 132

Ezekiel Tolten (Totten), 100, 798, 2000, 75, 482

Silas Fitzgonder (Fitzgouder), 10, 30, 160, 10, 337

Joseph Knox, -, -, -, 5, 21

J. McMason, 75, 75, 1500, 60, 346

Henry G. Sannden, -, -, -, 100, 220

John F. Sannden, -, -, -, 50, 300

Henry Lousey, -, -, -, -, 30

John D. Stall, 50, -, 500, 50, 282

Peter Mastin, 100, 100, 2000, 100, -

Joseph Wonderly, 100, 100, 700, 35, 409

William Love (Lowe), 200, 180, 3800, 100, 885

Thompson Low, -, -, -, -, 169

Henry Lowe, 20, 50, 426, 30, 208

Nathan A. Lowe, 30, 120, 900, 25, 162

Joseph King, 30, 50, 500, 50, 226

Hesteram King, 20, 80, 500, 25, 748

Isaac King, -, -, -, 5, 262

Mary More, 25, 25, 250, 25, 192

Johnson White, 100, 50, 1500, 80, 1168

James Kinter, 50, 50, 300, 10, 140

J. J. White, 60, 90, 1000, 50, 260

Ashton Biggs, 75, 75, 2000, 50, 297

Middleton S. Biggs, 50, 100, 800, 50, 127

George Riley, 800, 450, 2275, 35, 284

John Riley, -, -, -, 25, 240

Samuel Riley, -, -, -, 25, 123

Henry Thompson, -, -, -, -, 150

Thomas Riley, -, -, -, -, 150

Isaac Thompson, 533, 1066, 14400, 150, 2988

James Chisholm, 13, 137, 300, 610, 542

Elliott Tubb, 300, 300, 3600, 25, 734

Gough C. Tubb, 60, 120, 1080, 50, 275

John W. White, 85, 65, 900, 75, 312

Amos Stelby, 25, 25, 300, 5, 350

Grilch Pendleton, 200, 1202, 4200, 200, 1329

John Haze (Haye), 250, 650, 4500, 50, 440

Henry Knapp, 40, 148, 542, 25, 230

John Knepp, 40, 195, 470, 25, 375

John Bremble, 60, 70, 565, 100, 507

Joseph Fraizer, 15, 650, 5000, 100, 746

Jno. W. Abernethy, 20, 150, 212, 10, 138

William Waltz, 70, 158, 2000, 85, 210

Ezra Houser, 90, 10, 1200, 20, 232

Eve Lee, 50, 109, 159, 5, 120

J. N. Rhoth, 30, 20, 500, 40, 345

John Bean, 45, 121, 598, 25, 236

John G. Ganes, 80, 379, 3000, 100, 720

George N. Ganer (Ganes), 8, 77, 684, 42, 195

Samuel Guergey, 35, 81, 1160, 100, 252

John Roath, 22, 105, 750, 12, 208

George Routh, 20, 110, 150, 100, 344

John Girner, 65, 185, 2000, 50, 559

John A. Shaffer, 85, 193, 1000, 50, 530

Peter Miller, 50, 125, 1750, 100, 300

David Fike, 5, -, 200, 75, 115

John Phillips, 100, 600, 10000, 200, 525

John Kemp, 75, 25, 10000, 100, 300

William Rolf, 50, -, 500, 20, 201

D. S. Arnold, 95, 150, 1000, 25, 663

Thomas Moore, 75, 75, 900, 50, 170

John Riddle, 40, 110, 950, 60, 434

David Wilt, 120, 228, 3000, 125, 603

Henry Beckman, 60, 140, 1000, 25, 445

John G. Styer, 200, 247, 3576, 50, 829

Elisha Harvey, 50, 200, 750, 25, 370

James Wilson, 200, 700, 2500, 50, 1045

Jarret N. Moore, 75, 125, 2000, 12, 468

Isaac Moore (Moon), 75, 200, 1375, 12, 169

Benjamin Moon, 90, 3, 750, 12, 299

John B. Davis, 100, 200, 1500, 50, 510

Lloyd Kittsmuller, 25, 75, 500, 10, 150

Dudley Lee, 100, 50, 100, 30, 885

L. F. Klepstine, 40, 23, 1000, 30, 285

John Bray, 100, 100, 2800, 85, 627

Henry Hammell, 100, 100, 1000, 100, 516

Henry Tasker, 40, 60, 500, 16, 146

Hannah Havely, 100, 50, 1000, 50, 292

Nelson Baker, 150, 310, 1600, 150, 878

John Lewis, 50, -, 300, 10, 470

Daniel Lewis, 40, -, 200, 7, 105

Saml. Rodeham, 30, 20, 500, 25, 71

Joseph Friend, 75, 442, 4000, 25, 630

William Schooley, 30, 20, 600, 15, 232

Isaac Mail, -, -, -, 15, 289

John Cope, 40, 200, 1200, 10, 180

Nathan Casteel, 180, 320, 6000, 75, 1192

Colvin Savage, 100, 575, 3000, -, 141

George w. Griffith, 20, 480, 2500, 20, 215

Patrick Hamil, 200, 320, 600, 250, 1145

Richard Kight, 40, 140, 920, 25, 279

John Friend, 50, 300, 2000, 10, 216

E. E. Blackburn, 200, 665, 5000, 75, 500

Abraham Wilson, 50, 150, 1000, 100, 1185

Samuel Bosley, 20, 30, 150, 50, 163

John Paugh, 50, 50, 300, 20, 199

Jas. E. Paugh, 35, 65, 300, 25, 238

Benj. Moon, 25, 39, 222, 25, 208

Abraham Moon (Moore), 30, 970, 5000, 25, 75

Andrew F. McCrobie, 20, -, 100, 25, 319

Michael Garrett, 14, 86, 500, 5, 59

John Jenkins, 130, 70, 1600, 40, 3338

Samuel W. Friend, 200, 455, 2500, 75, 380

Henry Shanks, 15, 35, 400, 75, 160

Thomas Wilson, 75, 325, 1500, 20, 583

Josiah Friend, 15, 85, 800, 48, 219

Fletcher Friend, 15, 35, 500, 10, 32

W. H. Bernard, 100, 300, 800, 75, 690

J. B. Bernard, 100, 300, 800, 75, 699

Edward Barnard, 150, 50, 1000, 50, 660

Geo. Barnard, -, -, -, 20, 30

Wright Thayer, 78, 300, 2268, 200, 521

William Kennig (Kenney), 25, 25, 1000, 50, 248

Luke Mole, 35, 565, 1250, 10, 562

John A. Drogie, 125, 825, 12000, 200, 989

Stephen Merrill, 100, 100, 3000, 25, 652

Patrick Callan, -, -, -, 5, 271

Thomas Grimes, -, -, -, 20, 416

J. W. Deimett, 18, 100, 200, -, 124

M. Sharpless, 25, 25, 300, 50, 288

John Poland, 60, 1450, 1000, 25, 316

John Paugh, 100, 120, 1500, 100, 569

Henry Beckman, 50, 101, 1000, 25, 321

Herman Beckman, 25, 25, 600, 50, 366

David Bevans, 25, 125, 400, 10, 130

Saml. B. Harvey, 70, 150, 1000, 10, 196

Shadrack Harvey, 15, 3335, 400, 10, 144

Joseph Paugh, 4, 146, 400, -, 24

Francis Thomas, 60, 340, 2000, 50, 120

M. G. Harvey, 50, 100, 600, 10, 115

Michael Paugh, 50, 100, 700, 15, 84

John Beckman, 600, 200, 3000, 50, 1256

O. Harshbyer, 40, 46, 700, 40, 194

Jas. Biddinger, 45, 48, 500, 10, 203

Danl. Slaubough, 45, 195, 1300, 50, 195

George Starke, 45, 40, 1520, 10, 216

Barbarab Orndorff, 24, -, 300, -, 113

Peter Lohr, 85, -, 1000, 75, 295

Alexander Lohr, 75, 125, 1500, 10, 278

Solomon Biddinger, 50, 67, 600, 10, 305

Jac. Bramaman, 100, 150, 2000, 150, 560

Peter Starke, 25, 75, 250, -, 87

Wm. Cunningham, 146, 1650, 5000, 50, 740

John Brady, 100, 300, 1500, 100, 545

Henry Macgraff, 40, 40, 1000, 50, 189

John Hart, 75, 125, 2000, 30, 529

John Slicer, 150, 150, 4000, 50,3 80

James Rich, 110, 56, 1200, 50, 339

John Boyer, 75, 175, 2000, 125, 459

Andrew Deihl, 200, 400, 3000, 150, 735

John Opel, 50, 15, 900, 25, 160

Michael Zebnor, 20, 20, 400, -, 204

John Herman, 30, 20, 400, 25, 186

Adam Vestler, 20, 30, 200, 5, 73

Louis Hauft, 90, 85, 1500, 80, 273

Conrad Hauft, 40, 35, 1000, 10, 215

George Deihl, 22, 25, 300, -, 73

Jonas Harshberger, 40, 80, 620, 10, 221

Andrew Fisher, 40, 10, 600, 25, 158

John Sinearman, 100, 500, 1000, 50, 403

John Miller, 200, 57, 5000, 150, 568

John Simon, 150, 60, 3000, 100, 538

George Gerack, 75, 25, 1000, 100, 255

Elizabeth Daniels, 80, 120, 800, 50, 226

Martin Deihl, 100, 57, 2000, 150, 588

Danl. Fleckenger, 70, 30, 2000, 50, 239

Rudolph Grider, 90, 94, 2500, 50, 251

Alestine Spicker (Spiller), 50, 50, 1000, 50, 225

Michael Berchly (Buckly), 100, 50, 1800, 150, 551

Isaiah Buckly, 80, 120, 1200, 500, 336

Nicholas Mufaker, 75, 75, 2000, 50, 363

Phillip Goor (Goox), 35, 165, 1200, 50, 273

John Karling, 40, 160, 600, 50, 192

Henry Grosse, 40, 160, 1000, 15, 180

Andrew Fox, 20, 3, 150, 5, 82

John Frantz, 200, 100, 5000, 50, 448

Daniel Haubaugh, 150, 20, 2000, 100, 492

Christ. Bowman, 60, -, 600, 10, 111

Simeon Logsdon, 30, 70, 800, 30, 279

Eli McMillan, 12, 20, 400, 25, 156

Christ. Masser (Musser), 50, 144, 900, 10, 101

George Garvigle, 50, 30, 1000, 30, 201

James Hardin, 55, 95, 1000, 10, 215

George Ault, 25, -, 251, -, 131

Richard Farral, 150, 250, 7000, 300, 1155

Andrew Kernsold, 60, 220, 1000, 50, 190

Joseph Logdson, 40, 110, 1800, 125, 410

John J. Stine, 80, 141, 2400, 150, 308

John Camp, 18, 85, 300, 75, 200

David Johnson, 80, 1725, 4500, 400, 248

Hiram Mitchell, 102, 129, 1500, 300, 443

Frank Bitner, 100, 211, 1500, 60, 121

Jacob Bauser, 35, 15, 600, 50, 198

Andrew Codflesh (Cooflish), 10, 40, 500, -, 63

Sarah Benson, 60, 40, 400, 70, 323

Hiram Hileman, 15, 35, 300, 25, 50

Solomon Turney, 70, 190, 1500, 110, 233

Jonathan Glass, 70, 80, 2000, 75, 293

Adam Keller, -, -, -, 116, 240

David Spielman, 100, 205, 2400, 125, 187

George Hielman, 50, 50, 500, 75, 264

John M. Davis, 15, 185, 250, -, 5

William Hileman, 15, 35, 200, 50, 35

Samuel Fickman, 40, 60, 500, 100, 176

Susanna Turney, 50, 50, 500, 70, 80

Fredk. Bowman, 25, 75, 500, 75, 110

Daniel Bowman, 15, 10, 250, -, 61

Fredk., Snyder, 40, 180, 510, 150, 470

Christ. Slosnagle, 20, 80, 500, -, 118

George Ryley, 150, 54, 2700, 125, 454

Joseph Lette, 60, 65, 1000, 75, 135

John E. Frazee, 25, 75, 500, 125, 256

Phillip Shultz, 18, -, 1200, 75, 275

John Delfer, 18, -, 1200, 65, 150

Sophia Hughes, 30, 100, 1600, 50, 175

John Dennis, 200, 900, 8000, 100, 830

Sarah Elliott, 60, 37, 4000, 100, 500

John J. Rudemark, 180, 20, 1800, 350, 350

Andrew Rice, 60, 50, 2000, 1100, 800

Frederick Valentine, 60, 60, 1500, 100, 400

John Rice, 80, 87, 2500, 200, 700

Henry Rudemark, 150, 50, 10000, 500, 1400

Thomas Wilkinson, 60, 240, 5000, 150, 500

_. Savage Company, 80, 50, 600, 50, 200

Michael Naughton, 100, 300, 4000, 150, 400

_. Savage Company, 70, 530, 4000, 60, 300

Jno. M. Buchanan, 300, 200, 15000, 100, 875

Levi Staller, 90, 30, 6000, 75, 190

John Clarke, 200, 149, 8500, 200, 725

Morris Wincbrumer, 60, 58, 3000, 175, 600

John Winguts, 200, 30, 8000, 150, 450

Levi R. Ficktig, 52, -, 5000, 200, 385

John Valentine, 200, 100, 6000, 100, 600

W. & L. Leppold, 70, 60, 5000, 100, 350

Susan Brant, 90, 20, 3000, 50, 400

Andrew Stineline, 60, 50, 1200, 50, 275

Michael Rudemark, 100, 100, 5000, 150, 900

William Frantz, 160, 177, 6500, 300, 900

Henry D. Smouse, 238, 200, 6000, 150, 350

Henry Deakins, 40, 20, 2500, 150, 520

John Smouse, 140, 100, 8000, 200, 600

William Frantz, 300, 285, 8000, 300, 5300

Patonea O. Baker, 65, 20, 1800, 150, 200

Casper Close, 32, -, 900, 50, 175

Clabaugh Jones, 20, -, 1500, 50, 125

Andrew Stagmyer, 40, 40, 1300, 100, 200

Nathan Wilson, 50, 20, 2000, 50, 380

Amos Graso, 100, 75, 1800, 100, 350

John F. Hinkle, 90, 60, 4000, 150, 400

Caroline Hinkle, 60, 70, 1000, 150, 200

Hester Rabey (Raley), 80, 60, 4000, 100, 300

William Brant, 70, 30, 2000, 125, 350

A. B. Beall, 200, 100, 4000, 200, 1000

Alex. Beall, 60, 80, 1500, 100, 200
G. P. Johnson, 45, 75, 2000, 150, 170
S. C. Eastman, 100, 54, 2500, 150, 600
Henry Hardman, 150, 50, 3500, 200, 800
Daniel Folck, 140, 58, 5000, 100, 300
William Beall, 200, 80, 6000, 200, 600
Peter Smouse, 300, 200, 5000, 500, 1200
Peter Smouse, 125, 175, 3000, 50, 200
William H. Beall, 20, 20, 2000, 50, 70
John Henlde, 100, 170, 3200, 150, 350
Ashael Willson, 50, 40, 2000, 100, 300
Sarah Buel, 100, 100, 3000, 100, 400
Edward Drake, 100, 100, 2500, 150, 300
A. B. Rice, 60, 65, 1000, 125, 325
Jno. Henderson, 70, 200, 2000, 125, 450
Saml. Bucey, 60, 40, 2000, 50, 400
Saml. Bucey, 80, 40, 3000, 125, 700
George W. Durst, 65, 15, 800, 100, 500

John Fisher, 70, 80, 1500, 100, 450
J. M. N. Henderson, 80, 80, 2000, 100, 400
John R. Brooke, 300, 250, 8000, 10, 500
Lewis Fres (Tres), 300, 700, 15000, 50, 300
Jas. Kuykendall, 150, 150, 5000, 100, 600
Allaghany County, 80, 60, 8000, 200, 325
Margaret Adams, 45, 5, 2500, 50, 200
D. Shriver, 200, 200, 5000, 160, 850
Chas. H. McBlair, 450, 520, 19000, 150, 750
J. B. Keller, 50, 75, 750, 100, 225
John Humbard, 100, 60, 5000, 100, 330
A. P. Shephard, 400, 200, 8000, 200, 1300
A. L. Withers, 60, -, 3500, 200, 750
C. M. Thurston, 300, 150, 20000, 400, 1200
Wm. Knost, 40, -, 2000, 250, 200
F. M. Grombeck, 150, 50, 2000, 200, 700
B. P. Limpkin, 80, 70, 3000, 80, 225
John Shade, 40, 10, 1800, 100, 250
Q. Weiskettle, 25, -, 6000, 100, 250
Geo. Dornin, 20, -, 1500, 100, 200

Anne Arundel County Maryland
1860 Agricultural Census

The University of North Carolina Library under a grant from the National Science Foundation microfilmed agricultural Census records. Records were filmed at the University of North Carolina from original records at the Maryland State Library.

Columns 1, 2, 3, 4, 5, and 13 represent the following information on the census:
1. Name of Owner, Agent or Manager of Farm
2. Acres of Improved Land
3. Acres of Unimproved Land
4. Cash Value of the Farm
5. Value of Farming Implements and Machinery
13. Value of Livestock

Pages are out of sequence.

Joseph Howard, 75, 100, 4000, 150, -

Edward Jones, 50, 3, 5000, 100, 550

Isaac Jones, 350, 250, 6000, 100, 1000

Ephraim Story, 25, -, 600, -, 150

John S. Sellman, 520, 220, 14000, 250, 1800

James Fulton, 150, 100, 7000, 200, 700

Edward Harney, 260, 40, 18000, 700, 1890

S. A. Brown, 160, 4, 9800, 100, 650

T. K. Kent, 300, 52, 24800, 800, 2300

James A. Iglehart, 340, 21, 25000, 800, 1875

Thos. Davidson, 200, 15, 15000, 800, 1700

E. A. Ditty, 75, 55, 4000, 150, 600

John Iglehart, 270, 30, 21000, 800, 1480

John W. Iglehart, 230, 10, 16800, 800, 1450

John Hardesty, 90, -, 4000, 50, 500

James Larrimore, 11, -, 350, 75, 300

J. S. Nicholson, 80, 70, 1200, 100, -

Richard Harwood, 30, -, 300, 25, 100

Jeanette Z. Chaney, 150, 50, 10000, 300, 1000

P. W. Whitenight, 175, 20, 7800, 250, 650

Stephen Lee, 350, 100, 18000 150, 1600

James E. Nicholson, 75, 25, 1300, 25, 200

T. S. Iglehart, 161, 68, 19600, 900, 3300

Estate of J. H. Watkins, 270, -, 13500, -, -

Estate of J. W. Watkins, 200, 190, 3900, -, -

T. Talbott, 5, -, 3000, 200, 150

McCeney Howes, 9, -, 2000, 100, 100

S. H. Durett, 230, 71, 20000, 1000, 1700

John W. Williams, 200, 54, 15000, 800, 1000

Fayett, Ball, 200, 233, 9000, 500, 1000

Geroge Bell, 200, 25, 13000, 600, 1000

S. A. Clagett (Ten), 200, 100, 12000, 500, 1000

D. M. Brogden, 180, ½, 24000, 600, 1600

James Black, 40, 20, 2400, 200, 665

James Black, 60, 5, 2600, -, -

T. Sheppard, 200, 30, 6000, 300, 1400

Wm. G. Makall, 300, 36, 23500, 1000, 1900

M. S. Robinson (Mgr), 210, 60, 15600, 400, 1470

Harriett Dorsett, 150, 51, 12000, 500, 1100

James Chester, 450, 250, 49000, 1200, 2900

James Chester Jr. Ten, 250, 100, 21000, 200, 1050

Edward McCeney(McCerney), 200, 60, 13900, 500, 2100

Aby Clayton, 140, 60, 8000, 100, 725

Henry Bassford, 100, 70, 2120, 30, 600

Gassaway Owens, 100, 17, 5800, 100, 650

William Gardner, 80, 32, 1200, 50, 500

Lloyd Brown, 100, 60, 1900, 50, 100

Benjamin Phipps, 153, 93, 2400, 50, 300

Benjamin Streets, 100, 40, 7000, 75, 800

Robert Burkett Ten, 200, 35, 5800, 50, 700

Mary Clayton, 200, 45, 17000, 200, 1150

Jacob McCeney, 190, 30, 11000, 500, 1000

Ben Carrick, 25, 5, 900, 50, 150

Thos. McCeney, 175, 36, 10500, 200, 310

Saml. Purdy mgr., 2000, 500, 87500, 1200, 5325

R. S. Stuart, 1200, 400, 112000, 3000, 5250

Estate of Saml. Owens, 140, 84, 12200, 200, 850

Deborah Jones, 75, 25, 1000, 50, 175

John Brashears, 125, 75, 6000, 100, 600

Wm. Sims Ten, 150, 50, 10000, 100, 659

Lucy Mitchell, 260, 40, 18000, 1000, 1328

Charles H. Steel, 230, 70, 21000, 800, 1526

William Ghiselin, 130, 20, 9000, 400, 795

P. J. Drury, 90, 15, 6300, 500, 1020

P. J. Drury, 120, 64, 11400, -, -

P. J. Drury, 120, 51, 5130, 100, 300

Wm. H. Peake, 190, 25, 12900, 300, 100

James Bird, 200, 40, 12000, 500, 1000

Wm. Brogden, 300, 100, 24000, 500, 1925

Lucinda Sellman, 230, -, 16000, 600, 1075

J. B. Stockett, 120, 80, 8000, 400, 800

Geo. B. Stockett, 120, 19, 5500, 50, 600

Wm. F. Holliday, 150, 161, 10800, 100, 965

Sarah E. Stackett, 50, 20, 2100, 40, 100

Allen Marriott, 200, 140, 17200, 300, 150

Thos. H. Ditty, 140, 100, 4000, 200, 800

Lemuel Bird, 100, 10, 5000, 100, 600

Susan Boyle, 248, 12, 12400, 300, 1225

John Holliday, 150, 27, 8850, 300, 875

Saml. A. Hamilton, 826, 240, 67650, 1000, 1800

B. W. Marriott, 388, 200, 23800, 400, 1060

J. W. Bird, 162, 20, 10920, 300, 800

J. W. Bird Jr., 300, 125, 20000, 520, 1600

H. M. Dinall, 200, 63, 15780, 500, 1300

Wm. H. Tucker, 300, 150, 22500, 400, 2200

Chas. W Stockett, 200, 77, 19390, 200, 1300

Wm. Grimes Ten, 150, 100, 10500, 100, 550

Wm. P. Bird, 23, 12, 2100, 50, 300

James J. Gray, 200, 180, 22800, 300, 1987

John B. Owens, 175, 70, 12500, 300, 15000

Denis Orme, 125, 55, 9000, 200, 700

Marcia O'Hara, 175, 40, 7950, 200, 900

John Collinson, 300, 149, 22550, 500, 1570

R. Weims, 295, 59, 23280, 652, 1460

Eduard Collinson, 100, 33, 6650, 250, 676

Ed. Lee, 100, 36, 4800, 50, 700

V. U. Brewer, 100, 50, 6000, 40, 532

Sarah A. Brewer, 100, 50, 6000, 50, 477

Margaret Purdy, 75, 58, 2800, 25, 590

Thos. Purdy, 100, 44, 3500, 50, 600

Thos. Purdy, 100, 213, 2145, 50, -

R. P. Davis (mgr), 200, 60, 13000, 300, 162

A. J. Nichols, 150, 60, 6000, 25, 400

Wm. H. Dawson, 100, 25, 5000, 500, -

Joseph T. Dawson, 100, 25, 5000, -, -

Thos. E. Dawson, 200, 80, 6000, -, 380

James Harris Ten, 170, 30, 6000, 50, 416

John H. Linthicomb Ten, 100, 38, 4140, 50, 600

Wm. A. Journey Ten, 90, 70, 4800, 25, 570

Frank R. Brashears, 180, 40, 11000, 400, 1150

Kelly Mangum Ten, 100, 20, 600, 50, 625

Isaac Mayo, 400, 200, 36000, 2500, 2600

Richard Sellman, 300, 78, 18700, 400, 2355

Richard Sellman, 240, 46, 17160, 400, 1710

Richard Contee, 270, 114, 23000, 500, 1870

Charles Contee, 270, 90, 21600, 500, 720

B. O'Hara, 80, 16, 5760, 500, 1650

Matilda O'Hara, 100, 30, 5520, -, -

M. E. S. O'Hara, 100, 300, 6550, -, -

Virginia O'Hara, 100, 30, 6650, -, -

Charles Watkins, 25, -, 500, 20, 165

Anthony Baston(?), 25, -, 500, 30, -

Francis Waters, 240, 110, 21000, 1000, 194

Joseph Busey, 300, 130, 18360, 600, 1530

Wm. H. Busey, 140, 60, 8000, 200, 680

R. S. Mercer, 200, 70, 19320, 1000, 1300

H. M. Murray Ten, 450, 247, 41800, 250, 2680

Mary S. Mercer, 500, 300, 48000, 1000, 2250

Charles A. Stewart, 200, 147, 20840, 700, 1760

T. R. Anspach, 600, 200, 36000, 1000, 2848

Thos. Bird, 75, 25, 500, 50, 550

Rinaldo Pindel, 200, 150, 24500, 300, 165

Thos. Richardson, 175, 90, 18200, 400, 1220

Thos. H. Hopkins, 110, 29, 8340, 300, 1014

Gassaway Nintuser, 120, 45, 9900, 300, 1110

Benjamin Brown, 6, -, 150, 25, 100

Dolphe Rogers, 90, 34, 8750, 100, 523

Murray Thomas, 250, 150, 28000, 500, 2300

B. A. Welch, 225, 95, 22400, 800, 1705

Julia Carr, 150, 50, 10000, 300, 875

Henry Carr, 175, 63, 8750, 200, 695

Geo. B. Steuart, 200, 178, 26460, 500, 1180

Benj. Bird, 175, 35, 14700, 600, 1263

Eleonir C. Owens, 140, 60, 14000, 200, 1054

Ann Murray, 125, 35, 18200, 300, 963

Nathaniel Chew, 100, 63, 8150, 200, 488

Thos. T. Bird, 80, 97, 7850, 150, 1036

E. Burnell, 150, 25, 10500, 800, 1190

John Carr, 200, 60, 18200, 600, 1400

Henry Basford, 200, 100, 5000, 75, 420

Albertus Basford, 11, 11, 600, -, 150

N. C. Bowen, 125, 25, 7500, 100, 70

Nicholas Owens, 175, 60, 14000, 300, 1078

Ben. Hall Ten, 350, 50, 24000, 500, 1490

D. Sisk Ten, 90, 40, 7800, 40, 325

Wm. Sheppard, 175, 58, 14900, 400, 400

A. G. W. Owens, 100, 48, 7400, 50, -

Frank Deal, 210, 78, 17000, 600, 1000

Joseph Nicholson, 200, 11, 4000, 125, 500

R. F. Higgins, 150, 170, 1500, 100, 400

James Sanford, 225, 50, 6000, 200, 1520

Joshua Higgins, 160, 70, 2300, -, 400

F. Dunall, 350, 150, 10000, 500, 2000

John Linthicomb, 180, 110, 10000, 300, 1000

Theodore Linthicomb, 250, 65, 6000, 150, 1100

R. Anderson, 193, 3, 8000, 325, 1937

Saml. Hopkins, 350, 97, 13000, -, -,

B. D. Hal, 350, 207, 24000, 420, 1690

T. W. Hammond, 160, 20, 8000, 200, 900

Ann Barber, 162, 50, 5000, 75, 375

M. W. Durall, 250, 100, 16000, 275, 1300

Chas. V. Hammond, 170, 61, 11000, 450, 1140

Thos. Crandle, 65, 65, 4000, 50, 200

Jas. F. Balderson, 110, 20, 3000, 100, 400

Wm. H. Baldwin, 200, 170, 15000, 40, 500

R. Baldwin, 130, 56, 8000, 200, 800

John H. Brown, 70, 83, 1200, 400, 300

Matilda Savage, 60, 30, 1200, 50, 100

Thos. E. Watson, 90, 83, 2500, -, 245

Rachael Woodard, 2, -, 800, -, 75

William Rich, 150, 114, 600, 450, 436

R. D. Woodward, 350, 70, 16000, 600, 1800

B. Williams, 2, -, 300, -, 125

Ann Ganett, 45, 22, 800, 50, 25

Ann Hammond, 80, 10, 2500, 50, 150

Eliza Sewell, 70, 30, 2500, -, 100

Samuel Chaney, 125, 46, 3500, 250, 335

A. Jenkins, -, -, -, 20, 310

Wm. H. Carter, 175, 93, 10000, 137, 500

R. M. Dodson, 80, 70, 4500, 50, 400

R. H. Carr, 200, 50, 7000, 100, 150

P. Whitnight, 350, 350, 14000, 70, 785

George Cutler, 150, 250, 8000, -, -
Wm. E. Anderson, 423, 315, 14000, 200, 1000
John H. Jones, 100, 66, 4000, 100, 600
Wm. Hawkins, 60, 150, 2500, 100, 240
J. Cadel, 150, 50, 4000, 100, 450
R. Hodges, 120, 27, 5000, 225, 750
J. Foard, 60, 144, 2700, 50, 321
Eleanor Cooksey, 200, 70, 2500, 50, 600
Amelia Linthicomb, 190, 70, 5000, 75, 700
E. Magruder, 275, 205, 9600, 175, 1600
R. Hopkins, 200, 50, 10000, 90, 600
G. L. Hopkins, 100, 80, 4600, 110, 700
Thos. Linthicomb, 200, 62, 7800, 228, 1000
M. Linthicomb, 100, 10, 6000, 75, 160
Joshua Carter, 45, 23, 1700, -, 40
Dr. B. Watkins, 434, 153, 37000, 400, 1100
Thomas Welch, 460, 120, 20000, 700, 2200
R. Welsh, 200, 175, 9000, 300, 1100
H. Gaither, 70, 56, 1900, 30, 250
John Clagett, 100, 125, 9000, 100, 650
Wm. H. Turtor, 350, 250, 15000, 240, 1700
C. A. Waters, 200, 187, 7000, 350, 1500
J. W. Gray, 150, 75, 6000, 125, 650
C. Seigler, 15, 7, 440, 25, 100
Joseph F. Jones, 125, 8, 33000, 290, 685
Mary A. Sevier, 150, 163, 6000, 75, 382
B. J. Worthington, 150, 65, 8000, 150, 675
C. F. Worthington, 1200, 70, 10000, 150, 700

E. Lee Lusby, 250, 276, 16000, 200, 11000
B. Lusby, 160, 195, 10000, 200, 1000
D. Durall, 200, 95, 10000, 325, 1400
W. Anderson, 90, 24, 1800, 100, 1100
Geo. D. Clayton, 225, 135, 11000, 400, 1000
Armistead Owens, 100, 40, 8000, 50, 660
H. F. Owens, 100, 147, 6000, 200, 500
Johnathan Pinkney, 217, 20, 20000, 300, 6000
L. S. Hazzard, 175, 55, 10000, 175, 900
Chas. Hodges, 230, 120, 9000, 250, 1600
A. Clagett, 160, 225, 7000, 200, 935
John B. Fulton, 125, 25, 7000, 75, 800
T. R. Beard, 350, 293, 16100, 340, 1700
A. A. Hall, 400, 105, 20000, 500, 1900
John Tucker, 400, 150, 15000, 1000, 2000
Wm Prather, 180, 184, 7000, 300, 840
J. M. Durall, 150, 100, 7000, 200, 500
A. G. Walsh, 250, 150, 12000, 500, 800
G. S. Welch, 175, 165, 10000, 350, 728
Joseph Fowler, 500, 300, 16000, 250, 1300
E. R. Dorsey, 600, 350, 20000, 150, 1300
D. D. Kent, 160, 17, 7000, 400, 845
J. T. Hughes, 700, 125, 37000, 2000, 2700
J. H. Hodges, 200, 24, 13000, 1500, 1600

Stephen Beard, 275, 165, 13000, 200, 1100

John Beard, 200, 53, 8000, 500, 1800

Louisa Howard, 140, 134, 5000, -, -

Peter Miller, 125, 19, 3000, 200, 485

Wm. P. Paranett, 200, 40, 8000, 200, 875

T. J. Fongue, 200, 144, 13000, 200, 1500

H. H. Brown, 400, 200, 15000, 600, 1189

Maria Waters, 100, 100, 5000, -, -

Luther Giddings, 250, 300, -, 500, 1000

Judson Clark, 214, 148, 3000, 50, 375

R. Sellman, 100, 100, 8000, 200, 400

B. S. Nichols, 230, 100, 4000, 100, 500

John Bryan, 22, -, 800, -, 125

N. B. Worthington, 125, 75, 6000, 150, 450

J. Drelson, 50, 93, 3000, 75, 300

B. Meade, 90, 30, 3000, 100, 650

Julia Iglehart, 250, 250, 12000, 200, 700

S. E. Durall, 100, 58, 4000, 100, 135

T. Seifles, 70, -, 700, 10, 30

Ezra Sheckles, 280, 40, 7000, 400, 900

S. E. Carr, 100, 12, 2000, 100, 600

W. Brashears, -, -, 100, 50, 250

Heirs of Denton Durall, 100, 20, 2500, -, -

D. Claude (Cloud) Jr., 180, 95, 10000, 300, 400

J. W. Allen, 200, 141, 14000, 200, 1200

Wm. Wells, 120, 35, 3000, 50, 430

Heirs of W. Rawlings, 133, 130, 3500, 50, 425

Wm. Rawlings, 125, 60, 3700, -, -

P. Rawlings, 140, 75, 3500, -, -

L. H. Scott, 60, 147, 2500, -, -

Roger Tydings, 25, -, 60, -, -

Richard Tydings, 25, -, 600, -, -

Chapman Harwood, 350, 150, 25000, 1000, 1200

Frank H. Stockett, 250, 168, 10000, 500, 1310

Thos. H. Ceney, 100, 32, 8000, 100, 375

Mrs. Kent, 150, 20, 3000, -, -

A. Smith, 150, 75, 2000, -, 400

John Walter (Walten), 200, 60, 26000, 200, 1100

George Wells, 650, 1400, 4000, 500, 3200

E. C. Taylor, 100, 47, 5000, 200, 400

J. N. Steuart, 100, 50, 4000, -, 75

D. Hart, 300, 63, 8000, 400, 1200

R. McKublin, 150, 30, 15000, 2000, 1000

Lieut Wood, 140, 20, 8000, -, -

Barbers Legatus, 200,80, 9000, -, -

Estate of N. J. Worthington, 230, 25, 700, -, -

Heirs of Wm. Murdock, 75, 101, 7000, -, -

Estate of D. Claude (Cloud), 200, 200, 16000, 200, 1000

Hammond Cranston, 150, 113, 70900, -, -

J. R. Houison (Harison), 180, 308, 10000, 300, 1000

C. M. Ray (Roy), 100, 175, 4000, -, -

Henry Hopkins, 60, 75, 3000, 50, 200

T. S. Lockerman, 125, 45, 5000, -, -

Henry Bush, 71, -, 2800, 50, 200

George Franklin, 100, 125, 6700, 100, 400

R. W. Higgins, 150, 100, 2500, 100, 400

Wm. Watson Ten, 260, 44, 1200, 100, 400

Albert Cooksey, 150, 33, 10000, 75, 500

Joshua Linthicomb, 200, 100, 21000, 850, 1850

Alfred Parker, 60, 20, 1600, 50, -

_. J. Linthicomb, 150, 300, 13500, -, -

Landon Bakney (Rakney), 125, 75, 4000, 100, 200

Saml. Purdy, 100, 100, 6000, 40, 500

Saml. Anderson, 140, 135, 5000, 150, 600

Absalom Anderson, 250, 50, 9000, 150, 750

Rezin Saithers (Gaithers), 150, 150, 9000, 50, -

H. Ridout, 150, 50, 6000, 150, 1125

Thos. Arnold, 100, 200, 5000, 150, 750

Rebecca Arnold, 300, 200, 7500, 150, 500

J. W. Bourke, 140, 60, 2800, 50, 400

W. T. Derall, 50, 50, 800, 20, 250

John Ridout, 136, 64, 3000, 150, 1000

W. D. Basie, 60, 40, 3000, 100,3 50

Lem. Bourke, 100, 49, 3000, 100, 500

J. Bourke, 100, 85, 3000, 100, 500

Wm. V. Stinchcomb, 130, 70, 4000, 100, 490

T. Baldwin, 50, 27, 1000, 40, 110

L. Wheeler (mgr.), 70, 30, 5000, 100, 400

J. Iglehart Jr., 350, 150, 20000, 400, 1500

Susan Brice, 175, 125, 6000, 100, 850

John Miller, 150, 70, 4100, 100, 1000

J. Winchester, 350, 250, 18000, 200, 1900

M. Phipps, 30, 10, 600, 23, 150

Z. Durall, 100, 20, 4000, 50, 1300

E. Merrikin, 400, 100, 15000, 150, 1200

J. Ridout of H., 383, 320, 5000, 1000, 2800

J. C. Williamson, 50, -, 3000, 50, 650

J. Ridout (agt.) 50, 10, 3000, -, -

John Boone, 40, 10, 2000, 200, 495

P. Pettibone Jr., 100, 100, 3300, 100, 500

S. R. Richardson, 170, 80, 9000, 200, 1300

M. Williams, 100, 25, 2500, -, -

Julia Kent, 600, 300, 18000, 200, 1500

H. M. Zuell (Buell), 125, 25, 4000, 150, 950

Ann Rideout, 400, 350, 30000, 500, 1750

R. E. Moss, 225, 75, 10000, 250, 900

W. S. Weedon, 200, 100, 12000, 200, 1400

B. Mezzick, 250, 206, 17000, 300, 1900

J. T. Hyde, 200, 25, 7000, 200, 900

W. Ridout, 120, 26, 6000, 50, 875

P. Pettibone, 250, 50, 12000, 230, 1000

W. Stinchcomb, 240, 100, 4000, 100, 875

R. Adams (agt.), 33, -, 1000, 50, 175

A. Stinchcomb, 130, 30, 1700, 30,875

H. Tydings, 150, 148, 2500, 100, 600

J. B. Jones (agt.), 150, 150, 3600, 75, 400

T. Hoge, 300, 80, 11400, 300, 2000

Jas. Hoge, 100, 18, 2400, 50, 625

A. S. Cook, 33, -, 1200, 50, 350

S. T. Redgraves, 222, 222, 5000, 100, 1000

Chas. Boone, 150, 37, 4000, 100, 850

Jas. Freeman, 150, 150, 3600, 150, 200

Joshua Brown, 80, 40, 3600, 150, 216

Brown & Martin, 385, -, 7700, 300, 400

Sam Lucas, 100, 100, 6000, 150, 350

J. B. Nichols, 170, 66, 11800, 300, 1125

Wm. Hammond, 100, 4, 2100, 125, 400

Mace Hawkins, 16, -, 300, 40, 100

Isaac Pully, 55, -, 1000, 100, 200

J. J. Fate, 28, -, 1400, 60, 100

Eliza Fate, 80, 20, 3000, 50, 100

C. Rocklitz, 6, -, 125, -, 125

Eliza Kerby, 70, 30, 1000, 75, 250

T. Sherbert, 75, 75, 1200, 75, 300

Wm. Jones, 50, 100, 1800, 73, 225

Jas. Spriggs, 200, 212, 4000, 150, 600

Sally Saffle, 120, 20, 2000, 100, 325

Thos. Tolson, 150, 50, 4000, 75, 400

Wm. Kilbern, 80, 20, 200, 80, 225

John Pettibone, 80, 45, 4000, 175, 850

J. S. Wilson, 250, 15, 10000, 200, 650

E. Rockhold, 200, 128, 6500, 100, 550

E. R. Rockhold, 40, 110, 1600, 50, 350

John Hoge, 110, 49, 3000, 75, 495

W. Pumphrey, 250, 84, 2500, 100, 220

Levi Sheckell, 60, 44, 832, 50, 225

W. H. Freeman, 75, 75, 1500, 100, 500

E. R. Arnold, 200, 80, 14000, 150, 850

T. Tydings, 90, 153, 2500, 75, 350

Thos. Stinchcomb, 80, 20, 800, 50, 175

J. H. Stallings, 100, 31, 2000, 75, 325

R. Steuart, 10, -, 200, 20, 50

M. A. Brown, 100, 27, 1300, 75, 150

H. Tydings, 170, 130, 3000, 100, 850

C. E. Gray, 100, 25, 300, 40, 275

R. Freeman, 90, 25, 300, 30, 150

S. Brittenhouse, 3, -, 150, 15, 75

Ann Stallings, 180, 20, 2400, 75, 500

Owen Johnson, 50, 20, 800, 40, 100

J. Robinson, 300, 500, 16000, 200, 750

G. R. Hillyard, 150, 150, 3600, 100, 538

J. J. Johnson, 60, 40, 700, 40, 150

C. R. Steuart, 250, 1350, 13000, 300, 1100

C. G. Pumphrey, 70, 110, 1500, 75, 300

L. N. Hill, 40, 70, 850, 50, 750

C. Rhoades, 120, 30, 4500, 200, 351

C. C. Prather, 30, 20, 350, 40, 125

R. Prather, 150, 627, 4500, 100, 260

Thos. Drdem, 40, -, 200, 30, 200

G. W. Wilson, 70, 30, 1000, 40, 100

Thos. Wade, 40, 20, 420, 50, 200

Jas. Menshaw, 30, 70, 500, 40, 150

P. Holland, 100, 100, 2000, 50, 200

D. Menshaw, 40, -, 200, 20, 100

D. M. Benson, 27, -, 100, -, 200

A. Mullican, 150, 150, 2000, -, -

A. Robinson, 275, 700, 14500, 400, 700

O. Pumphrey, 200, 150, 4500, 150, 300

J. Pumphrey, 200, 700, 9000, 150, 900

C. Robinson, 150, 300, 4500, 100, 500

D. R. Williams, 300, 200, 8000, 200, 600

J. Johnson, 200, 200, 2500, 100, 400

A. Johnson, 400, 400, 5000, 2500, 800

Susan Koflinsk, 100, 178, 1600, 50, 200

Wm. Burton, 100, 34, 1400, 40, 250

Wm. Cobbin, 60, 20, 800, 28, 120

J. H. Lark, 100, 87, 1200, 75, 325

Zach Gray, 60, 90, 750, 50, 150

J. W. Hunter, 130, 80, 4200, 125, 850

H. Joice, 450, 100, 8250, 400, 600

T. Boone, 60, 31, 1800, 100, 400

M. Linstead, 140, 44, 3600, 100, 250

R. Tydings, 50, 79, 2000, 100, 240

T. W. Williams, 230, 270, 3500, 150, 600

J. E. Williams, 400, 800, 12000, 300, 850

George Crane, 70, 650, 8400, -, 300

A. Franklin, 16, 800, 160, 50, 150

L. Freeman, 150, 630, 2500, 150, 350

Z. Pumphrey, 100, 100, 1800, 50, 125

E. Robinson, 75, 75, 1100, 50, 225

A. Osborne, 100, 80, 1080, 40, 80

L. Linstead, 200, 500, 4900, 200, 500

P. Blunt, 100, 170, 1700, 100, 250

J. D. Jacobs, 100, 180, 1700, 100, 51

F. Brown, 150, 250, 2400, 75, 300

C. Linstead, 50, 230, 1700, 40, 125

W. Robinson, 100, 180, 1700, 75, 250

S. Lark, 160, 173, 3300, 150, 750

E. A. Heath, 75, 139, 2200, 100, 400

W. T. Armiger, 50, 50, 1000, 50, 200

A. Johnson, 150, 121, 5500, 100, 400

R. Fowler, 100, 200, 1800, 75, 550

J. Henshaw, 60, 40, 700, 50, 224

W. Linthicomb, 150, 300, 4500, 150, 350

W. Stall, 150, 180, 7000, 150, 750

W. H. Cook, 75, 125, 2000, 150, 400

H. Dunbar, 300, 60, 10800, 300, 1050

J. L. Webster, 100, 63, 1600, 100, 400

M. S. Welch, 100, 196, 4400, 450, 600

Ed Hines, 100, 96, 1200, 100, 300

R. Chord, 100, 127, 1400, 100, 275

J. H. Meckins, 75, 76, 1500, 100, 280

J. H. Barnes, 100, 163, 3150, 150, 350

G. W. McCublin, 60, 120, 1800, 150, 250

J. Meckins, 40, -, 1000, 50, 150

J. Arminger, 200, 220, 5000, 150, 450

D. Jacob, 100, 83, 4000, 100, 350

L. Freeman, 80, 100, 3000, -, 150

O. Williams, 180, 160, 6820, 160, 225

John Jabb, 150, 100, 2500, 150, 125

S. Gray, 100, 100, 1600, 100, 350

A. Winfield, 80, -, 800, 100, 200

H. F. Fowler, 120, 30, 1600, 100, 400

J. Fowler, 120, 30, 1700, 100, 250

E. S. Powell, 200, 100, 1600, 50, 250

Jos. Osborn, 9, 22, 1500, 100, 150

Saml. Lynch, 520, 600, 20000, 200, 1100

Henry Chairs, 75, 25, 1500, 10, 150

Ed. Chairs, 100, 100, 2000, 100, 300

J. L. Tydings, 110, 10, 2400, 125, 375

W. R. Osborne, 100, 200, 3000, 100, 250

F. F. Hancock, 135, 100, 2500, 200, 475

L. W. Hancock, 350, 650, 10000, 150, 150

Wm. Marrs, 30, -, 360, 50, 80

Chas. Stepney, 20, -, 240, 40, 75

Jas. Cole, 70, 30, 2000, 110, 275

Jas. Daiger, 70, 30, 2000, 100, 600

Benj. Osborne, 60, 130, 2300, 150, 250

H. A. Hancock, 80, 20, 1200, 140, 225

Wm. D. Chord, 60, 140, 1600, 50, 100

John Boyd, 60, 40, 1600, 100, 300

Richd. Todd, 40, -, 1000, 75, 150

H. B. Wood, 160, 200, 2880, 100, 125

Phil Wales, 130, 295, 4500, 100, 300

L. Foreman, 100, 100, 1600, 95, 200

H. Johnson, 200, 480, 10880, 100, 250

John Peters, 12, -, 500, 50, 125

Robt. Pringlass, 6, -, 500, 50, 25

Elizabeth Hancock, 275, 200, 9000, 150, 400

Benj. Young, 130, 100, 2800, 150, 450

Lemuel Hancock, 200, 325, 5200, 200, 500

J. B. Turner, 87, -, 1000, 50, 200

Thos. Sollers, 200, 450, 6500, 100, 500

J. K. Garey, 50, 100, 3000, 75, 200

Jos. Gray, 75, 125, 2000, 100, 350

Jos. Wright, 60, 40, 1000, 50, 150

J. Zealdhall, 125, 175, 3600, 150, 500

H. R. Watson, 95, 65, 2210, 100, 400

R. J. Robinson, 80, 240, 3200, 100, 200

F. Chairs, 100, 135, 2800, 150, 300

G. Simpton, 200, 400, 1200, 300, 700

R. Hall, 130, 40, 2500, 150, 250

Jos. Hodges, 250, 410, 6600, 300, 475

Ben. Brown, 25, -, 250, 25, 200

Jos. Spencer, 60, 20, 600, 50, 250

Wm. Howard, 27, -, 175, 25, 80

E. Franklin, 30, -, 250, 30, 70

A. Franklin, 24, -, 210, 28, 125

Robt. Curry, 24, -, 205, 26, 100

Nathan Owens, 60, 10, 500, 50, 200

N. Brewer, 120, 20, 10000, 100, 200

Sarah Elder, 40, 10, 1000, 50, 120

J. Clark, 100, 105, 2000, 50, 200

B. F. Clark, 40, 80, 1000, 25, 100

B. Clark, 100, 180, 2500, 50, 100

H. C. Gray, 40, 55, 2000, 50, 400

H. Baldwin, 30, 70, 1000, 20, 100

Al___ House, 80, 70, 8000, 50, 100

Enos, Shepley, 40, 30, 1500, 100, 300

Larkin Shepley, 15, 35, 1000, 50, 100

J. M. Webb, 100, 364, 5000, 100, -,

J. J. Clark, 80, 50, 500, 25, 100

R. Miller, 60, 20, 2000, 25, 200

Jno. Loroman, 150, 70, 5000, 100, 600

Eliza Jackson, 40, 60, 2000, 50, 100

Wm. Turner, 50, 62, 1200, 25, 100

Eliza Bryan, 100, 20, 1200, 50, 200

T. S. Disney, 100, 200, 3000, 100, 400

A. H. Disney, 50, 150, 2000, 50, 150

Saml. Bordler, 81, 31, 1200, 25, 100

John Raul, 80, 50, 1000, 50, 200

Jos. Clark, 30, 20, 600, 25, 200

Jesse A. Watts, 90, 155, 2000, 50, 100

Saml. Goodwin, 70, 40, 1200, 100, 150

Saml. Watts, 80, 5, 1000, 50, 200

Jno. Gardner, 150, 70, 5000, 100, 400

T. J. Webster, 60, 70, 5000, 50, 300

Julia Jenkins, 30, 20, 1000, 50, 50

Geo. Gardner, 125, 100, 4500, 100, 300

Wm. Simpson, 50, 35, 6000, 100, 400

Benj. Ray (Roy), 150, 250, 8000, 100, 400

Jno. Boyer, 50, 25, 700, 25, 200

Matilda Pope, 20, 23, 4000, 25, 100

Joshua Owens, 150, 50, 1000, 50, 100

Phil Disney, 150, 50, 1000, 100, 400

R. Jones, 50, 100, 1500, 50, 200

Wm. Disney, 75, 73, 1500, 25, 200

Jno. H. Herman, 100, 40, 1500, 50, 152

Ma__ Disney, 50, 50, 1000, 25, 300

R. Gainer, 100, 132, 5000, 50, 500

_. McGill, 20, 130, 1500, -, -

R. Isaacs, 150, 176, 9000, 50, 500

J. Turner, 100, 112, 7000, 100, 500

R. Hart, 60, 30, 900, 10, 100

J. Harman, 130, 80, 1200, 40, 225

Wm. T. Anderson, 100, 300, 3000, 25, 150

Wm. Philips, 70, 90, 1000, 25, 100

Wm. Anderson, 31, -, 900, 50, 100

Jno. Redmiles, 60, 41, 1000, 20, 150

T. W. Cormer (Conner), 70, 55, 3000, 50, 250

R. Donaldson, 75, 37, 900, 20, 100
Nathan Jones, 70, 130, 1400, 25, 120
Thomas Duvall, 150, 350, 12000, 100, 400
R. Jones, 50, 22, 600, -, -
C. Donaldson, 40, 44, 800, 25, 200
N. Naylor, 75, 128, 3600, 25, 500
J. Shucks, 30, 20, 400, 20, 100
Noah Donaldson, 100, 12, 2000, 50, 100
Jno. Spears, 15, 50, 4000, 50, 500
Jno. Mullineau, 50, 50, 2000, 40, 250
Jas. Hurles, 80, 20, 600, 30, 280
M. E. Hurles, 75, 75, 1500, 25, 125
Reson Snowden, 250, 650, 12000, 350, 1000
E. S. Lee, 27, 130, 1200, 50, 200
Walter Smith, 600, 515, 40000, 500, 1000
S. Middleton, 90, 30, 5000, 100, 400
Wm. Merser, 200, 300, 5000, 50, 400
A. Merser, 20, 200, 2000, -, -
J. Lloyd, 100, 400, 5000, 50, 250
L. Welsh, 140, 71, 4000, 100, 700
N. Wells, 150, 25, 3000, 100, 250
J. H. Waters, 15, 75, 4000, 100, 500
Victor Shales, 150, 450, 18000, 50, 250
Luke Shields, 60, 40, 800, 25, 150
Geo. Jacobs, 300, 200, 5000, 100, 600
E. Hardon, 100, 25, 2500, 50, 420
J. Beallmear, 75, 25, 1000, 50, 250
A. Beallmear, 150, 180, 3500, 50, 600
R. C. Hardesty, 250, 110, 12000, 40, 1000
{Joseph Coroman, 215, 52, 5000, 200, 900
{J. J. Coroman, 200, 200, 500, -, -
Frank Sheckels, 180, 100, 10000, 100, 400
R. Anderson, 140, 40, 3600, 15, 250
Beall Duvall, 300, 200, 6000, 100, 800

Marshal Hopkins, 120, 50, 3000, 100, 600
Isaac Chaney, 100, 300, 4000, 50, 400
Lot Mayfield (Marfield), 400, 108, 20000, 500, 1200
Osbourn Belt, 80, 115, 2000, 50, 200
William Jones, 60, 8, 3500, 50, 400
Geo. Hammdon, 80, 113, 3000, 50, 400
James Rawlings, 100, 46, 4000, 50, 300
Jane Sappington, 3, 3, 600, 25, 200
R. G. Morgan, 30, 62, 1400, 25, 250
James Parker, 100, 600, 4000, 50, 400
Rachael Watts, 60, 25, 600, 10, 100
Robt. Humphrey, 6, -, 120, 10, 100
Wm. Wall, 20, 25, 500, 20, 150
Joshua Stewart, 75, 300, 3000, 50, -
Wm. Mewshaw(Menshaw), 30, -, 300, 10, 100
Rufus Benson, 107, 20, 8000, 100, 600
Rufus Benson, 80, 80, -, -, -
Rufus Benson, 100, -, -, -, -
Rufus Benson, 70, -, -, -, -
James Martin, 25, 75, 1500, 25, 150
Philip Myers, 40, 20, 500, 40, -
John V. Gray, 30, 57, 800, 10, 400
Othias Mackber, 75, 180, 2000, 50, 225
Waters Beallmore, 100, 117, 1500, 150, 300
Charles Beallmore (Beallman), 100, 112, 2200, 90, 400
_. F. White, 250, 139, 4600, 300, 800
R. Phelps, 50, 150, 2000, 20, 100
B. C. Mullikin, 330, 330, 13000, 400, 1000
Thomas Queen, 25, 50, 300, -, -
S. B. Talbott, 225, 275, 10000, 270, 700
A. G. Woodward, 200, 140, 6800, 300, 750

Wm. T. Anderson, 275, 150, 8500, 200, 700
P. D. Carr, 150, 38, 4700, 30, 400
George Miller, 28, -, 5000, 50, 400
Jno. W. Waugh, 40, -, 2000, -, -
Eleanor Aisquith, 175, 75, 10000, 50, 400
R. G. Duvall, 200, 160, 10800, 500, 600
George McNeal, 100, 200, 400, 25, 200
Samuel Smith, 5, 45, 40, 10, -
N. Day, 175, 51, 2000, 50, 250
Elizabeth Gambrill, 100, 140, 7000, 50, 300
Henry Saffles, 100, 70, 2500, 50, 400
Mary A. Hammond, 150, 558, 10000, 100, 500
Julia A. Hammond, 100, 250, 7000, 75, 400
Cecilia Hammond, 75, 171, 4800, -, 200
John Howard, 50, 170, -, -, -
Catherine Change, 150, 250, 4000, 50, 200
Ann S. Kirby, 75, 84, 1500, 25, 200
John R. Bell, 100, 150, 5000, 25, 200
P. Newborn, 300, 300, 9000, 100, 400
Joseph Hopkins, 75, 38, 2000, 20, 150
E. J. Butler, 100, 70, 3000, 100, 600
R. Webb, 150, 110, 8000, 50, 200
S. Pinkney, 200, 160, 6600, 50, 300
N. Wheat, 80, 137, 2000, 40, 300
D. Griffith, 80, 78, 2000, 25, 150
J. Wheat, 200, 200, 4000, 100, 400
G. Valentine, 20, -, 200, -, 125
Thomas Upton, 50, 190, 2000, 40, 300
Charles Griffith, 100, 152, 2500, 20, 150
H. Mojers (Majors), 50, 108, 600, -, -
W. Upton, 25, 25, 250, 25, 100
Chas. Wood, 20, 5, 250, 50, 125

John Wood, 25, 25, 600, 50, 100
John Hatter (Hatten), 50, 100, 900, 25, 200
Louis Wood, 20, 20, 300, 30, 100
G. Sappington, 100, 32, 2500, 100, 300
W. Queen, 9, -, 200, 25, 100
Seth Bonads, 25, 23, 400, 50, 200
James Dumer, 20, 8, 200, 20, 160
Bailey Smith, 300, 300, 12000, 100, 300
Ezekiel Oliver, 20, 6, 400, 25, 125
John Scott, 26, -, 400, 25, 100
Mary Benson, 60, 98, 2000, 50, 150
M. Stinchcomb, 100, 150, 4000, 100, 350
P. Newburn, 80, 30, 1500, 50, 100
Hickman Morris, 160, 20, 2000, 100, 400
A. M. Bennet, 125, 175, 5000, 100, 600
Wm. Bennet, 25, 25 500, 25, 100
J. K. Bennet, 20, 15, 350, 50, 100
J. McMullan, 6, -, 400, 40, 100
M. Shoals, 15, 2, 400, 50, 150
E. S. Kilbourn, 250, 200, 20000, 1000, 2000
J. W Biggs, 30, 47, 4000, 50, 300
John Williams, 150, 410, 1000, 100, 600
Thos. Whitle, 20, -, -, -, -
Wm. Harrison, 75, 95, 12000, 200, 200
Noah Dorsey, 100, 227, 10000, 100, 1000
W. Shaler, 520, 150, 30000, 200, 6000
R. Jacobs, 150, 95, 5000, 100, 300
Mathias Shaw, 75, 25, 2000, 100, 300
Anthony Murry, 100, 100, 2000, 75, 300
Wm. Downs, 150, 100, 3500, 100, 400
D. M. F. Stewart, 350, 50, 16000, 300, 1000

Eli Gaither, 160, 45, 800, 100, 400
Zadock Duvall, 225, 225, 4000, 50, 400
Wm. Disney, 75, 16, 1500, 50, 310
W. Donaldson, 250, 88, 6000, 100, 400
T. R. Woodward, 150, 50, 1000, 100, 400
Ellen Henderson, 100, 70, 500, 20, 300
Edward Clark, 70, 35, 1200, 50, 400
George Thomas, 50, 38, 1000, 50, 200
W. J. Sappington, 150, 80, 2000, 100, 400
Mathias Lourmer, 30, 26, 300, 50, 150
Thos. Chaney, 13, -, 200, 20, 50
Nathan Watts, 125, 50, 5000, -, 200
Thos. Phelps, 60, 140, 1800, 50, 400
Saml. Shoemaker, 4, -, 30, 30, 100
H. H. Turner, 58, 20, 600, 50, 100
E. Turner, 60, 40, 8000, 100, 200
J. W. Turner, 60, 50, 600, 50, 200
A. Warfield, 50, 50, 700, 40, 200
S. R. Gaither, 75, 175, 2500, 100, 200
Ann Warfield, 100, 75, 2500, 100, 400
Chas. Disney, 40, 20, 800, 50, 200
John Jackson, 80, 20, 1000, 100, 300
N. Needhopper, 120, 20, 1700, 50, -
L. Manold (Manald), 10, -, 100, 20, 100
J. H. Gairy, 70, 40, 1200, 50, 200
J. C. Biggs, 150, 60, 4500, 100, 500
W. W. Elliott, 70, 12, 500, 50, 300
J. A. Bruel, 75, 47, 6000, 100, 400
W. L. Gardner, 75, 57, 2000, 50, 200
R. Harman, 75, 35, 2000, 100, 300
G. Diggens, 25, 25, 500, 40, 200
Wm. Bond, 200, 120, 16000, 150, 700
Elias Harman, 60, 20, 2500, 100, 400
Wm. Jenkins, 200, 200, 4000, 150, 400

Jos. Clark, 100, 50, 2000, 50, 300
A. L. Lynthicum, 90, 20, 10000, 200, 400
Wm. Anderson, 500, 162, 25000, 1500, 1500
Amos Clark, 50, 80, 6000, 100, 100
Wm. Gardner, 80, 40, 4000, 50, 250
Chas. Hance, 30, 10, 1000, 25, 100
Denton Hance, 60, 40, 3000, 100, 300
Chas. Warfield, 125, 25, 3000, 40, 150
R. Owens, 100, 40, 1000, 50, 100
F. Disney, 140, 48, 3000, 200, 400
H. Fairall (Farrall), 10, 40, 500, 50, 200
Saml. Watts, 100, 30, 1300, 100, 200
Sarah Elder, 50, 50, 2000, 50, 50
Susan Anderson, 75, 37, 1200, 50, 200
Chas. Duvall, 300, 228, 20000, 200, 800
Emma Dorsey, 50, 16, 3000, 50, 200
N. P. Benson, 75, -, 1500, 75, 200
D. Wesley, 197, -, 1600, 75, 200
James Phelps, 300, 750, 15735, 300, 600
Giles Cromwell, 10, 23, 350, 75, 150
Thos. Baldwin, 100, 62, 3300, 300, 400
H. R. Banker (Barker), 30, -, 450, -, -
Aaron Hawkins, 120, 30, 6000, 400, 775
T. H. Merritt, 50, 250, 14000, 200, 60
L. Thielman, 20, 100, 3600, 100, 200
Wm. Thornton, 100, 56, 4680, 300, 300
Charles Hodges, 110, 130, 7200, 150, 350
S. S. Tracey, 40, 4500, 136200, 150, 200
Geo. Williams, 40, 15, 1650, 100, 250
E. H. Stansbury, 115, 55, 6800, 150, 400

Geo. Williams, 40, 10, 2040, 75, 300
Wm. Taylor, 40, 10, 2040, 75, 300
Geo. Grimage, 25, -, 100, 75, 200
S. Cromwell, 100, 80, 7000, 150, 250
J. S. Winfield, 170, 19, 100, 225, 950
John Merritt, 100, 150, 7500, 250, 600
G. W. Merritt, 30, 20, 200, 150, 250
Thos. Crogan, 40, 15, 2200, 150, 300
E. H. Gardner, 112, 3, 10200, 400, 750
J. R. Thomas, 75, 73, 12000, 200, 250
J. M. Lucas, 100, 50, 150000, 300, 600
Jas. Steuart, 100, 120, 13200, 250, 650
Eduard Brooks, 60, -, 3000, 100, 300
J. T. Redish, 10, -, 400, 30, 125
F. G. Crisp, 90, 10, 10000, 200, 550
R. F. Crisp, 90, 45, 10800, 100, 200
R. O. Crisp, 100, 70, 11900, 500, 650
E. R. Bales, 100, 40, 20000, 100, 150
James Ench, 35, -, 2500, 50, 175
R. Cromwell, 250, 190, 52400, 600, 1000
Nicholas Crisp, 130, 92, 17900, 1200, 750
Thomas Jones, 13, -, 400, 100, 200
Geo. Riley, 10, -, 300, 50, 125
P. Reddish, 100, 50, 7500, 100, 300
A. Kernelshue, 20, -, 600, 75, 200
Jesse Conway, 40, 40, 1600, 40, 200
John Smith, 100, 33, 8000, 300, 200
John Frill, 30, 10, 1600, 150, 175
E. J. Henkle, 83, -, 8300, 500, 600
Rezin Hammond, 560, 200, 38000, 2000, 2100
Joseph Brian, 400, 170, 27500, 800, 2100
H. H. Thomas, 150, 18, 8400, 1000, 700
S. Linthicomb, 40, 26, 3300, 200, 350

Hugh McCall, 50, 58, 3240, 200, 250
Clara Peters, 30, 10, 2000, 75, 150
Wm. Linthicomb, 400, 100, 25000, 1000, 1700
Geo. Waters, 40, 15, 2360, 200, 200
H. Linthicomb, 140, 23, 8300, 700, 600
Wm. H. A. Brian, 100, 37, 6850, 250, 375
H. Zealdhall, 82, 23, 4600, 250, 375
Jos. Hood, 17, 1, 720, 100, 150
Wm. Shipley, 540, 399, 37560, 1000, 1130
S. Hambleton, 182, 30, 12720, 500, 500
J. H. Caples (Casels), 75, 26, 5050, 225, 500
D. Warfield, 80, 13, 2790, 200, 250
H. B. Smith, 60, 10, 4200, 75, 200
Jethro Smith, 75, 29, 4200, 150, 450
Thos. Cole, 175, 85, 7800, 150, 350
Jas. Benson, 225, 70, 14750, 1000, 1200
A. S. Cromwell, 30, 7, 2040, 100, 350
Jos. Hawkins, 140, 142, 14140, 250, 600
Jos. Steuart, 125, 25, 7500, 500, 500
R. Cromwell, 200, 100, 15000, 500, 750
Thomas Pumphrey, 300, 300, 24000, 600, 1400
Charles Pumphrey, 70, 63, 5000, 300, 400
Wm. Pumphrey, 50, 30, 400, 250, 400
Wm. H. Downs, 50, 50, 4000, 300, 200
Ada Wesley, 100, 140, 4700, 250, 200
R. Wilhelm, 60, 240, 4500, 200, 250
Lewis Kelley, 140, 121, 7850, 400, 550
John Smith, 140, 56, 10800, 400, 600

Basil Benson, 200, 200, 20000, 500, 800

Geo. Rider, 90, 82, 1720, 100, 300

Jos. Smith, 90, 86, 1769, 100, 350

A. Warfield, 75, 32, 5350, 300, 200

Jos. Cole, 675, 25, 20580, 300, 700

Wm. Cole, 60, 28, 5281, 150, 300

John Martin, 100, 43, 8580, 200, 250

B. Plummer, 20, -, 1200, 150, 200

Susan Gaither, 50, 50, 500, 100, 125

Nelson Phelps, 110, 120, 6900, 200, 350

H. Kelly, 80, 113, 5790, 300, 550

J. W. Hawkins, 60, 77, 4010, 50, 30

R. Harmon, 70, 40, 3300, 150, 250

Peter Harman, 70, 40, 3300, 200, 275

J. Q. A. Jones, 30, 82, 1640, 150, 1200

Enos Jeffers, 15, 44, 590, 100, 140

R. Phelps, 75, 71, 4380, 400, 350

Wm. F. Boyer, 20, 60, 2400, 150, 150

Isaac Boyer, 30, 11, 1200, 150, 150

Peter Griffin, 335, 10, 900, 100, 125

John Hammond, 60, 10, 1400, 150, 450

R. A. Shipley, 65, 42, 2675, 100, 450

Jesse Boyer, 39, 1, 1200, 125, 175

Thos. Shipley, 20, 12, 960, 100, 160

Rezin Boyer, 40, 10, 1500, 80, 275

Eliz. Harmon, 80, 55, 2700, 150, 200

Julia Disney, 100, 100, 2000, 125, 200

F. M. Smith, 80, 100, 7600, 200, 300

Q. A. Elliott, 90, 40, 6500, 100, 100

G. T. Dyson, 32, -, 640, 100, 125

M. J. Dyson, 22, -, 440, -, 125

Philip Harman 80, 20, 5000, 200, 250

Saml. Sailor (Gailor), 40, -, 2000, 125, 250

J. L. Merchant, 60, 24, 4200, 200, 350

Chas. Crook, 150, 160, 6200, 300, 700

H. J. Williams, 150, 125, 8190, 400, 1450

Wm. H. Dorsey, 150, 80, 10200, 600, 1000

J. H. Craggs, 130, 100, 10200, 200, 600

M. Steuart, 100, -, 4000, 250, 225

J. Richardson, 80, 47, 2340, 200, 600

Wm. Turner, 40, -, 400, 125, 200

R. Randall, 60, -, 600, 25, 100

R. T. Wood, 50, 40, 900, 50, 400

T. Weems, 250, 97, 9400, 200, 1300

T. Weems, 90, 62, 4500, 25, -

J. Chubb, 100, 34, 1600, 40, 450

C. Watson, 100, -, 1000, 20, 400

D. Blunt, 50, 14, 600, 20, 400

T. Norris, 80, 60, 1680, 50, 500

J. Aisquith, 100, 70, 1700, 25, 200

F. Crandall, 50, -, 600, 25, 400

Wm. Crandall, 40, -, 400, 30, 150

J. Atwell, 70, 30, 1000, 50, 600

T. G. Phiffs, 30, -, 300, 25, 200

A. Scott, 5, -, 50, 20, 200

G. Slimamom, 40, -, 400, 25, 300

J. Dennis, 100, 20, 1200, 50, 400

Wm. White, 140, 100, 5000, 50, 500

J. Crandall, 40, 40, 800, 25, 250

Wm. T. Crandall, 65, 40, 1050, 50, 350

Ezra Parrish, 100, 100, 4000, 50, 400

J. W. Parrish, 125, 75, 5000, 50, 650

J. Good, 65, 35, 2000, 25, 600

G. W. Hyde, 200, 70, 8000, 400, 1300

B. F. Wells, 55, 16, 2000, 20, 575

N. J. Glover, 400, 228, 22500, 1000, 1600

D. C. Thomas, 120, 40, 9700, 1000, 1250

T. Franklin, 250, 50, 1500, 300, 1300

J. C. Weems, 450, 130, 29000, 800, 3400

T. Anderson, 250, 20, 3600, 100, 500

Anna Connor, 120, -, 6000, 100, 900

R. Perry, 80, 43, 4920, 100, 1300

T. Perry, 80, 16, 3500, 53, 500

T. H. Phipps, 80, 26, 1300, 50, 200
Thos. Sears, 150, 125, 8000, 100, 500
J. Frott, 5, -, 150, 25, 100
J. Petherbridge, 125, 25, 7500, 100, -
Wm. F. Prout, 240, 150, 9600, 125, 700
Ann Ganner (Conner), 70, 34, 5200, 200, 400
A. Lerch, 180, 133, 7000, 100, 1100
A. Hartze, 175, 107, 7000, 200, 980
W. R. Norman, 40, 40, 1200, 100, 875
Mary Norman, 40, -, 800, 50, 250
A. H. Hall, 230, 70, 2100, 600, 1250
Mary Hall, 80, 20, 7000, -, -
M. Hardesty, 200, 50, 10000, 600, 600
T. Crandall, 40, 20, 2400, 100, 800
Wm. DaShields, -, -, -, -, -
E. A. Weems, 150, 50, 2000, 50, -
W. DaShields, 90, 17, 1500, 50, 600
T. Sheppard, 200, 102, 4500, 150, 840
W. Bevans, 140, 60, 8000, 100, 700
R. Estep, 160, 45, 12000, 350, 1200
D. Owens, 220, 70, 16200, 500, 1180
Mary Meekins, 275, 126, 6000, 400, 1200
G. W. Nutmell, 400, 160, 33600, 1000, 3580
Enoch Owens, 62, 12, 5000, 150, 500
Joseph Owens, 175, 40, 13000, 500, 1330
G. W. Welch, 130, 70, 12000, 500, 1000
Ann Gardener, 100, 50, 2000, 30, 350
H. Deale, 60, 56, 1600, 50, 220
Deborah Weems, -, 60, 3000, 50, -
Wm. B. Smith, 100, 80, 7600, 100, 578
Wm. Hall, 315, 315, 18900, 200, 2471
Wm. Hall, 110, 110, 6000, 50, 780

Wm. Hall, 185, 150, 9450, 100, 640
Wm. Hall, 103, 103, 2600, 25, 600
A. L. Hall, 200, 200, 16000, 300, 1400
Augustus Hall, 200, 200, 20000, 300, 500
Martin Fenwick, 30, 152, 21000, 300, 1500
T. J. Franklin, 278, 158, 2600, 700, 3350
R. Simmons, 175, 49, 4480, 50, 1000
Wm. Crandall, 150, 50, 6000, 50, 800
B. Tongue, 200, 90, 19500, 300, 1700
A. Gardener, 200, 40, 9600, 100, -
J. M. Deal, 160, 40, 6000, 200, 1100
James Deal, 40, 24, 800, 25, 250
Robt. Franklin, 150, 90, 7000, 100, 1400
B. Davis, 200, 60, 3000, 50, 500
Charlotte Lietch, 125, 71, 16000, 200, 800
Thos. Wells, 40, 12, 1000, 50, 150
Robt. Atwell, 20, 10, 900, 25, 100
Clay Sparrow, 220, 4, 15000, 250, 1000
John Randoll, 75, 25, 5000, 50, 500
S. Scrivener, 300, 70, 18000, 300, 1000
Joseph Childs, 175, 50, 12000, 150, 600
Sprigg Harwood, 200, 50, 15000, 200, -
J. M. Green, 20, -, 2000, -, -
G. W. Hughes, 600, 250, 85000, 1000, 2000
John Davis, 50, 50, 500, 25, 2000
N. L. Darnell, 200, 10, 16000, 500, 800
H. B. Darnell, 140, -, 10500, 400, 700
T. Wood, 30, 27, 400, 50, 250
B. Darnell, 100, 25, 7000, 200, 1200
James Owens, 40, 27, 2000, 50, 80
G. Pindel, 50, -, 2000, 25, 250

R. S. Pindel, 100, 25, 6000, 100, 500
S. Sheppard, 100, 36, 5400, 50, 400
S. Plummer, 175, 50, 11000, 100, 700
Mary Childs, 50, 20, 35000, 100, 500
J. Armiger, 10, 36, 5000, 75, 500
Sarah Drury, 60, -, 3000, 50, 400
E. E. Telyard, 140, 60, 14000, 200, 900
James Wayson, 40, 10, 2500, 75, 400
Samuel Brook, 250, 50, 24000, 300, 1000
E. Gott, 200, 100, 24000, 300, 1000
E. Gott, 150, 50, 12000, 50, 700
T. Owens, 100, 3, 7800, 200, 500
T. Owens Jr., 400, 138, 40000, 250, 1900
T. Owens Sr., 100, 37, 10000, 100, 7500
J. Armiger, 50, 22, 2000, 50, 500
Jos. Armiger, 150, 80, 9000, 100, 400
H. C. Welch, 150, 75, 16000, 50, 1100
H. C. Drury, 200, 100, 21000, 300, 1200
F. H. Gary, 600, 500, 55000, 1000, 2300
Alex. Owens, 90, 40, 6500, 700, 700
Wm. Hopkins, 200, 50, 48000, 500, 1300
T. L. Webb, 72, -, 2160, 200, 200
S. E. Wood, 175, 125, 7000, 50, 1000
Samuel Dove, 120, 160, 3600, 100, 500
Marshall Carr, 150, 110, 7500, 50, 225
Mortimer Carr, 103, 50, 9000, 50, 200
Arthur Carr, 196, 140, 8000, 200, 600
Eliza Carr, 140, 190, 5600, 50, 500
N. S. Knighter, 8, -, 3000, 20, 380
B. A. Carr, 28, -, 2200, -, 300

F. Ward, 132, 120, 6600, 100, 500
Geo. W. Jones, 100, 34, 5300, 50, 400
Thos. Owens, 440, -, 300, 13000, 150, 1000
F. Scrivener, 102, 500, 7610, 100, 1000
D. E. Jacobs, 50, 250, 15000, 50, 2000
E. Laurence, 53, 45, 2600, 100, 700
Wm. Scrivener, 350, 140, 25000, 300, 300
J. R. Dowell, 60, 20, 5000, 25, 350
E. F. Childs, 120, 50, 8500, 100, 600
R. Estep, 60, 9, 2100, 30, 400
M. V. Whittington, 90, 10, 5000, 50, 400
S. W. Carr, 75, 37, 4400, 30, 510
S. Gover, 200, 100, 15000, 50, 1000
John Ward, -, 240, 25, 150
P. H. Smith, 10, -, 500, 20, 225
Eduard Blake, 60, 60, 5000, 50, 400
Robert Queen, 18, -, 540, 20, 200
Wm. T. Clayton, 200, 40, 12000, 150, 1400
T. Sheppard, 225, 38, 18500, 300, 700
Eliza Sheppard Ten, 100, 67, 5000, -, -
J. Sellman, 140, 90, 13800, 100, 800
J. Nutmell, 50, -, 3000, 150, 800
G. W. Starlings, 80, 19, 5000, 75, 600
F. T. Owens, 90, 13, 9180, 100, 700
F. R. Owens, 200, 3, 12180, 300, 1000
V. Compton, 116, 25, 5640, 100, 800
M. Dillon, 60, -, 3000, 25, 200
J. F. Wilson, 400, 150, 2200, 500, 2000
E. Hall, 150, 40, 13000, 300, 1000
Alice King, 20, -, 4000, 20, 100
Horace Owens, 70, 13, 4000, 50, 500
E. O. Gardner, 65, 29, 2800, 50, 650
R. Drury, 200, 100, 15000, 150, 700

E. O. Gardner, 65, 29, 2800, 50, 650

R. Drury, 200, 100, 15000, 150, 700

F. H. Drury, 250, 150, 20000, 200, 2000

Susan Drury, 200, 100, 15000, 500, 1200

B. C. Ness, 200, 75, 13700, 300, 900

Emily Hill, 300, 150, 12000, 100, 600

Wm. C. Hopkins, 75, 55, 3000, 50, 500

Gilbert Nowell, 60, 18, 900, 25, 200

Louisa Jones, 70, 30, 1000 25, 200

R. Suit, 35, 25, 600, 50, 225

R. A. Harwood, 100, 70, 3000, 50, 250

Jane Duvall, 100, 80, 3000, 50, 300

James Wells, 80, 20, 16000, 100, 300

E. Atwell, 300, 260, 1400, 150, 1000

R. Jones, 75, 29, 3500, 100, 500

A. Gibbs, 140, 60, 1000, 200, 1000

Thos. J. Hall, 350, 92, 15000, 200, 1000

Thos. J. Hall, 150, 100, 10000, 100, 700

Thos. J. Hall, 300, 194, 19000, 200, 1000

Baltimore County Maryland
1860 Agricultural Census

The University of North Carolina Library under a grant from the National Science Foundation microfilmed agricultural Census records. Records were filmed at the University of North Carolina from original records at the Maryland State Library.

Columns 1, 2, 3, 4, 5, and 13 represent the following information on the census:
1. Name of Owner, Agent or Manager of Farm
2. Acres of Improved Land
3. Acres of Unimproved Land
4. Cash Value of the Farm
5. Value of Farming Implements and Machinery
13. Value of Livestock

Pages in this county are out of sequence.

Thos. B. Sumwalt, 100, 50, 10000, 300, 600
Ephraim Pierce, 35, 40, 18000, 300, 400
T. S. Boggs, 75, -, 20000, 300, 300
A. P. Perry, 27, -, 10000, 100, 400
James White, 26, -, 8000, 100, 300
Eveline Comegys, 67, -, 20000, 100, 500
Levi Hofman, 42, -, 15000, 100, 200
M. Swan, 200, 100, 90000, 200, 500
C. Heal, 100, 25, 50000, 200, 1000
C. B. Wainbaugh, 50, -, 25000, 100, 300
Peter Nesbeck, 5, -, 1000, 100, 100
S. Richmond, 4, -, 1000, 100, 300
Hugh Gelston, 200, 50, 120000, 700, 1200
R. Johnson, 300, -, 12000, 700, 600
T. Monmonier, 41, -, 16000, 200, 300
W. Taylor, 64, -, 15000, 100, 200
Mrs. Carter, 100, -, 40000, 100, 200
C. Stinchcomb, 10, -, 4000, 100, 100
J. Zimmerman, 52, -, 15000, 200, 900
Thomas Winans, 90, 100, 140000, 1200, 2000
George Kelly, 35, -, 15000, 200, 500

James Read, 30, -, 15000, 100, 300
Elias Read, 10, -, 5000, 50, 100
J. T. Hutson, 32, -, 15000, 200, 300
T. Z. Canby, 62, -, 30000, 200, 400
John Burke, 12, -, 4000, 100, 100
James Steel, 300, -, 10000, 100, 100
John Andrews, 13, -, 4000, 100, 100
Geo. W. Riley, 70, 70, 16000, 300, 1000
Paul Rust, 30, -, 9000, 50, 50
Wm. H. Freeman, 165, 200, 65000, 150, 500
A. R. T. Heald (Hand), 14, -, 2800, 50, 300
Adam Seak (Scak), 14, -, 2800, 50, 300
T. S Alexander, 115, 115, 20000, 200, 800
Joseph Crosby, 200, 75, 60000, 300, 860
John Dunn, 80, 19, 4000, 100, 300
John Pierce, 5, -, 1000, 50, 100
Julius Lader, 74, -, 5000, 150, 400
Wm. B. Stinchcomb, 74, -, 7000, 150, 200
T. B. Wooten, 10, -, 1000, 100, 200
G. Ellis, 23, -, 2300, 50, 50
J. Helnig, 15, -, 4000, 100, 50
Joshua Barnes, 17, -, 2500, 50, 150

J. Foreman, 30, 50, 6000, 100, 350
John Calk, 30, -, 2000, 150, 400
A. Pollack, 100, 66, 24000, 400, 2000
Robert Dorsey, 270, 200, 47000, 400, 900
John Bevan, 36, 15, 5000, 100, 200
S. K. Crosby, 200, 25, 35000, 300, 700
F. Risener, 180, 170, 25000, 300, 1500
Phillip Thomas, 15, -, 3000, 100, 150
Gasper Riply, 26, -, 5000, 100, 150
A. Stinchcomb, 15, -, 2000, 50, 50
B. Billmyer, 40, -, 4000, 100, 300
Wm. Lee, 50, -, 2000, -, 50
John Brickman, 22, -, 1000, 21, 150
Mary Boony, 30, -, 1500, 50, 200
John Busack, 75, 25, 8000, 50, 400
Henry Jones, 40, -, 8000, 50, 401
Wm. Price, 33, -, 7000, 50, 100
Humphrey Moore, 40, -, 7000, 100, 300
John Upton, 34, -, 4000, 100, 400
C. Harrison, 57, -, 10000, 100, 400
H. Repler, 70, 14, 4000, 100, 600
A. C. Thomas, 7, -, 1500, 50, 100
Thomas Gamble, 25, -, 2500, 100, 300
F. Hahn, 50, 16, 6000, 150, 600
T. Brasman, 10, -, 2060, 50, 150
N. Warfield, 3, -, 1500, 50, -
Wm. Devers, 100, 50, 23000, 200, 1000
John Blume, 25, -, 2000, 50, 100
H. Kiser, 10, -, 3000, 50, 50
J. Smith, 100, -, 1000, 50, 100
E. Powers, 40, 7, 6000, 100, 150
M. Blume, 18, -, 2000, 100, 100
S. House, 21, -, 2000, 100, 100
J. S. Wilson, 25, -, 3000, 50, 100
C. Miller, 100, 38, 10000, 100, 150
A. Wherling, 24, -, 3000, 100, 150
H. Williams (Ten), 20, -, 2000, 50, 50

James Linthicum, 40, -, 6000, 200, 200
W. Gamble, 80, -, 12000, 400, 200
J. H. Hiser (Hiner), 40, -, 6000, 150, 400
J. Clemons, 26, -, 3000, 100, 300
Wm. Kelly, 14, -, 2500, 50, 100
J. Bell, 26, -, 3500, 100, -
P. W. Gibbons, 70, -, 16000, 200, 800
Wm. Graham, 51, -, 7000, 100, 200
J. Watson, 21, -, 3000, 50, 100
W. Pierpont, 27, -, 6000, 100, 300
W. Haycock, 26, -, 5000, 100, 800
J. A. Thomas, 30, -, 15000, 100, 500
H. Simm, 22, -, 8000, 100, 150
S. Roberson, 16, -, 500, 50, 100
Wm. Taylor, 16, -, 7000, 50, 100
Wm. Gover, 12, -, 7000, 50, 100
Wm. McPherson, 42, -, 12000, 200, 300
John McDowell, 25, 5, 6000, 50, 100
David Patterson, 36, 30, 10000, 50, 50
J. C. Holland, 66, -, 13000, 100, 150
B. F. Hawes (Ten), 40, -, 8000, 100, 240
A. A. Lynch, 75, 25, 20000, 200, 700
J. K. Smith, 6, -, 4000, 50, 50
J. P. Furting (Fusting), 60, -, 2400, 400, 800
Henry Wise, 22, -, 5000, -, 100
Parnell Rich, 10, -, 4000, 50, 100
M. Bensinger, 87, 36, 26000, 200, 900
James Battee, 165, 36, 4000, 300, 800
R. Sewell, 25, -, 18000, 200, 400
J. E. Dorsey, 170, 30, 60000, 300, 600
M. A. Zell, 43, -, 3000, 100, 400
S. Schwartze, 100, 25, 45000, 200, 400
Wm. A. Mullikin, 80, -, 40000, 300, 400
J. Herbert, 93, -, 25000, 400, 400

J. Q. Hewlett, 132, -, 13000, 400, 1200

D. C. Howard, 20, -, 20000, 50, 1000

F. Smith, 20, -, 4000, 50, 40

Asa Needham, 25, -, 10000, 100, 200

Seth Reed, 100, -, 10000, 300, 600

T. C. Rutten, 13, -, 7000, 50, 200

W. Taylor, 133, -, 40000, 200, 700

R. P. Jones, 13, -, 6000, 100, 300

C. Cooper, 120, -, 24000, 150, 60

Wm. E. Shiply, 100, -, 5000, 150, 300

M. Morrison, 58, -, 10000, 200, 400

G. Femanus, 100, 70, 25000, 200, 1400

S. W. Waring, 100, -, 25000, 200, 1000

Wm. A. Purdy, 10, -, 5000, 50, 1000

Fred P. Bensinger, 15, 6, 6000, 50, 400

Wm. Thompson (Ten), 100, 100, 40000, 400, 700

A. Shiply, 100, -, 15000, 200, 800

J. C. Smith, 200, 144, 20000, 300, 1200

John Glenn, 500, 600, 100000, 1000, 4000

G. Richstien, 100, 80, 12000, 300, 800

G. Reineker, 100, 32, 50000, 300, 800

Benjamin Bell, 53, 11, 7000, 100, 360

D. Swoke (agent), 200, 300, 40000, 200, 1000

D. Borman, 51, 18, 15000, 150, 400

Wm. J. Albert, 42, -, 25000, 20, 1000

P. M. Huse, 62, -, 10000, 100, 200

Wm. Marr, 25, -, 1000, 50, 100

Wm. Baynes, 50, -, 12000, 150, 300

N. R. Smith, 100, 69, 35000, 200, 2000

G. W. Lenmare (Lermase, Lenmase), 350, 130, 75000, 100, 5000

T. C. Yearly, 56, -, 15000, 100, 300

Henry Beamer, 70, 14, 25000, 400, 600

J. B. Thompson, 50, -, 20000, 200, 500

J. Eschbaugh, 35, -, 10000, 200, 400

A. Seemuller, 32, -, 25000, 100, 700

H. Towson, 100, -, 15000, 50, 150

M. Taylor, 100, 50, 200600, 200, 700

J. B. Brinkly, 25, -, 15000, 50, 100

Ross Campbell, 50, 30, 20000, 100, 1000

James Gibson, 100, 32, 20000, 400, 1000

L. Van Bokelyn, 65, -, 40000, 400, 1400

Joseph Hopkins, 4, -, 4000, 50, 100

John Love, 18, -, 3600, 50, 500

Chew VanBibber, 14, -, 2500, 50, 300

James Glanville (Ten), 14, -, 3000, 100, 200

B. Winters, 20, -, 4000, 150, 250

J. Midline, 36, -, 10000, 200, 300

Thos. Miller, 100, 25, 10000, 250, 600

J. Carter, 12, -, 5000, 50, 180

C. Wolfington, 22, -, 8000, 100, 250

H. Simon, 25, -, 2600, 50, 150

A. Miller, 22, -, 5000, -, 200

J. Roff, 50, -, 5000, 100, 250

L. Rhinehardt, 12, -, 2500, 50, 200

Theodore Boseman (Bauman), 60, 15, 3700, 200, 500

Nat. Carnay (Comay), 100, 84, 4000, 200, 600

V. B. Boseman, 200, 25, 10000, 500, 1000

Herod Gosnell, 100, 15, 2500, 100, 400

Henry C. Bennett, 125, 30, 9000, 200, 600

E. H. Bennett, 200, 30, 7000, 200, 400

Joseph M. Allen, 80, 20, 4000, 100, 700

John P. Righam, 23, 6, 1000, 50, 50
Lewis Allen, 75, 25, 3000, 50, 500
Perry Gosnell, 100, 50, 5000, 154, 300
Wesly Boseman (Bauman), 150, 100, 8000, 200, 800
Edwin Saffle(Soffle), 150, 50, 8000, 200, 600
G. W. Aler, 92, 48, 6000, 200, 600
Eduard Triplet, 100, -, 1000, 51, 300
Charles Ware, 80, 10, 5000, 200, 400
Zebedee Gosnell, 100, 47, 3000, 100, 200
John B. Armstrong, 270, 130, 25000, 400, 2000
Walter J. Odel, 100, 30, 7000, 200, 700
William Beam, 30, 9, 10000, 200, 50
William Randal, 70, 30, 2000, 100, 150
Lloyd Randall, 20, 12, 1500, 50, 100
John Stansfield, 24, 7, 1000, 50, 300
John Stansfield, 20, 20, 1200, 50, 100
James Stansfield, 12, 6, 1000, 50, 200
William Odel, 40, 10, 1000, 50, 200
E. P. Robinson, 200, 50, 10000, 200, 700
Wm. Black, 100, 70, 5000, 100, 200
Vernon W. Dorsey, 167, 50, 10000, 150, 500
William Kane, 150, 50, 8000, 200, 400
George E. Odel, 130, 20, 15000, 400, 400
Henry C. Lettyinger, 15, -, 2000, 50, 200
George W. Bailey, 75, 10, 7000, 100, 400
Samuel Ward, 100, 45, 10000, 200, 600
T. C. Worthington, 300, 200, 10000, 50, 750
John Miller, 40, -, 4000, 100, 700

George W. Bourns (Bowers), 100, 35, 6500, 200, 700
Wm. Chapman, 300, 150, 2000, 300, 1000
James Baker, 25, -, 2000, 100, 400
James Harker, 140, 75, 10000, 400, 600
Richard Heiland, 60, 42, 4500, 300, 250
Samuel Connery, 100, 31, 3000, 150, 250
Richard Baker, 54, -, 2000, 100, 200
George Bennison, 120, 25, 6000, 200, 500
C. H. Owings, 210, 40, 12000, 300, 600
Urias Bailey, 100, 200, 15000, 300, 600
James Sargeant, 90, 16, 2500, 1000, 250
Joshua Sumwalt, 107, 50, 10000, 200, 400
C. D. Owings, 175, 50, 10000, 400, 700
N. H. Worthington, 186, 200, 20000, 400, 600
Wm. A. Mansfield, 60, -, 3500, 100, 500
John K. Harvey, 150, 110, 8000, 200, 800
Louis Ehler, 145, 10, 10000, 200, 600
G. Hamilton, 70, -, 1200, 100, 100
Thos. G. Offutt, 100, 16, 5000, 150, 700
Atwood Blunt, 268, 40, 18000, 500, 1700
Lemuel Offutt, 150, 50, 10000, 500, 1000
S. T. Shipley, 175, 25, 10000, 200, 700
George Diger, 75, -, 4000, 100, 200
C. A. Worthington, 86, 20, 9000, 100, 300
Richard Choate, 140, 60, 7000, 200, 200

Benjamin Stansfield, 32, 8, 2500, 100, 300

Geo. T. Whitney, 155, 35, 4000, 150, 300

Nicholas Hate, 93, 35, 6000, 300, 300

William Cusler, 100, 20, 6000, 200, 300

John C. Brooks, 56, 40, 8000, 200, 250

Jesse Carr, 135, 75, 7000, 150, 200

James Carr, 135, 15, 7000, 250, 500

William Fete, 220, 40, 10000, 1000, 1400

Samuel Walden, 140, 59, 8000, 400, 700

Basil J. Dorsey, 90, 10, 4500, 250, 300

John Worthington, 230, 70, 15000, 300, 1000

Noah Worthington, 150, 120, 10000, 300, 600

T. M. Spencer, 220, 50, 16000, 1500, 1000

David Steuard, 120, 40, 8000, 150, 500

Anna Edelen, 250, 50, 15000, 500, 1000

Noah Worthington Jr., 500, 400, 36000, 1000, 2000

Susan Dorsey, 100, 24, 6500, 300, 400

Marand Duval, 15, -, 600, 50, 100

Jabez Hamilton, 60, 40, 1800, 100, 300

George Allen, 38, -, 800, 50, 200

Joseph H. Wright, 60, 35, 800, 300, 400

Wm. P. Hartly, 52, 8, 2500, 200, 300

Samuel Hartly, 68, 20, 4500, 300, 600

John P. Haycock, 42, 3, 2500, 20, 300

Phineas Hartly, 36, 6, 2500, 150, 300

Thomas Hartly, 45, 4, 4000, 400, 400

Elias P. Hartly, 42, -, 3500, 300, 600

Edm. Iglehart, 100, 40, 6000, 200, 300

John Blackburn, 100, 70, 6000, 400, 600

R. H. Worthington, 500, 200, 35000, 600, 3000

Henry Fryfogle, 75, 6, 6000, 200, 400

Philip Foshender, 40, -, 200, 50, 100

Thomas Ritter, 20, -, 1400, 50, 150

Alfred Crook, 55, 15, 2500, 50, 200

John Fryfogle, 30, -, 3000, 150, 200

James B. Waters, 26, -, 700, 51, 50

Ada Clay, 56, 10, 4000, 100, 500

David Jean, 720, 30, 11000, 600, 1500

Henriette Randle, 40, 10, 2500, 100, 700

Amy Phillips, 200, 50, 15000, 400, 300

John Crooks, 40, -, 1000, 40, 300

Luther Tomanus, 40, -, 1500, 100, 200

Anna W. Rogers, 90, 30, 6000, 200, 400

William Lowry, 60, -, 600, 200, 300

George Kraft, 70, -, 1100, 200, 400

Henry Owings, 66, -, 3500, 100, 100

Mathias Homes, 17, -, 1200, 50, 100

Henry Bread, 15, 5, 1500, 50, 100

William Steakle, 15, -, 1600, 50, 100

G. Zimmerman, 50, 50, 3500, 50, 100

Joseph Hask, 40, 26, 2500, 50, 150

Wm. B. Monmonier, 30, 22, 4000, 100, 200

Joseph Wideman, 50, 10, 4000, 200, 300

Thomas Ely, 106, 75, 11000, 200, 800

Gerrett Emmons, 200, 50, 10000, 500, 1000

Joseph France, 80, 90, 8500, 500, 400

George Wideman, 10, -, 1200, 50, 100

John Russell, 80, 24, 9000, 200, 400

T. A. Muller, 100, 18, 4500, 200, 800

Joseph Gardner, 100, 18, 6000, 200, 800

Ann Wemer, 80, 10, 4500, 100, 500

Rob Oliver, 500, 318, 40000, 2000, 5500

Owings Griffith, 150, 45, 10000, 200, 360

R. F. Maynard, 200, 50, 10000, 400, 600

James A. Walters, 130, 60, 4000, 200, 600

S. W. Crooks, 100, 50, 6000, 100, 400

Wm. C. Haney, 80, 20, 3000, 100, 200

Thomas Burke, 80, 20, 4000, 400, 200

Joseph Pauly, 5, 6, 1800, 100, 400

Wm. H. Pupert (Rupert), 40, -, 1000, 100, 100

George Latisth, 148, 50, 10000, 200, 600

Solomon Fogle, 20, -, 800, 50, 20

Israel Owings, 80, 20, 6000, 100, 200

Henry Ripley, 10, 4, 1500, 100, 200

Achsah Owings, 80, 15, 7000, 180, 300

George Lynch, 90, -, 6000, 200, 500

John Davies, 150, 60, 20000, 400, 1000

Mary A. Stinchcomb, 50, -, 3000, 150, 50

John Hiel, 12, -, 2000, 50, 300

Herod Clark, 31, 6, 3000, 51, 200

Jacob Hinkly, 140, 6, 4000, 50, 300

John H. Hines, 40, -, 2500, 50, 150

B. H. D. Ball, 100, -, 20000, 500, 600

Nathaniel W. Huff, 12, -, 1500, -, -

Alexander Burdock, 17, -, 25000, -, 1500

Robert Whitelock, 2 ½, -, 20000, -, 100

Ephraim Huffman, 2, -, 1500, -, 300

John A. Bunseman, 12, -, 1600, -, 400

Brice Shipley, 30, -, 3700, 50, 100

Orphy Shipley, 12, -, 3200, 50, 300

Cornelius Hartzell, 5 ½, -, 3000, 50, 250

Joshua Cowan, 3 ¼, -, 2000, 25, 150

John Counselman, 52, -, 10000, 300, 300

Diretee (Dactee, Deirdre) Fox, 5, -, 2500, 100, 30

John Beacham, 6, -, 2000, 25, 100

Henry Bean, 35, 20, 3000, 50, 150

John Ruppert, 42, 42, 10000, 100, 200

Gallaway Chiston, 37, -, 23000, 500, 300

Robert W. Belt, 9, -, 5000, 30, 150

Jasper N. Berry, 8, -, 8000, 100, 320

Godleib Kane, 80, -, 16000, 150, 400

George Emert, 12, -, 1600, 50, 150

James R. Weagley, 100, -, 6000, 100, 700

Rebecca Smith, 100, 64, 8000, 100, 300

Michael Wilson, 120, 57, 7000, 100, 100

George Smith, 20, 158, 1000, 25, 200

Elizabeth Coates, 40, 10, 2500, 100, 100

William Taylor, 9, 9, 1800, 25, 75

Rachel A. Dorsey 3, -, 700, 25, -

Robert Wylie, 180, 40, 35000, 200, 1100

Louisa Merryman, 100, 33, 14000, 20, 800

Albert J. Boyer, 45, -, 15000, 300, 400

Jno. E. Kirkpatrick, 20, -, 2000, 50, 500

John Fattle, 30, 58, 5000, 50, 200
Edward Harvey, 110, -, 7000, 200, 6000
John Numson, 20, 33, 10000, 150, 600
John Brady, 20, -, 10000, 100, 600
Caleb Horsus, 40, 12, 6500, 400, 1200
William J. Bland, 60, 10, 5000, 150, 600
Joseph Edge, 75, 38, 28000, 200, 500
Joseph Smith, 226, 60, 15000, 500, 2000
Martin Coleman, 3, 10, 800, 50, 100
James Barns, 76, -, -, 100, 100
Charles Thomas, 100, 25, 5000, 50, 100
Luke J. Pierce, 300, -, -, 300, 1200
Aaron Tucker, 67, 10, 4000, 200, 300
Noah Walker, 250, 200, 22000, 500, 2000
Cardiff Taggert, 325, -, 30000, 500, 500
Robert Dennison, 200, 215, 20000, 2000, 1000
Oliver P. McGill, 90, 12, 15000, 500, 1000
G. B. Milligan, 100, 20, 15000, 500, 1500
Thomas B. Cocksey, 330, 120, 22500, 500, 1200
Francis B. Lawrenson, 450, 150, 50000, 5000, 3000
John S. Turner, 150, 40, 8000, 300, 400
John Fisher, 215, 68, 25000, 800, 2000
Richard Brown, 70, -, -, 150, 250
Elizabeth Flint, 60, -, 9000, 300, 500
Eliza Wise, 70, 130, 30000, 200, 500
John McKeen, 44, 16, 20000, 50, 300
Robert Holliday, 21, 4, -, 100, 500
James Hooper, 10, -, 1000, 20, 3000

Hanson P. Rutter, 57, 3, 25000, 200, 200
Joshua Smith, 22, -, 6000, 100, 300
Edward Keen, 35, -, 7500, 200, 500
John P. Fisher, 30, -, 10000, 150, 250
Wm. Bucklie, 25, -, 15000, 150, 500
Canney Brooks, 80, -, 80000, 200, 1000
William Martin, 5, -, 5000, -, 700
John W. Quick, 75, 20, 20000, 300, 900
Thomas Matthews, 100, -, 40000, 400, 400
W. Wright, 90, 61, 6000, 300, 700
Thomas H. Moore, 150, 81, 9000, 200, 100
Wm. F. Johnson, 300, 150, 40000, 600, 1500
Mary Stinglofp (Stinglass), 130, -, 20000, 300, 600
Uris Rogers, 120, 75, 25000, 300, 1600
Reubin Stump, 125, 25, 20000, 500, 500
Henry Onderdink, 134, 41, 20000, 50, 600
David Graham, 11, -, 3000, 50,, 250
Joseph Art, 80, -, 6000, 100, 350
Charles G. Lyon, 175, 25, 20000, 400, 1000
Charles L. Rogers, 160, 26, 25000, 400, 100
James Manor, 180, 70, 40000, 400, 1100
William Derris (Davies), 42, -, 4500, 200, 500
James B. Councilman, 180, 50, 20000, 500, 1200
Alexander Ridell, 180, 20, 20000, 200, 1300
Benjamin Howard, 50, -, 15000, 500, 800
Jas. H. McHenry, 500, 300, -, 17500, 13600

Thomas Cradock, 275, 60, 30000, 400, 600

Geo. H. Wittnacht, 107, -, 15000, 300, 500

William Lewis, 40, 26, 5000, 100, 500

Debra C. Thomas, 18, -, 16000, 150, 400

Henry Miles, 300, -, 300000, 200, 2100

John B. Morris, 50, -, 25000, 200, 400

Thomas Goss, 50, -, 25000, 150, 303

Wm. P. Griffith, 36, -, 36000, 200, 700

John Lewis, 180, 20, 40000, 500, 500

John F. Thomas, 20, 36, 10000, 20, 500

H. H. Williams, 50, -, 10000, 200, 800

Martin Shraly, 200, 60, 60000, 500, 2000

N. P. Hayward, 75, 25, 25000, 600, 750

Jacob Smith, 30, 7, 4000, 200, 400

Thos. Sanderson, 106, 60, 30000, 500, 400

Jahera Parsons, 10, 11, 4000, 200, 200

John A. Craig, 52, -, 40000, 75, 1800

John M. Brierly, 25, -, 15000, 100, 200

John Dervan, 33, -, 16000, 50, 500

James Owings, 20, 60, 10000, 100, 100

James Hamilton, 40, 35, 15000, 250, 250

John Grason, 100, -, 20000, 200, 600

Ferdinand Myers, 68, -, 20000, 300, 200

William Shirly, 35, -, 20000, 500, 500

Charles B. Keywrite, 50, -, 25000, 300, 400

Wesly Paine, 100, -, 30000, 200, 500

Richard W. Hook, 30, 17, 10000, 200, 300

Levi Buffington, 80, 20, 25000, 50, 250

Caspar Harman, 12, -, 10000, 100, 600

Hamilton Custine, 26, -, 26000, 200, 700

John Miller, 25, -, 20000, 100, 400

Tobias Avrey, 25, -, 20000, 200, 500

John D. Smith, 48, -, 40000, 200, 400

Jesse Stingluff, 150, 200, 50000, 500, 1000

John Corner, 35, -, 10000, 200, 500

Joseph Pearson, 105, 38, 50000, 500, 1000

Joseph Pearson, 31, -, 25000, 300, -

John Smith of D, 56, -, 30000, 100, 500

Isabella Brown, 80, -, 80000, 500, 800

John M. Oram, 30, -, 50000, 308, 600

Lloyd M. Rogers, 250, 175, 400000, 400, -

Jacob Olir, 40, 26, 7000, 200, 300

Samuel Hook, 30, 41, 3000, 50, 300

Rudolph Hook, 100, 40, 4000, 40, 300

Adolphus C. Schaffer, 100, 35, 3000, 700, 1000

Mary Pearson, 31, -, 30000, 150, 500

Oliver Perigo, 15, -, 2000, 50, 150

Levi R. Pickens, 17, 113, 20000, 100, 400

Joseph Stiger, 55, -, 3000, 50, 700

Frederick Hoffman, 35, 20, 15000, 200, 1500

John S. Gettings, 400, 700, 100000, 200, 2000

Edward Butler, -, -, -, 150, 400

Catherine Cusack, 16, -, 300, 150, 100

Henry Lentz, 25, 15, 2500, 150, 100

John W. Ridgley, 40, 62, 5000, 100, 120

Samuel Stone, 400, 100, 40000, 400, 1500

Basil S. Bennett, 150, 112, 18000, 300, 1000

Henry Sevirson, 175, 39, 25000, 400, 900

Burlington Carlisle, 200, 100, 30000, 400, 1100

Joshua Gent, 175, 25, 25000, 300, 800

George H. Elder, 400, 200, 30000, 200, 1600

Isaac Sumwalt, 70, 41, 4000, 100, 500

John _. Bowen, 140, 860, 17000, 300, 800

Joshua Young, 10, -, 6000, 100, 200

Benjamin W. Cox, 65, 5, 15000, 250, 600

Thomas Watts, 60, 40, 5000, 100, 300

Wm. Williamson, 250, 150, 40000, 400, 1000

John Armstrong, 100, -, 20000, 200, 700

Andrew Huff, 200, 150, 30000, 300, 1000

New Mount Hope Insane Asylum, 100, 100, 40000, 400, 900

John H. Cole, 90, 27, 9500, 500, 1000

Nicholas Hatten, 125, 36, 6000, 200, 600

Jesse Stockdale, 175, 175, 10000, 200, 500

Samuel Wright, 45, 21, 2500, 250, 300

Micajah Martin, 20, 6, 1000, 75, 50

Wm. R. Deaver, 20, 40, 1000, 103, -

John W. Triflet, 70, 20, 3000, 75, 300

Nancy Hervitt, 6, 54, 2000, 54, -

Kitty A. Fort, 200, 100, 12000, 100, 60

William J. Cook, 150, 40, 7000, 40, 600

William E. Cook, 30, 20, 1000, 20, 30

Greenbury Cook, 35, 20, 1000, 20, 200

Aquilla McComas, 50, 100, 700, 100, 700

Silas H. P. Fuller, 3, 7, 550, 10, 120

Wm. R. Haser, 50, 73, 3000, 100, 300

Mordecai C. Stockdale, 130, 24, 5000, 200, 500

Jesse L. Burnett, 150, 20, 10000, 250, 700

Wm. Addison, 140, 50, 2000, 25, 150

Thomas T. Stockdale, 7, -, 1000, 50, 100

Alexander Walters, 80, 20, 4000, 100, 400

Franklin Anderson, 350, 150, 40000, 1000, 1500

Conroud Uhlers, 103, 20, 2000, 100, 500

Charles Blizzard, 90, 35, 3000, 100, 500

Charlotte Polglose, 100, 38, 2000, 50, 100

John Elley, 50, -, 6000, 50, 100

George Crawford, 125, 45, 8000, 100, 500

Alfred Gore, 40, 10, 300, 100, 100

Henry H. Derckson, 130, 130, 8000, 200, 20

Bial Owings, 182, 100, 10000, 200, 700

John Benson, 130, 85, 6000, 200, 600

John D. Smith, 30, 30, 1000, 50, 100

David Uhler, 20, 12, 2000, -, 50

Jacob Bickley, 33, 8, 3000, 25, 300

Jacob Uhler, 33, 7, 4000, 100, 200

Susan T. Norris, 300, 200, 20000, 3000, 3000

William Gore, 40, 10, 1000, 100, 100

Ephraim Triplet (Triflet), 16, 2, 1000, 30, 120

John B. Slade, 200, 100, 1100, 300, 1000

Supther Fry, 25, 5, 1000, 30, 200

Thomas Wooden, 50, 10, 4000, 50, 100

Elijah Wooden, 100, 20, 5000, 200, 1000

Thomas Coakley, 56, 6, 3000, 50, 500

Nicholas Gardiner, 40, 5, 3000, 100, 250

Eli Henkle, 70, 20, 3000, 200, 400

John Jones of Ch., 110, 25, 5000, 100, 700

John F. Richards, -, -, -, 20, 300

Isaac E. VanBibber, 66, 30, 3000, -, 100

George W. Barker, 250, 50, 10000, 200, 600

Jacob Algier, 130, 80, 8000, 150, 600

Noah B. Stockdale, 75, 25, 2500, 50, 150

Andrew Burke, 150, 50, 6000, 300, 500

John T. Gill, 50, 50, 2000, 200, 400

Frederick Fowke, 205, 40, 4000, 300, 400

John Warren, 46, 4, 3000, 100, 300

Henry L. Frazier, 65, 15, 4000, 500, 500

Bonnet Garritson, 115, 14, 4000, 100, 200

David Abbott, 65, 24, 3000, 100, 400

John Houck Jr., 51, 15, 3000, 200, 400

Daniel Helms, 57, 20, 3000, 200, 200

Thomas Upperco, 100, 20, 4000, 200, 400

Jacob Upperco, 70, 28, 3000, 100, 400

Henry Helms, 67, 10, 3000, 100, 400

Nicholas Emrick, 16, -, 2500, 200, 300

Andrew Uhler, 130, 32, 6000, 200, 400

Hugh Neal, 120, 55, 10000, 200, 400

Edward G. Penny, 100, 50, 6000, 300, 500

Henry Lockard, 60, 45, 3000, 200, 400

Danl. Trout, 85, 30, 6000, 200, 500

McArthur Turner, -, -, -, 100, 300

Alfred Low, 180, 50, 10000, 200, 900

John Ford, 180, 30, 4000, 300, 800

John T. Ford, 75, 25, 4000, 50, 400

Henry Baily, 160, 40, 7000, 150, 400

Herod Choate, 80, 20, 5000, 100, 600

William Geary, 30, -, 2000, 100, 100

Richard W. White, 53, 10, 3000, 150, 500

Lewis Tarno, -, -, -, 200, 600

David Owings, 225, 25, 7000, 100, 500

Samuel T. Martin, 90, 10, 8000, 200, 700

Milton Pointer, 300, 100, 40000, 500, 1500

Samuel H. Blakely, 207, 111, 20000, 500, 700

J. Bennington, -, -, -, 300, 200

Francis B. Groff, 100, 100, 10000, 300, 1000

Jacob B. Groff, 40, -, 8000, 100, 200

John R. Reese, 200, 75, 15000, 300, 700

George F. Wagly, 100, 25, 15000, -, 500

Joseph Ritter, 30, 10, 5000, -, 100

George Stirey, 111, 170, 12000, 100, 400

Resin Brown, 50, 10, 4000, 200, 200

John Morrow, 19, 30, 20000, 300, 800

William Hagy, 100, 40, 8000, -, 4000

Ann Moran, 62, 2, 5000, 100, 100

Silas W. Carr (Cann, Carn), 86, 12, 8000, 200, 600

Seth Hocke, 20, -, 3000, 100, 250

Lewis Trittle, 32, -, 5000, 300, 300

John Kissett, 90, -, 10000, 300, 700

Wm. Berryman, 75, 20, 6000, 50, 500

John Bickley, 31, 4, 5000, 100, 100

H. C. Berryman, 60, 4, 6000, 100, 100

Amos A. Hayman, 50, 6, 6000, 100, 400

Solomon Choate, 43, 45, 6000, 300, 500

Nehemiah Berryman, 42, -, 4000, 200, 300

Thomas Haughy, 88, 12, 6000, 300, 400

Matilda Beckley, 110, 20, 6500, 200, 400

Jacob L. Torney, 200, 175, 2000, 700, 1000

Jeremiah F. Duckey, 150, 130, 6000, 100, 500

Shelby Reister, 100, 30, 3000, 100, 200

Silas Sersh, 26, 8, 2000, -, 100

Christoper Gill, 160, 75, 5000, 300, 700

Gotlieb Hester, 90, 14, 4000, 100, 500

Wm. H. Bushey, 90, 90, 4500, 100, 200

Michl. Shamloffie, 17, -, 1000, 100, 200

Wm. H. Musselman, 85, 20, 3000, 100, 400

John Duce, 36, -, 1000, 50, 300

Edward Gill, 130, 70, 8000, 300, 700

Harry Brown, 30, 7, 1000, 50, 300

Thomas Byerly, 150, 50, 8500, 300, 800

Thomas Watts, 120, 50, 7000, 200, 600

John Myers, 120, 50, 5000, 300, 600

Philly Myers, 120, 10, 2000, 50, 100

Mary A. Brown, 60, 10, 2000, 50, 50

Henry Nelta, 90, 60, 4000, 200, 500

David Osborn, 150, 10, 2000, 200, 500

Richard Belt, 80, 25, 2000, 200, 400

Geo. W. Ambrose (tenant), -, -, -, 300, 600

Geo. Derrick, 40, -, 1000, 100, 300

Benjamin Gill, 100, 95, 5000, 500, 800

David Longsucker, 158, 25, 14000, 200, 900

John Kemp, 150, 80, 4000, 150, 800

Geo. W. Norris, 150, 200, 17000, 1000, 1800

Danl. Shugars, 25, 25, 4000, 200, 300

Abraham Groff, 90, 20, 6000, 100, 100

Francis Kindig, 100, 400, 10000, 500, 500

William T. Gorin, 144, 11, 15000, 300, 1000

John S. Givin, 90, 10, 10000, 200, 800

James J. Gover, -, -, -, 150, 400

Mary C. Worthington, 600, 900, 70000, 1000, 1500

John R. Cockey, 300, 140, 17200, 1000, 1500

Richd. J. Worthington, 400, 800, 44000, 1000, 3000

Edward Worthington, 600, 400, 15000, 400, 1000

Kinsay Worthington, 400, 400, 40000, 200, 800

Julia Duncan, 40, -, 2000, 100, 100

Samuel Maden, 20, -, 800, 50, 50

Nicholas Waters (tenant), -, -, -, 100, 450

James Bruster, 200, 50, 12000, 500, 1000

Benj. J. Worthington, 150, 125, 15000, 500, 1200

G. G. W. Johnson, 97, 40, 7000, 200, 300

Frank Rogers, -, -, -, 400, 1200

E. F. Johnson, 105, 75, 800, 300, 1000

Richard Johns, 450, 450, 30000, 600, 2500

Thomas Lucas, 202, 34, 14000, 1000, 1000

Charles H. Cole (tenant), -, -, -, -, 400

Edward Philpot, 200, -, 20000, 1500, 1500

John H. Johns, 500, 90, 35000, 300, 1200

Henry Long, 25, -, 1500, 100, 200

John Long, 20, -, 1000, 100, 200

Richard Scott, 85, 15, 4000, 300, 700

Henry Swatts (S. Watts), 200, 100, 12000, 400, 1000

Henry Cromhart, 25, -, 2000, 200, 400

Wm. C. Gent, 100, 50, 5000, 100, 400

S. Howard, 80, 20, 4000, 200, 300

Jacob Wigler, 100, 30, 7800, 200, 300

Mary Wilderson, 200, 100, 6000, 200, 400

C. T. Cockey, 225, 225, 18000, 200, 1200

Daniel Cummings, 100, 15, 9000, 200, 700

G. W. Norris, 100, 130, 15000, 300, 800

Jesse Gore, 80, -, 3500, 100, 400

James R. Manly, 60, 20, 4000, 102, 560

Francis Humphries, 200, -, 15000, 300, 600

Christian Wissanple, 125, 25, 10000, 200, 300

Samuel Gersuch, 200, 90, 14000, 200, 700

D. Vondersmith, 100, 30, 700, 200, 400

Isaac Snavely, 200, 50, 10000, 200, 500

George B. Gingrich, 30, 50, 5000, 200, 200

Job Maden, 13, -, 1000, 50, 250

William Gardiner, 100, 25, 400, 100, 400

George Gore, 100, 30, 5000, 100, 30

Johnsey Gore, 88, 20, 600, 200, 300

Elizabeth Gore, 100, 20, 7000, 400, 700

Washington Gore, 68, 20, 6000, 100, 200

William Mahan, 80, 20, 4000, 100, 100

William Pridle, 200, 53, 15000, 100, 500

John R. Cox, 40, -, 2500, 100, 200

Salathis Cole, 100, 20, 4800, 100, 500

John H. Harman, 100, 39, 10000, 100, 500

Mary B. Patterson, 400, 600, 35000, 1500, 3000

E. Brackenridge, 35, 5, 1600, 200, 300

Wm. Stockbridge, 40, -, 2400, 100, 100

John Duckman, 15, -, 2500, 100, 100

Stephen Griffith, 400, 400, 33000, 400, 800

Jacob Criager, 400, 400, 32000, 300, 1000

Benjamin Goodwin, 500, 600, 55000, 300, 500

George Gephardt, 190, 60, 16000, 400, 500

Jacob Grim, 25, 24, 800, 25, 125

George Banblitz, 30, 10, 800, -, 75

Charles Banblitz, farms, for, others, -, 200

Barbary Cooper, 50, 25, 2000, 50, 300

David Panther, 70, 34, 2500, 100, 400

Harrison Fowble, 25, 27, 1500, 50, 178

Abraham Panther, 27, -, 500, 50, 200

Margaret Hare, 20, 40, 1000, -, 20

Ivanna Hale, farms, for, others, 30, 200

Peter Fowble, 100, 37, 3000, 100, 300

Jacob Hare, 30, 30,700, 50, 400

James Price, 75, 29, 2000, 150, 350

Jacob Rish, 50, 40, 1500, 100, 300

Eli Crowther, farms, for, others, -, 100

Jabez Armcast (Armcost), 26, -, 1000, 15, 350

Jacob Folk, 40, 15, 1500, 75, 200

Julian Jane, 20, 20, 1000, -, -

Chas. W. Merryman, 25, 5, 1000, 40, 130

Thomas Wheeler, 58, 58, 1500, 50, 150

Samuel Fan (Fare), 35, 24, 1000, 50, 150

Francis Hernt (Hemt, Herrit, Hunt), 50, 42, 1400, 50, 200

Caleb Martin, 100, 44, 2500, 60, 300

Luther Martin, 100, 27, 4000, 200, 300

Mordecai Martin, 200, 90, 4000, 200, 600

Elisha Merryman, 50, 23, 1500, 75, 250

William Rence, 50, 40, 1500, 75, 300

Shadrick Kemp, 250, 108, 4000, 250, 600

Edmund Myers, 100, 25, 2000, 100, 150

John Shultz, 50, 26, 1500, 75, 150

George Ports, 100, 25, 2000, 200, 600

Elizabeth Kelly, 50, 30, 2000, 50, 150

Edward Lawson, 100, 38, 3000, 125, 500

Staple Armacast, 50, 50, 1700, 50, 400

John Armacast, 85, 15, 3000, 100, 425

William Curtis, 100, 75, 2000, 100, 200

Thomas Curtis, 100, 75, 2000, 100, 150

Jno. Patterson, 40, 7, 1000, 20, 150

John B. Hand (Hare), 40, 20, 800, 30, 250

Jacob Bull, 100, 33, 3000, 140, 500

Ambrose Bull, 40, 25, 2000, 100, 300

Solomon Wolfgang, 45, 80, 2000, 100, 200

John Wilhelm, 100, 23, 2000, 100, 400

Parnel Bickley, 200, 68, 6000, 200,700

Henry Zenker, 130, 60, 5000, 200, 600

John Hare, 40, 20, 1000, 100, 400

David Wilhelm, 90, 34, 3000, 200, 300

Caspar Milidero, 126, 60, 2000, 100, 300

Abraham Wilhelm, 30, 50, 1500, 100, 350

Peter Hoover, 100, 45, 4000, 100, 400

Robet Royston, 50, 27, 2000, 100, 250

Rachel Kilbaugh, 35, 7, 1000, 40, 200

Elisha Wheeler, 200, 100, 4000, 200, 500

Daniel Price, 50, -, 1000, 100, 200

Asa Armacast, 50, 48, 2000, 100, 200

Caleb Price, 85, 70, 2000, 200, 400

John Armacast, 141, 80, 4000, 300, 400

John W. Armacast, 25, 10, 1800, 95, 125

Darby Belt, 21, 8, 800, 20, 150

Isaiah Wheeler, 200, 50, 3000, 400, 850

John C. Zonck, 200, 55, 10000, 700, 1200

Isaac Marfit, 46, -, 1500, 100, 150

Luther Martin, 100,75, 3000, 300, 600

Greenbury Borig, 50, 50, 2000, 100, 300

John P. Kelly, 50, 45, 2000, 100, 350

William Sparks, 60, 60, 20000, 200, 400

Morgan Cox, 60, -, 800, 50, 250

Henry Storm, 15, 20, 2000, 50, 250

Mordecai Gest, 136, 93, 1500, 200, 300

Bryant Wheeler, 100, 100, 3500, 100, 200

Michael Hare (Hand), 63, -, 1000, 100, 200

Nicholas Ensor, 80, 20, 4000, 150, 500

Mordecai Price, 70, 70, 2000, 100, 400

Hannah Alforn, 80, 42, 4000, 150, 700

David Wilhelm, 80, 33, 3500, 150, 350

Isaiah Dyhoff, 40, 27, 1500, 50, 300

John L. Price, 155, 27, 4500, 75, 500

John Zonck, 300, 137, 4000, 390, 1200

Henry Peregoy, 25, 33, 600, 23, 100

Christian Kilbaugh, 75, 26, 2000, 60, 400

Abraham Rosson (Bosson), 100, 100, 3000, 100, 400

Charles Ash, -, -, -, 100, 300

Daniel Wilhelm, 107, 15, 1500, 60, 300

Michael Armacast, 550, 100,7000, 300, 1050

Jos. Tracy, 70, 30, 4000, 150, 300

Noah Wisner, 200, 150, 5000, 300, 800

Richard Tracy, 100, 35, 4000, 200, 350

John Marfoot, 115, 55, 4000, 100, 425

John Peregoy, 43, 1200, 25, 150

Mous Cullison, 100, 45, 1800, 100, 150

Joshua Fowble, 250, 150, 9000, 400, 1000

John M.Wheeler, 250, 50, 6000, 300, 600

J. P. Snyder, 100, 28, 4000, 150, 250

Benjamin Rush, 70, 84, 2000, 75, 375

George Cullison, 164, 50, 2000, 100, 400

John W. Fowble, 100, 37, 3000, 150, 800

Thomas Cole, 80, 80, 3500, 100, 300

George Algire, 65, 15, 2000, 100, 250

Richard Gill, 150, 150, 4000, 200, 400

Elisha Brace, 75, 25, 4000, 200, 600

Elizabeth Birchy, 130, 30, 5000, 200, 400

Henry Zouck, 200, 325, 10000, 400, 800

Thomas Belt, -, -, -, 80, 175

Peter Fowble, 280, 190, 7000, 300, 700

Isaac Howard, 8, 8, 200, 20, 80

Thomas Osborn, 108, 10, 2500, 150, 500

Noah Serfield, 124, 120, 5000, 300, 730

George Alburn, 34, -, 1000, 25, 200

Richard Fowble, 140, 60, 2500, 100, 300

Jacob, L. Cople_, 75, 25, 3000, 100, 500

Johnas Kemp, 230, 100, 10000, 300,700

Shadrack Kemp, 140, 70, 2500, 150, 520

Hosea Kemp, 100, 30, 4000, 40, 380

Didimus Gile, 80, 28, 2000, 100, 500

Lewis Pitts, 100, 23, 6000, 200, 700

Samuel S. Cole, 80, 20, 3000, 150, 300

Samuel Price, 60, 40, 4000, 150, 450

Samuel Shawl, 150, 75, 6000, 100, 125

Evan Davis, 100, -, 10000, 1000, 2000

Thomas Scott, 225, 75, 4500, 150, 400

Alijah Cole, 70, 30, 5000, 250, 600

Isaac Griffith, 100, 100,7000, 200, 300

Jacob Armacast (Armacost), 150, 50, 6000, 100, 440

Michl. Crowther (Crouther), 45, 15, 2000, 250, 500

Jemma Cole, 100, 30, 3000, 200, 500

Jense Benson, 50, 7, 2500, 100, 350

Mi__ Fowble, 140, 60, 10000, 150, 650

Roseanna Tracy, 150, 50, 4000, 150, 500

Benjamin Gersuch (Gorsuch), 180, 100, 9000, 500, 500

Anna Trowbridge, 25, 25, 1000, 50, 350

Abraham Mays, 55, 15, 1500, 75, 250

Elisha Lovell, 200, 70, 3000, 100, 400

Mary Cole, 100, 75, 2500, 100, 500

John Turnbaugh, 38, -, 500, 50, 150

Robert Miller, 370, 100, 5000, 200, 300

Geroge Thompson, 30, 7, 600, 50, 150

Thos. Thompson, 20, 8, 800, 50, 200

John Devin, -, -, -, 100,600

Thomas Hale, 100, 70, 1800, 100, 300

John Dehoff, 100, 65, 4000, 200, 400

Elerewis Guin, 100, 74, 3000, 100, 400

Bayn Kelly, 96, 14, 2000, 100, 175

John Hale, 150, 50, 1500, 100, 200

Elisha Jackson, 82, 15, 1200, 100, 400

Emanuel Guin, 34, 34, 900, 50, 300

Margaret Benson, 80, 50, 4500, 200, 400

Christa Wisner, 95, 34, 3500, 125, 450

Mathias Wisner, 50, 50, 1500, 50, 200

Stephen Wisner, 40, 50, 1800, 50, 200

Ab__ Wisner, 80, 27, 1500, 80, 200

Stephen Fowble, 40, 144, 2000,75, 300

Abraham Wisner, 160, 40, 1500, 30, 200

Isaac Wisner, 50, 10, 1500, 25, 150

Richard Shawl, 200, 200, 3000, 100, 600

Thomas Wilhelm, 75, 25, 2000, 125, 400

Benj. Benson, 138, 100, 5000, 300, 800

Janna James, 100, 12, 2000, 20, 125

John Benson, -, -, -, 40, 300

John Royston, 90, 10, 1500, 50, 300

Thomas Thompson, 40, 10, 1000, 50, 300

Daniel Stump, 60, 20, 2000, 100, 200

Mary Stump, 9, -, 200, 40, 25

Joseph Walker, 42, -, 1000, 100, 200

John Royston, 40, 17, 1000, 100, 300

George Walker, 35, 9, 1500, 75, 500

Henty Coltritter, 30, 18, 4000, -, 86

William Rhule, 100, 40, 3000, 100, 250

Peter Rhule, 10, -, 500, 50, 200

John Dick, 90, 22, 1000, 125, 100

Wm. T. Matthews, 40, 10, 1000, 125, 100

John Palmer, 60, 12, 2000, 500, 150

Archibald Sterling, 60, 34, 3000, 200, 300

Jacob Schafer, 80, 90, 5000, 225, 800

Charles Leader 120, 55, 6500, 200, 600

Zacharia Jacobs, 175, 125, 6000, 225, 800

John Leroni, 170, 250, 7000, 300, 1000

William Hill, 160, 38, 7500, 400, 800

Geo. Hampshire, -, -, 3000, 100, 175

Joseph Keens, 25, 29, 2000, 50, 200

George Nonemaker, 10, 23, 500, 10, 30

Charles Wentz, 65, 22, 2000, 150, 500

Charles Ruling, 62, -, 2000, 50, 300

Jarlnis Ruling, 42, -, 2000, 75, 400

Edward Ruling, 42, 19, 2500, 50, 400

Francis Munday, 50, 60, 2000, 75, 300

John Carr, 24, -, 700, 50, 50

William Rhule, 25, 10, 600, 100, 75

John Shupert, 60, 15, 200, 75, 150

Edmond Walsh, 50, 60, 2000, 50, 400

Wm. McAbee, 16, 16, 1000, 50, 200

Elisha Durey, 60, 21, 3000, 150, 400

Jacob Hampshire, 350, 56, 15000, 300, 1200

Wilson Houseman, 37, 15, 1500, 50, 300

George Keller, 25, 5, 1000, 40, 250

Edward Hoskal, 80, 96, 3400, 50, 328

Nicholas Haskal, 150,75, 4500, 130, 800

Dan__ Stabler, 75, 65, 2000, 100, 400

Jeremiah Baily, 14, 30, 800, 20, 100

John Day, 90, 50, 400, 100, 400

Luther Williams, 50, 26, 4000, 50, 300

James Gisford, 36, 5, 1800, 50, 300

George Baker, 25, 20, 1600, 20, 150

Jacob Pankey, 41, 41, 2000, 50, 200

Mary Sampson, 75, 26, 3000, 75, 400

Carrander (Canada) Eaton, 50, 47, 1000, -, 50

Samuel Kline, 75, 25, 2000, 100, 300

Charles Little, 30, 53, 2500, 50, 300

Samuel Williams, 15, 15, 2000, -, 100

Matilda Keyes, 100, 110, 2000, 50, 2000

Alicanth Keys, -, -, -, 50, 300

Stephen Freeland, 170, 105, 7000, 250, 500

Henry Baker, 40, 45, 1000, 60, 200

John Shock, 20, 48, 1000, 30, 300

Charles Gore, 4, 32, 2000, -, 70

William Hare, 50, 46, 3000, 10, 175

George Marion, 20, 85, 2000, 60, 400

Jacob Palmer, 100, 40, 6000, 300,500

Joseph Palmer, 100, 30,7000, 200, 400

Edward Amos, 35, 32, 1000, -, 90

Enoch Dorsey, 100, 50, 6000, 60, 300

Abraham Oligrath, 15, -, 300, 75, 225

John Tarmor, 65, 20, 1800, 80, 200

Nathaniel Spicer, 20, -, 1000, -, 30

Thomas Wantland, 18, -, 1800, 100, 300

William Davis, 30, -, 4000, 300, 278

Jacob Stephens, 13, -, 1500, -, 200

Joseph Cooper, 50, 34, 3500, 30, 200

Grifton Marshal, 50, 67, 3000, 75, 350

Theodore Masemore, 35, 17, 1500, 60, 182

Samuel Young, 50, 40, 1200, 65, 225

Susanna Taylor, 40, 20, 400, 40, 300

Lydne Shafer, 25, 33, 1000, 50, 200

William Fisher, 5, 20, 1250, 150, 550

Edward Matthews, 50, 50, 3000, 150, 500

Valentine Price, 25, 79, 2000, 26, 150

Daniel Kish, 20, 60, 1000, -, 200

Walter Walker, 100, 98, 5000, 60, 400

George Palmer, 20, 25, 1000, 20, 150

Thomas Hornt (Hanet, Hermet, Hunt), 15, 193, 2500, 60, 306

William Morris, 50, 25, 2000, 50, 150

Ezra Morris, 31, -, 800, 50, 100

Luther Amos, 60, 18, 1800, 60, 100

James Leach, 26, 26, 1000, 30, 200

Nelson Cullings, 150, 104, 4000, 200, 600

William McDonough, 30, 14, 1000, 20, 60

Thomas Thompson, 10, -, 500, 20, 61

Jesse Hoshall, 250, 250, 6000, 200, 151

John Mayes, 50, 40, 1000, 30, 300

Caleb Hoshall, 22, 100, 1000, 60, 400

Jacob Hoshell, 100, 50, 4000, 400, 800

Jacob Makey, 129, 100, 5000, 100, 600

Jacob Shamlager, 160, 80, 7000, 50, 800

Henry Hoffacker, 70, 197, 1400, 150, 650

Nelson Hoshall, 100, 90, 2000, 30, 250

John Hoshall, 25, 20, 1500, 80, 200

Ephraim Hoshall, 50, 50, 1500, 50, 150

Elisha Jones, 50, 50, 2000, 60, 200

Micajah Cooper, 150, 150, 2500, 60, 400

Joseph Pope, 58, 22, 1800, 150, 250

Amos Cross, 50, 64, 2000, 50, 300

John Spicer, 42, 12, 900, 20, 100

Christopher Bull, 100, 30, 2500, 125, 500

John W. Liperman, 30, 32, 1000, 25, 100

George Alburn, 82, 40, 2000, 100, 250

John Wisner, 40, 45, 3000, 100, 400

Jacob Amos, 75, 25, 3500, 100, 300

Samuel Gore, 115, 75, 6000, 400, 600

James W. McCullough, -, -, -, 75, 350

William Ensor, 75, 12, 3500, 200, 400

Daniel Wilhelm, 140, 100, 2500, 125, 200

John Ingham, 100, 70, 4000, 200, 600

Thomas Mays, 30, 10, 1300, 50, 100

John Ensor, 65, 9, 1500, 75, 350

Samuel Alburn, 14, -, 1000, 50, 200

Nich___ Bull, 50, -, 1500, 75, 30

Elizabeth Bering, 37, -, 600, -, 50

John Sanble (Sauble), 75, 75, 4500, 300, 700

John SS. Sanble (Serble), 30, 25, 1500, -, 100

Daniel Baublitz (Banblitz), 50, 50, 800, 25, 50

Zachariah Alburn, 100, 50, 3000, 150, 600

Charles Burke, 125, 100, 8000, 200, 600

John Bublets, 125, 100, 8000, 200, 600

Samuel Baublitz, 100, 36, 1500, 100, 200

John Holenbaugh, 50, 42, 1500, 100, 275

Charles Ender, 20, 10, 600, 75, 150

Abraam Hase (Hare), 80, 63, 2000, 75, 200

Daniel Hase, 70, 50, 3600, 250, 600

John Cooper, 80, 60, 2500, 125, 300

Peter Wilhelm, 150, 100, 8000, 200,700

William Young, 150, 45, 3000, 100, 250

Sophia Paist, 10, 30, 1000, -, 60

Levi Rust, 80, 50, 5000, 150, 300

Valentine Rust, 100, 38, 1500, 60, 200

Mary Seigman, 50, 68, 1500, 75, 100

Mordecai Alburn, 39, 14, 2000, 100, 150

Benjamin Gooding, 116, 117, 2500, 100, 600

Christian Baublitz, 10, 44, 800, 10, 60

John Zimmerman, 50, 50, 1500, 50, 250

Mils Davis, 60, 35, 3000, 100, 280

William McCullough, 100, 108, 5000, 200, 600

Christian Gore, 100, 40, 3000, 150, 300

Joseph Boilinger, 100, 100, 4000, 200, 300

Henry Hase, 110, 100, 4000, 200, 300

Thomas Alburn, 84, 25, 3000, 150, 400

Henry Hofacker, 100, 85, 5000, 200, 600

Joseph Hare (Hase), 50, 45, 1500, 100, 300

William Hoffman, 100, 150, 3000, 300, 500

George Miller, 150, 100, 4000, 150, 600

Jesse Hampshire, 150, -, 2000, 100, 450

John Hampshire, 100, 80, 2000, 100, 350

Jacob Sharer, 100, 80, 2000, 100, 600

John Cross, 150, 80, 3000, 200, 175

William D. Hoffman, 25, 26, 1500, -, 100

William H. Hoffman, 360, 1050, 5000, 500, 400

George Henry, 12, -, 500, 15, 100

Frederick Bailey, 100, 25, 2000, 50, 250

Levi Emmie (Emmric), 32, -, 900, 25, 100

James Shire (Shere), 43, 20, 1200, 30, 200

Andrew Hick, 80, 39, 2000, 30, 200

Samuel Foreman, 25, 25, 500, -, 80

Henry Hilker, 70, 23, 1800, 50, 400

Noah Miller, 33, -, 1000, -, 60

Onrad Waltman, 135, 135, 10000, 200, 800

William Masmore (Masman), 25, -, 1000, -, 50

Margaret List, 25, 5, 7700, -, 4000

Henry Rhule, 30, 20, 1800, 30, 120

Nicholas Hile, 60, 20, 1500, 40, 150

Christopher Hoff, 22, -, 800, 40, 150

Henry Baker, 100, 75, 2000, 150, 60

Jacob Osler, 75, 25, 2000, 100, 800

Isaac Raines, 43, 11, 2000, 200, 300

Joshua Frederick, 50, 60, 1000, 100, 150

John Bailey, 50, 50, 2000, 80, 200

Absalom Baker, 40, 42, 1500, 25, 100

Rachel Wilson, 40, 40, 800, 10, 75

William Kidd, 50, 30, 2000, 200, 275

John W. Cooper, 25, 50, 900, 75, 250

Daniel Wilson, 20, 9, 800, -, 60

Lysander McCullough, 60, 24, 500, 150, 400

John Whistler, 100, 120, 4000, 80, 300

Jacob Gosnell, 100, 38, 5000, 125, 370

Eli Kidd, 53, 10, 1000, 50, 350

Edward Kelly, 75, 40, 2500, 40, 300

Rebecca Dorsey, 50, 61, 1500, 40, 150

Samuel Morris, 65, 15, 1500, 75, 225

Abraham Downs, 90, 34, 2000, 75, 325

Lewis Morris, 80 25, 1800, 50, 250

Tigo Cooper, 95, 35, 4000, 200, 400

William Miller, 10, 50, 3500, 50, 300

John Bull, 90, 10, 2000, 75, 600

Henry Wilson, 200, 20, 4000, 50, 230

Aquilla Wilson, 150, 50, 4000, 50, 350

William Craft, 60, 41, 2300, -, 200

Adam Bailey, 70, 20, 2000, 40, 350

Josiah Wheeler, 80, 50, 2000, 40, 280

Jemima Foster, 240, 70, 2000, 200, 350

Joshua Spindler, 25, 8, 1000, 20, 200

Enoch Spindler, 70, 30, 2000, 20, 200

William Hamilton, 80, 18, 1500, 20, 195

William Johnston, 63, 60, 3500, 100, 630

Conrad Turnbaugh, 110, 30, 2000, 75, 700

John O. Mays, 200, 100, 3000, 100, 500

Milcher Armacost(Armacast), 246, 212, 8000, 400, 1500

Thomas Cooper, 75, 25, 3000, 100, 325

Lusamia Jones, 170, 170, 2000, 200, 650

Maria Cox, 100, 60, 4000, 50, 300

Jacob Vance, 40, 7, 800, 50, 275

Wm. Turnbaugh, 40, 14, 1500, 200, 300

Thomas Teft (Taft), 25, 9, 1200, 50, 350

John Knight, 12, -, 800, 25, 200

George Childs, 200, 167, 5000, 500, 1000

Joseph Nash, 50, 14, 200, 100, 280

Henry Merryman, 140, 59, 5900, -, 650

Martin Conn, 140, 59, 5900, 200, 600

John P. Mays, 150, 35, 7000, 200, 1500

Abraham Hicks, 60, 50, 2000, 100, 250

John R. Butt (Batt), 100, 60, 1800, 60, 345

Thomas Satie (Satio, Latio), 65, 60, 1500, 50, 300

Welsey Rosnic, 200, 50, 4000, 100, 400

Charlott, Withers, 450, 800, 15000, -, 250

Salem Tracy, 43, 3, 1500, 100, 51

Ann Kauffman, 16, -, 1500, 100, 200

John McCabbin, 64, 20, 2000, 100, 600

Joshua Standiford, 110, 50, 3000, 100, 300

William Plowman, 100, 20, 2500, 50, 200

David Sampson, 150, 50, 6000, 200, 600

Albert Woodrough, 60, 5, 5500, 100, 500

Ephraim Bell, 100, 35, 5500, 150, 500

James A. Mandiford, 135, 25, 6000, 400, 650

Rebecca Kennedy, 85, 8, 2500, 35, 375

John Hendric, 35, -, 1200, 100, 200

David Crawmer, 60, 20, 2000, 50, 150

Abraham McDonald, 200, 81, 5000, 300, 880

Matthew Hunt, 100, 70, 1500, 150, 300

Benjamin Koontz, 25, 72, 1000, 50,1000

Maranda Little, 110, 20, 4000, 50, 200

John Roser, 80, 50, 3000, 50, 500

Abijah Roser, 88, 30, 3000, 100, 500

Belinda Rutledge, 50, 10, 700, 50, 500

John Dalmer (Palmer), 75, 23, 1200, 30, 200

William Zatty, 40, 16, 1000, 50, 400

Benson Sullivan, 30, 69, 2000, 50, 250

Aaron Freeland, 60, 17, 1800, 80, 300

Daniel Keller, 75, 30, 2500, 200, 640

Isaac Sampson, 60, 20, 2500, 50, 275

James Harris, 160, 100, 6500, 300, 800

John Gemmell, 125, 80, 6200, 150, 500

Sims Whitcraft, 187, 30, 6000, 200, 400

Thomas Harkness, 110, 40, 4000, 200, 400

Stephen Gusuch (Gorsuch), 125, 25, 3000, 300, 360

William Gusuch, 80, 50, 3000, 100, 360

Luther Ward, 74, 50, 3000, 125, 330

Samuel Cornelius, 9, -, 1600, 30, 200

Joseph Seitz, 92, 30, 1600, 150, 250

Daniel Treadway, 90, 10, 1080, 100, 425

Mary Andrew, 100, 62, 1600, 50, 375

Benjamin Norris, 105, 50, 2800, 50, 700

Lysias Smith, 36, 8, 1800, 40, 300

Wesly Greer (Green), 80, 60, 5500, 250, 500

Elijah Chilcat (Chilcot), 100, 69, 2000, 200, 500

Samuel Billingsley, 70, 20, 1500, 100, 400

Silas Slade, 140, 7, 2500, 150, 450

Ross Bond, 100, 33, 3000, 200, 400

Elisha Bond, 131, 70, 5000, 100, 400

Pricilla Rutledge, 68, 100, 1500, 50, 350

Julia D. Gordon, 100, 80, 7500, 200, 300

Anne E. Almoney, 100, 90, 500, 20, 500

Robert Raknin, 100, 46, 5000, -, 280

Benjamin Burk, 90, 20, 4000, 100, 100

Thomas Wilson, 75, 30, 2000, 75, 125

Michael Minety, 80, 20, 2000, 50, 100

Clement Busey, 110, 40, 5000, 100, 250

Joseph Walker, 160, 380, 7000, 100, 1500

George Little, 200, 66, 7000, 150, 800

Sarah Stabler, 100, 40, 3500, 100, 340

Daniel Stabler, 125, 139, 6000, 200, 800

Wesly Cuddy, 63, 30, 1800, 100, 390

Madison Slade, 100, 80, 3600, 200, 750

David Wilson, 80, 20, 2000, 50, 400

John Slade, 60, 28, 1800, 200, 700

Thomas Slade, 200, 156, 5000, 100, 500

Reuben McCullough, 100, 38, 2500, 75, 400

Margaret Wright, 60, 24, 1500, 50, 140

Arthur Shane, 58, 20, 2000, 50, 300

Wm. Kirkwood, 100, 13, 3000, 250, 300

Robert Stirling, 150, 150, 8000, 300, 800

Joshua Shipley, 100, 66, 3500, 50, 300

John Shipley, 85, 86, 2500, 150, 300

Thomas Meredith, 75, 25, 2000, 50, 250

Buhna Bell, 100, 50, 3000, 100, 400

Micajah Meredith, 80, 20, 2500, 250, 350

Samuel Meredith, 250, 100, 3000, 200, 500

Matthew Birmingham, 100, 44, 4000, 175, 400

Sophrana Kella, 100, 69, 6000, -, 220

Franklin Anderson, 120, 42, 8000, 500, 620

Kean Curry, 96, 12, 2500, 20, 225

James Turner, 200, 900, 30000, 500, 400

James Anderson, 125, 45, 11000, 150, 525

James Price, 80, 50, 6500, 60, 540

Peter G. Hunter, 72, 40, 3000, 40, 530

Margaret Park, 200, 100, 6000, 55, 430

William Bull, 75, 60, 4000, 100, 500

Harth Frederick, 60, 139, 3000, 35, 2000

Nicholas Bull, 50, 80, 3000, 100, 400

Morris Frederick, 141, 25, 6000, 150, 530

Samuel Bull, 100, 25, 2000, 100, 300

William Collut, 50, 50, 3000, 100, 400

John Wise, 100, 50, 4000, 125, 500

Charles Burmeda, 170, 30, 11000, 100, 535

William Burns, 100, 108, 5000, 50, 400

Jos. Fitzpatrick, 40, 40, 4000, 50, 300

William Rowe, 50, 135, 4500, 150, 280

Mary Riley, 22, -, 700, 25, 65

Pleasant Hunter, 144, 24, 7000, 200, 1000

Joshua Pearce, 43, 20, 1260, 100, 300

Richard Nelson, 28, 10, 2500, 100, 225

William Wise, 44, 6, 2500, 50, 130

William Nelson, 40, 30, 1200, 20, 150

Thomas Hunter, 113, 10, 3000, 50, 400

John H. Nelson, 28, 2, 750, 100, 150

John Hunter, 100, 28, 3000, 100, 450

Phillip Stran, 30, 10, 1200, 50, 300

Janet Nelson, 80, 16, 4000, 100, 250

William Slade, 90, 25, 3000, 225, 385

James Elliott, 80, 20, 4000, 150, 250

Benjamin Garret, 80, 20, 2000, 200, 685

John Almoney, 77, 13, 4000, 100, 200

James Almoney, 112, 13, 2000, 100, 200

Dorcas Almoney, 115, 15, 1700, 100, 75

Rachel Almoney, 125, 25, 1600, 100, 240

Jefferson Almoney, 100, 50, 1500, 100, 900

Harrison Almoney, 70, 20, 3000, 50, 298

Chrisopher Slade, 80, 80, 3000, 400, 600

Norris Parish, 100, 50, 4000, 300, 600

Henry Wise, 62, 18, 4000, 100, 600

John Burrus (Barnes), 275, 275, 6000, 150, 600

John Mays, 200, 100, 8000, 200, 1000

John Rowe, 100, 70, 4000, 150, 400

Joseph Rowe, 50, 17, 1500, 100, 270

William Foster, 40, 47, 2000, 100, 340

James Kerr, 20, 30, 2000, 50, 180

Dorsey Mitchell, 60, 20, 2200, 50, 450

Thomas Wheeler, 63, 25, 2000, 50, 350

William Pearce, 70, 40, 1700, 50, 250

Daniel Matthews, 200, 115, 8000, 150, 500

John Bacon, 360, 40, 35000, 1000, 3000

James McBride, 75, 25, 5000, 200, 500

Richard Remare, 170, 50, 5000, 500, 1200

Thomas Gilbert, 75, 100, 5000, 100, 350

John Litzinger, 60, 10, 4000, 100, 300

Amos Stiltz, 37, 3, 1000, 50, 300

Ambrose Bull, 100, 80, 4000, 40, 100

John Miller, 467, 233, 4000, 200, 685

Stephen Miller, 34, 16, 2000, 30, 100

James Bull, 75, 25, 6000, 250, 600

Abraham Worthington, 80, 24, 52000, 500, 400

Charles Rogers, 18, 8, 40000, 150, 350

George Appold & Co. 19, -, 20000, 50, 200

George M. Lamby, 35, -, 36000, 100, 2500

John Ogiver, 19, -, 17000, 25, 300

JamesM. Taggert, 12, -, 7200, 50, 200

Jones & Bros., 27, -, 8100, 150, 400

Charles Melcher, 18, -, 7000, 200, 430

Mary Gales, 2, -, 1200, 25, 50

Amos King, 48, -, 38400, 100, 200

George Ward, 4, -, 1600, 12, 1120

John Miller, 10, -, 3000, 150, 325

Peter Stewart, 13, -, 6500, 60, 320

Christopher Fisher, 10, -, 10000, 150, 240

Alexander McBlaser (M. Blaser), 9, -, 9000, 200, 250

Michael Bullinger, 4, -, 4000, 200, 150

John Burtin, 15 ½, -, 3000, 25, 400

John Whiteford, 50, 20, 6000, 30, 300

Augustus Dabaugh, 75, -, -, 75, 100

Thomas Galloway, 120, 40, 11000, 200, 300

Joseph Tosborne (T. Osborne), 100, 69, -, 60, 100

Joshua T C. Talbott, 110, 30, 10000, 200, 800

Richard Padgran, 90, 10, 3000, 50, 60

Aquilla Galloway, 60, -, 5000, 40, 300

Eliza Bosley, 160, 7, 23000, 100, 20

James Jackson, 140, -, 14000, 40, 150

Robert F. Rice, 40, -, 8000, 25, 325

James A. Miller, -, -, -, 50, 400

Thomas B. Cockey, 70, 30, 6500, 100, 500

Edward D. Goodwin, 300, 200, 15000, 50, 400

Thomas D. Cockey, 135, 7, 16000, 150, 1000

John Bosly of W, 77, 50, 8000, 200, 1000

John Parks, 170, 68, 9000, 200, 800

John Galloway, 73, 7, 10000, 200, 700

Richd. Peddiarid, 150, 28, -, -, 500

Augustus Webster, 8, -, 3000, 40, 240

Temperance Talbott, 130, 15, 6500, 100, 400

Francis Corcoran, 21, -, 15000, 100, 400

William O. Wilson, 140, 36, 14000, 200, 1000

John Clark, 350, 150, 7000, 400, 1500

Joshua Talbott, 600, 425, 160000, 400, 2200

Jno. F. Fitzsimmons, 105, -, 10000, 150, 400

Milton Bosley, 120, 20, 6000, 25, 500

Joshua Bosley, 200, 50, 15000, 400, 1800

Milton Dance (David), 200, 32, 13000, 400, 2400

George Smith, 300, -, 25000, 200, 1000

John Farlett, 288, -, 14000, 300, 1000

Joseph Kelly, 350, 19, 30000, 400, 800

Elisha Backs, 96, -, 7500, 400, 800

William Price, 100, 20, 5000, 300,700

Charle Bosley, 710, 10, 9000, 400, 1300

Aquilla Talbott, 273, 8, 18000, 400, 2500

Catherine _. Todd, 167, 50, 8000, 300, 800

George Harneyman, 130, 70, 6000, 300, 1700

George Cole, 130, 20, 8000, 400, 1300

Geo. Merryman, 300, 100, 20000, 600, 709

John Justice, 30, -, 500, 30, 125

William Worden (Wooden), 15, 10, 500, 25, 150

Stephen Musgrove, 60, 140, 6000, 100, 750

Susanna Jones, 70, 6, 25000, 50, 175

Thomas Kenny, 10, 5, 300, 25, 100

Matthew Caspar, 16, -, 400, 50, 150

John M. Roff, 30, 5, 1500, 50, 200

John Whiteford, 50, 130, 2000, 100, 500

Aquilla Jones, te, na, nt, 100, 400

Joshua Tipton, 170, 30, 4000, 100, 500

George Eckart, 20, -, 600, 50, 120

Frederick, Kramble, 20, 5, 600, 50, 120

Henry S. Hipsley (Shipley), 130, 30, 5000, 100, 500

Andrew Kleintint, 9, -, 3000, 50, 250

Thomas C. Tracy, 45, -, 2500, 150, 250

Edwin R. Griffith, 60, 60, 3000, 30, 152

William Price, 150, 50, 9500, 100, 600

Joshua Price, 100, 50, 9000, 100, 300

William Gent, te, na, nt, 100, 600

John D. Matthews, 160, 40, 8000, 50, 650

John Merryman, 300, 250, 40000, 500, 6000

Charles McLean, 55, -, 8000, 75, 700

Thomas Bond, 63, -, 4000, 150, 1400

Mordecai Matthews, 150, -, 15000, 300, 900

Thomas Price, 140, -, 5600, 200, 400

George Ensor, 100, -, 500, 100, 300

William 0. Ensor, 31, -, 2500, 100, 600

Thomas L. Hall (Hale), 87, 10, 9000, 250, 400

John E. Bosley, 180, 20, 15000, 500, 1200

Richard D. Wheeler, 113, -, 4500, 150, 150

John F. Shipley, 200, 52, 20000, 400, 1100

Vincent Shipley, 120, 30, 20000, 50, 600

William Bosley, 200, 1017, 40000, 200, 1000

George Wisner, 100, -, 4000, 5, 300

Thomas F. Matthews, 200, 60, 15000, 200, 2000

Darby S. Ensor, 94, -, 7000, 100, 500

John Price, 250, 70, 20000, 600, 1600

Simone Fast, 130, 70, 10000, 150, 375

George Ensor, 140, 85, 8275, 500, 865

Peter Parks, -, -, -, 50, 160

Peter Parks, 43, -, 1500, 50, 160

John Thomas, 13, -, 500, 100, 200

Elias P. Thomas, 3, 30, -, -, 75, 200

Jane Morgan, 100, -, 10000, 100, 500

Thomas J. Lee, 200, 100, 45000, 100, 800

Joshua F. Cockey, 250, 24, 25000, 1000, 800

Thomas Love, 500, 70, 40000, 1000, 4500

Solomon Armstrong, 80, -, 8000, 100, 350

Zephana Polet, 225, 175, 20000, 250, 1900

William Carmary, 20, 10, 8500, 25, 70

Geo. R. Royston, 90, 30, 5000, 150, 600

Benj. Johnson, 200, 40, -, 100, 400

Isaac Webster, 250, 150, 4000, 1000, 600

{Charles Gorsuch, 300, 100, 12000, 100, 500

{Noah Gorsuch, Jointly, -, -, -, 100, 400

Saml. Manglelin, 30, -, 1400, 50, 100

John Beatty, 200, 20, 15000, 500, 2500

Josiah Price, 120, 50, 10000, 200, 900

M. Matthews, 70, 30, 4000, 100, 400

Benjamin R. Price, 130, 37, 6000, 400, 700

Benjamin Matthews, 100, 12, 4500, 200, 450

Joshua M. Gorsuch 65, 21, 5300, 25, 650

Thomas L. Gorsuch, 150, 113, 13300, 300, 1000

Thomas T. Price, 90, 50, 9600, 300, 1100

John Merryman, 30, -, 1100, 100, 300

Thomas C. Price, 90, 16, 5500, 300, 650

Mahlon Price, 12, -, 4000, 200, 125

Joel Price, 100, 31, 6000, 250, 720

Arin Wheeler, 40, 50, 2000, 25, 75

John Matthews, 260, 70, 11000, 500, 1100

Lewis R. Cole, 300, 244, 12000, 500, 1050

Abraham Cole, 250, 40, -, -, 500

George Jessop, 150, 200, 11000, 500, 1100

Jesse B. Gore, te, na, nt, 50, 700

John G. Rider, 135, 30, 10000, 300, 600

John Cockey, 150, 450, 29000, 150, 800

Robert Jones, 10, -, 600, 50, 300

Edward Rider, 500, 100, 80000, 1000, 3200

Harry Watson, 200, 100, 25000, 50, 700

Thomas B. Cockey, 67, 71, 5000, 200, 900

Mary A. Steenlagner, 65, 25, 3000, 100, 600

Jeremiah Mays, 94, 62, 3000, 75, 330

William Parks, 100, 20, 1800, 75, 300

Phil Brittenburger, 75, 30, 2500, 20, 450

Nanen Factory Farm, 100, 158, 7500, 250, 800

Faron Green, 55, 20, 3000, 100, 300

Louch E. Hoffman, 170, 30, 10000, 600, 2100

{William Jessop, 100, 122, 19000, 500, 2200

{William Jessop, 80, 53, 11000, -, -

(Richard Green, 30, -, 3000, 50, 4000

{Richard Green, 60, -, 6000, 20, 3000

Peter F. Cockey, 130, 40, 10000, 100, 800

Gest T. Cockey, 40, 50, 4000, -, 500

Abram Johnson, 92, -, 10000, 200, 600

Nelson Gorsuch, 90, 30, 5000, 100, 300

John Royston, 60, 40, 5000, 200, 300

Elisha Marg, 100, 20, 6000, 200, 400
William R. Mays, 75, 15, 5000, 100, 300
Benjamin P. Matthews, 30, -, 3000, 50, 200
Levi Bruce (Price), 83, 3, 7000, 200, 800
{Evan Matthews, 120, 15, 10000, 300, 700
{Evan Matthews, 118, 12, 13000, -, 300
{Elias Matthews, 47, -, 5000, 500, 500
{Elias Matthews, 120, 10, 10000, -, 300
Aquilla Matthews, 120, 18, 11000, 509, 1100
Dickinson Gorsuch, 240, 50, 25000, 600, 2000
Caleb Pearce, 175, -, 10000, 300, 400
Daniel Orrick, 90, 10, 8000, -, 300
Zelitha Matthews, 140, -, 10000, 400, 60
John Musselman, 200, 12, 15000, 400, 1000
George Matthews, 150, 50, 1500, 500, 800
George Chilcoat, 32, -, 4000, 20, 400
Naomi Ensor, 330, 20, 16000, 400, 1000
Charles Brooks, 110, 7, 6000, 300, 400
John Chilcoat, 70, 13, 5000, 200, 500
John B. Ensor, 130, 20, 9000, 300, 1700
William C. Ensor, 80, -, 6000, 200, 500
{John S. Ensor, 130, 20, 7500, 300, 1800
{John S. Ensor, 90, 60, 3700, -, -,
John Ensor of A, 120, 4, 7000, 300, -
John Henson, 150, 50, 10000, 300, 800

Salathiel Cole, 250, 30, 20000, 400, 1500
Joseph Fowler, 130, 75, 15000, 300, 900
Peter Bosly, 350, -, 15000, 400, 1200
William Brook, 90, 3, 7000, 300, 1500
Charles G. Whalen, 150, -, 10000, 400, 800
Aquilla Chilcoat, 85, 15, 7000, 300, 500
George W. Gill, 160, 20, 12000, 400, 1300
Amelia Gill, 40, -, 2000, 100, 1100
Edward Gill, 110, 30, 12000, 500, 1500
John T. Johns, 200, -, 16000, 500, 2900
George Chilcoat, 150, 40, 15000, 300, 1200
Alexander Brown, 800, 300, 50000, 600, 2500
Orie A. Belt, 40, 12, 4000, 100, 150
Henry Wilson, 50, 60, 4000, 50, 100
Edward Barnum, 50, 50, 4000, 100, 250
John Kelley, 30, -, 1000, 100, 200
Josiah Gill, 30, 51, 1000, 100, 200
Eliza A. Lee, 50, 65, 12000, 50, 150
Thomas Kelley, 100, 107, 5000, 200, 600
James Chalk, 60, 50, 4000, 100, 400
Thomas Ward, 95, 10, 8000, 200, 600
Matthew Casper, 16, -, 1000, 500, 200
Conrad Nolinberger, 34, -, 1500, 50, 150
William Gerot, 115, 10, 10000, 200, 700
Matthias Zenk (Zerk), 23, -, 2000, 100, 250
Joshua Marg (Mary), 50, 62, 8000, 100, 300
Harper Hoffman, 60, -, 4000, 100, 150

William Conver, 40, 5, 2000, 100, 200

James Warden, 20, -, 1200, 100, 250

John Knapp, 18, -, 1000, 50, 150

Henry Darver (Parver), 43, -, 2000, 50, 200

Robert Hutton, 60, 7, 2000, 50, 300

William Harris, 48, -, 3000, 100, 300

Ephraim Harris, 100, 90, 12000, 200, 600

Henry Hearhauls, 25, -, 1000, 100, 200

Uriah Cox, 90, 100, 10000, 200, 500

Geo. Dearhauls, 24, -, 1000, 50, 150

Geo. B. Richardson, 140, 50, 10000, 200, 700

Lewis A. Dehoff, 65, -, 5000, 100, 400

Samuel Worth, 400, 500, 60000, 400, 2000

Sophia Worthington, 575, 200, 70000, 1000, 3000

William C. Gent, 300, 100, 24000, 1000, 2200

Henry Merryman, 90, 6, 3500, 150, 300

Thomas M. Scott, 90, 16, 5000, 200, 450

Michael Merryman, 130, 20, 9000, 150, 500

Abraham Scott, 120, 40, 10000, 250, 800

Jacob Geist (Gust, Gist), 150, -, 10000, 300, 1000

Thomas T. Griffith, 120, 30, 10000, 200, 1000

Samuel H. Matthews, 50, -, 5000, 200, 300

Jas. R. J. Price, 104, -, 10000, 300, 600

Stephen Gill, 90, 15, 10000, 200, 700

Kesiah Ceele (Cole), 100, 57, 15000, 400, 800

John Scott, 100, 30, 15000, 400, 1200

Edwin Scott, 100, 6, 12000, 300, 1100

Mary Scott, 50, -, 3000, 100, 150

Rebecca Griffith, 200, 100, 25000, 400, 800

Geo. S. Ensor, 120, 40, 12000, 400, 900

Jos. Bosley, 220, 40, 20000, 400, 1000

Thos. C. Bosley, 180, 20, 20000, 400, 600

Thos. Ingham, 85, 5, 5000, -, 800

John Thomas, 13, -, 232, 50, 200

Elias Parks, 30, -, 900, 75, 200

Phil Ballenburger, 75, 30, 2500, 20, 450

Sylvester Gruser, 25, 15, 800, 50, 100

William Dawson, 40, 55, 500, 100, 400

James Wainwright, 104, 20, 10000, 200, 500

Thomas B. Coresly, 140, 100, 10000, 500, 1000

David Blacklock, 40, 9, 4000, 400, 400

Andrew Henies, 4, -, 400, 15, 40

James Hubbard, 6, -, 1500, 20, 60

Charles Masters, 1, -, 700, -, 6

J. W. Shanklin, 3, -, 500, -, -

J. W. Shanklin, 3, -, 500, -, -

Wm. Fuller, 2, -, 400, -, -

J. W. Shanklin, 250, 330, 20000, 250, 1200

Henry Maglat, 27, 3, 1200, 25, 50

John Letchthaler, 30, 5, 3500, 25, 300

Lewis Zarig, 20, 14, 2000, 100, 150

W. J. Cole, 75, 25, 10000, -, 650

Jno. A. Kilenshu, 21, 1, 4000, 600, 450

David M. Perine, 260, 90, 70000, 500, 1600

M. S. Watkins, 65, 25, 16000, 300, 1280

H. A. Ockerhanson, 21, -, 10000, -, 120

George H. Miles, 75, 50, 5000, 150, 200

Jno. H. Miller, 35, 55, 4000, 150, 250

Henry Wilberger, 10, 10, 1500, 75, 50

Charles Briscoe, 4, -, 800, 50, 300

Luke Magory, 36, 12, 9000, 25, 500

Joshua Kidd, 60, 40, 20000, 25, 250

George Hodinott, 5, -, 2000, 6, 100

John T. Thompson, 11, -, 8000, 25, 350

Samuel H. Ellis, 10, -, 6000, 20, 30

George Highthait, 10, -, 2000, 25, 150

Robert Moore, 50, 15, 20000, 200, 2000

John B. Hax, 20, 10, 10000, 40, 400

Crhstian Hax, 20, 10, 10000, 40, 350

Edward A. Slicer, 23, -, 7000, 150, 600

Andrew Riring (Riving), 14, -, 700, 10, 75

Van Ames, 9, 1, 500, -, 150

Phillip Ensor, 23, 3, 1400, 40, 250

George Brickman, 5, -, 300, 10, -

John Simms, 3, -, 700, 50, -

Henry Tritel (Triter), 6, -, 3000, -, 200

Gustave Banger (Ranger), 14, -, 4000, 100, 700

John Moffit, 60, 20, 16000, 100, 250

Thos. Johnson, 6, 5, 2300, 20, 75

John Reed, 120, 74, 19000, 50, 200

Wm. Dougherty, 4, 8, 4800, 100, 250

Thomas Robinson, 42, 6, 5000, 100, 300

George Botwell, 9, -, 4000, 125, 200

Shirden Guilew (Guilen), 40, -, 16000, 300, 300

Elizabeth Macklin, 40, -, 16000, 300, 250

Cas C. Graham, 27, -, 12000, -, 200

Archibald Stirling, 86, -, 2000, 500, 1200

John Baringer, 40, -, 4000, 200, 500

James Stevenson, 9, -, 10000, 100, 700

Ernest Hoen, 6, -, 10000, 200, 150

Henry Hoen, 2, -, 4000, 150, 350

T. J. Gestings, 19, -, 20000, 200, 500

August Hoen (Hour), 12, -, 15000, 300, 150

George B. Stine, 1, -, 4000, 100, 250

William Kennedy, 15, -, 20000, 100, 1000

Andrew Rutherford, 12, -, 15000, 100, 200

Levi Kesler, 80, 80, 24000, 200, 400

August Kohler, 18, -, 10000, 200, 500

Eugene Hunault, 50, 16, 4000, 200, 300

Wm. Williams, 35, 20, 8000, 100, 200

Thomas Brace, 8, 2, 600, 15, 60

Saml. McCormick, 3, -, 400, 10, 60

William S. Brown, 8, 4, 600, 10, 50

John Bayne, 16, 10, 2500, 20, 200

William Bayne, 10, 1, 1500, 20, 300

William P. Preston, 320, 500, 50000, 1000, 1000

Wm. E. Stansbury, 140, 55, 25000, 200, 600

Wilkinson Taylor, 120, 16, 25000, 200, 1000

Samuel Buckman, 45, 5, 7000, 100, 500

Isaac Anderson, 53, 2, 8000, 100, 450

Davis Sendall (Tindall), 80, 91, 10000, 150, 300

Patrina (Patricia) Wheeler, 60, 55, 8000, 150, 100

Peter Epps, 9, 3, 700, 10, -

Mary A. Foster, 10, 6, 6000, 100, 1000

Samuel H. Green, 32, -, 4000, 20, 300

John Erdman, 22, -, 8000, 130, 350
August Hemler, 5, -, 2000, 100, 150
Frederick Reuter, 6, -, 2000, 100, 150
H. Shroeder, 5, -, 1500, 100, 100
Richard Cromwelt (Cromwell), 25, 15, 12000, 10, 100
Mary A. Griswold, 12, 4, 3000, 50, 400
John List, 19, -, 3000, 100, 250
John Ray, 13, -, 2000, 100, 1000
Elijah Taylor, 150, 125, 25000, 500, 800
Henry Nous, 160, 5, 30000, 5000, 100
James Gethart, 136, 40, 50000, 1000, 2500
Ths. J. Lovesrowe, 40, 10, 15000, 250, 500
H. R. Curley, 30, -, 1700, 200, 500
Andrew Essig, 16, -, 3000, 50, 200
Isaac F. Lamley, 14, 1, 3000, 120, 180
Aled Brodie, 25, 4, 4500, 600, 490
Jno. Imwold, 18, 2, 3000, 200, 300
Allen Talbott, 12, -, 5000, 600, 1500
John Brodie, 26, 2, 5000, 200, 500
Ann M. Richard, 100, -, 10000, 200, 400
H. C. Grabenhunt, 70, 200, 100000, 1000, 200
Saml. G. Tucker, 50, 40, 15000, 300, 650
Obadiah Kemp, 25, -, 4000, -, 500
Endracia Stansbury, 140, 80, 22000, 100, 600
Jos. Stevenson, 200, 75, 25000, 200, 800
O. P. Merryman, 70, 31, 12000, 100, 600
Lewis Merryman, 60, 70, 14000, 100, 500
John Porter, 70, 12, 12000, 100, 300
Lewis Bustee, 70, 30, 12000, 200, 1000

Isaac Taylor, 100, 120, 20000, 400, 750
Mary Jones, 14, -, 16000, 75, 180
Thos. Harget, 8, -, 18000, 75, -
Robert Gale, 6, -, 7000, 50, 150
An__ Carothers, 20, -, 10000, 50, 200
N. G. W. Teackle, 73, -, 20000, -, 700
Thos. Neiser, 34, 6, 15000, -, 200
Gotleib Engle, 7, -, 7000, 50, 50
Chas. R. Taylor, 120, 13, 70000, 50, 400
Martha Gibson, 130, -, 70000, -, 600
Jas. McKellar, 13, -, 10000, 25, 200
Daniel Chase, 12, -, 8000, 100, 250
Wm. R. Constable, 25, -, 12000, 200, 800
C. D. Burnitz, 24, -, 4800, 50, 150
Edward Taylor, 23, -, 10000, 400, 250
J. Stevenson, 98, 10, 15000, -, 450
Robert P. Brown, 85, 40, 37000, 400, 400
F. Harrison, 75, 45, 35000, 400, 400
Francis H. Mery, 26, -, 15000, 200, 500
Ed. S. Myers, 100, 50, 35000, 400, 600
Abel C. Hopkins, 100, -, 40000, 300, 100
James N. Brown, 180, -, 60000, 500, 600
Martha A. Hill, 23, -, 8000, 200, 300
James J. Fisher, 43, -, 25000, 200, 2000
Benj. Vanhorn, 304, 40, 60000, 300, 800
Benj. Bowen, 80, 40, 4800, 200, 700
Ed. C. Lucas, 89, -, 20000, 20, 200
Henry C. Turnbull, 120, 25, 35000, 50, 850
James W. Curley, 60, -, 18000, 200, 800
William Fisher, 25, -, 2000, 400, 800

Edward Niring, 180, -, 60000, 400, 700

Wm. D. Breckinridge, 27, -, 16000, 300, 200

Mary Steuart, 80, 20, 8000, 300, 400

Isaac Darling, 90, 10, 6000, 100, 400

Edward Liberry, 75, 65, 7000, 200, 500

John Tool, 63, 10, 5000, 250, 450

John Plaskett, 160, 40, 8000, 400, 800

Stephen Tagg, 20, 5, 750, 30, 150

Thos. C. Ristum (Ristain), 100, 200, 40000, 300, 1000

Kirkpatrick Ewing, 100, 86, 13000, 120, 400

Jacob Wisner, 200, 80, 25000, 500, 1500

Joseph Parker, 115, 19, 10000, 500, 1200

Thomas R. Crane, 170, 107, 15000, 1000, 3000

Wm. Andrews, 151, 35, 9000, 300, 700

Uria Carter, 6, 5, 1000, -, 30

Wm. Corbin, 10, 3, 1200, -, 50

John Torbet, 18, 3, 600, 10, 70

Isaac Francis, 200, 200, 12000, 35, 400

Joshua Proctor, 40, 60, 2000, 60, 200

Edward Scovans, 3, -, 200, 25, 50

George Bayne, 12, 4, 800, 50, 60

Nathan H. Ward, 65, 2, 25000, 300, 1000

Amis Matthews, 170, 75, 20000, 400, 800

Grafton M. Bosley, 140, 24, 25000, 200, 1200

Sarah Boards, 17, -, 7500, 50, 300

Wm. Shealey, 28, 27, 8000, 300, 400

George Preston, 38, -, 20000, -, 1000

George M. Hirs, 18, -, 10000, 100, 400

Jno. C. Zinck, 3, -, 3000, 50, 12

Jno. Neagle, 8, -, 40000, 200, 300

Wm. C. Wilson, 146, 40, 150000, 500, 1200

Thomas Brown, 53, -, 50000, 100, 250

N. Schumacher, 55, -, 60000, 200, 1500

Edwin L. Parker, 20, -, 18000, 150, 650

Michael Aldin, 130, 40, 25000, 400, 900

Elizabeth Ward, 120, 17, 25000, 200, 700

Agnes Bradford, 125, -, 75000, 500, 1200

Jas. Reynolds, 12, -, 15000, 200, 250

Wm. H. Ward, 16, -, 15000, 100, 500

Benj. Kurtz, 60, -, 50000, 200, 500

Robert Moore, 50, 30, 60000, 200, 700

Chas. A. Buchanan, 100, -, 20000, 200, 200

Frederick Waters, 13, -, 10000, 50, 450

Geo. B. Clark, 38, -, 16000, 250, 125

Victor Sirata, 40, -, 12000, 200, 200

A. F. W. James, 150, 200, 80000, 15, 250

Thos. Keyworth, 20, -, 40000, 100, 900

G. G. Williams, 15, -, 30000, 300, 300

Andrew Harvey, 35, -, 60000, 200, 1200

J. M. Lamdin, 60, -, 60000, 200, 300

Francis J. Keller, 4, -, 4000, 100, 100

Geo. W. Arney, 11, -, 20000, 100, 300

John Gibson, 21, -, 11000, 200, 250

Sarah Jones, 22, -, 11000, 100, 100

Jno. P. McCormick, 25, -, 5000, 200, 130

Wm. T. Fendall, 80, 40, 25000, 150, 1200

Benj. Tipton, 17, -, 6000, 200, 300

Grayson Hoe, 12, -, 3600, 50, 100

Joney Riser (Reyer), 25, -, 12000, 50, 200

Stephen Bradbery, 25, -, 20000, 200, 1000

Jas. Malcome, 25, -, 2500, 100, 400

Henry Kemp, 60, 20, 40000, 150, 700

J__ Merryman, 80, 55, 75000, 150, 300

Wm. McDonald, 140, 110, 200000, 300, 30000

Mary L. Patterson, 10, -, 20000, 100, 500

Saml. Wyme (Wynn), 50, 85, 100000, 300, 800

Wm. Holmes, 40, -, 30000, 150, 400

Wm. S. Whiteley, 40, 730, 70000, 150, 700

John Lowback, 30, 80, 3000, 100, 300

Jno. A. Craig, 52, -, 40000, 75, 1800

Joshua Miles, 100, 15, 2500, 50, 300

John McComas, 120, 120, 8500, 300, 850

Thomas Elliot, 110, 25, 3000, 75, 700

John Bosly, 150, 50, 18000, 300, 700

George Norris, 100, 25, 1500, 150, 600

Nasson Hughes, 40, 120, 2500, 50, 460

James Uorris (Norris), 110, 40, 3000, 250, 600

Columbus League, 92, 50, 3800, 150, 430

Thomas Syttle (Lyttle), 70, 30, 1000, 100, 600

Nicholas Syttle, 70, 30, 1000, 100, 600

William Carland, 80, 20, 2500, 100, 400

William Carland, 100, 80, 4500, 100, 400

Martin Fugate, 50, 50, 3500, 200, 125

John P. Cuddy, 51, 57, 3500, 200, 470

Edward Terrell, 28, -, 1500, 250, 675

William Pearce, 400, 800, 20000, 500, 1600

John B. Holmes, 250, 56, 10000, 400, 200

Matthew Sparks, 295, 24, 18000, 300, 2000

Franklin Sparks, 100, 8, 6000, 150, 600

John Perdue, 290, 10, 18000, 280, 2000

Hezekiah Shorborough, 70, -, 2000, 60, 200

Sallie M. Graw (McGraw), 400, 225, 14500, 300, 2000

William Haskins, 270, 100, 15000, 500, 2000

Victor Holmes, 190, 20, 15000, 500, 1000

Nicholas Stanford, 100, 50, 6000, 200, 700

William Vance, 100, 25, 4000, 100, 550

Joshua Starrett, 100, 25, 4000, 100, 200

Robert Elliot, 82, 41, 2500, 100, 450

Francis Gelley, 20, 20, 3000, 30, 100

Thomas Matthews, 75, 66, 12000, 100, 600

James Rowe, 100, 171, 6000, 300, 600

James Wilson, 200, 200, 6000, 100, 800

Solomon Tipton, 160, 18, 8000, 200, 800

Benjamin Bosley, 88, -, 5000, 150, 600

Ira Tipton, 61, -, 4000, 100, 350

John Merryman, 163, 50, 10000, 80, 441

Elizabeth Sparks, 260, 25, 16000, 500, 1200

William Anderson, 55, 6, 3700, 300, 600

Ira Anderson, 150, 50, 5000, 300, 550

Caleb Hunt, 155, 20, 10500, 400, 950

Levi Curtis, 75, 5, 4800, 150, 500

Josiah Shepperd, 60, 30, 2700, 100, 325

Elias Sheppard, 60, 30, 3600, 70, 250

William Curtis, 140, 60, 8000, 100, 700

William Bacon, 170, 30, 8000, 200, 900

Thomas Anderson, 100, 10, 5000, 200, 300

Nathan Griffin, 145, 20, 15000, 100, 575

John B. Pearce, 260, 204, 12000, 400, 1150

Joshua Hutchens, 480, 20, 30000, 400, 4000

Mary D. Hutchens, 130, 40, 10000, 250, 1500

Richard Hutchens, 89, 7, 4800, 100, 600

John Smith, 40, 10, 4000, 100, 600

Elisha Gallaway, 95, 155, 4500, 75, 550

William Hutchens, 90, 10, 4000, 300,700

John Huckins, 90, 10, 4000, 50, 350

Nathan Nelson, 120, 5, 10000, 1000, 700

Mary Slade, 80, 4, 5000, 51, 600

Lewis Parsons, 90, 10, 8000, 300, 200

Samuel Miller, 22, -, 3000, 100, 400

John Scott, 45, 35, 10000, 100, 200

Joseph Parson, 562, 18, 3000, 100, 300

Labon Sparks, 138, 12, 8000,150, 1000

Nelson Miles, 36, 6, 2000, 150, 150

Matilda Miles, 80, 20, 5000, 150, 425

Jemina Johnston, 160, 40, 10000, 150, 775

Jacob Shock, 140, 30, 9000, 200, 1500

George Austin, 120, 20, 15000, 5000, 1500

Warwick Price, 80, 40, 6000, 200, 680

Peter Mowell (Morrell), 150, 70, 12000,700, 1200

Charles Evans, 400, 100, 16000, 150, 875

Henry Carroll, 1000, 1000, 36000, 1200, 3550

Nicholas Dusey, 400, 100, 16000, 200, 400

Jacob Johnston, 130, 25, 5000, 200, 800

Levi A. Made, 136, 24, 8000, 500, 1480

Thomas Griffith, 140, 40, 8000, 200, 800

Joshua Richardson, 85, 30, 4000, 150, 275

William Richardson, 110, 30, 5000, 300, 700

Edward Price, 160, 40, 8000, 250, 1100

Zekial Masumae (Masumue), 75, 15, 4000, 100, 600

Thomas Richardson, 100, 33, 5000, 200, 1000

Thomas Richardson, 95, 20, 4000, 150, 250

John Pierce, 200, 47, 9000, 200, 1000

John P. Pierce, 140, 40, 7200, 200, 900

James Richardson, 107, 8, 4000, 200, 635

Samuel Moore, 116, 52, 8000, 250, 6500

William Wilson, 40, 28, 4600, 100, 333

Samuel Merryman, 50, 20, 3000, 100, 300

Thomas Anderson, 90, 15, 4000, 10, 400

John Brown, 50, 28, 3000, 60, 320

John B. Pursell, 120, 70, 8000, 300, 900

William Beard, 56, 20, 1500, 50, 300

Micajah Pearce, 100, 50, 8000, 200, 800

Isaac King, 57, 17, 3500, 120, 315

Joshua Price, 120, 121, 7250, 100, 700

Joshua Price, 100, 50,7500, 100, 900

Jackson Wilson, 150, 120, 7000, 500, 700

William Curtis, 365, 125, 6200, 500, 850

Alexander Gother, 60, 18, 2500, 75, 200

Rebecca Griffin, 80, 24, 4000, 150, 400

William Slade, 80, 10, 7000, 100, 700

Robert Royston, 120, 30, 5000, 300, 800

Ann Emory, 800, 250, 60000, 700, 4300

Edward Stansbury, 100, 293, 11800, 600, 1081

Catherine Johnston, 90, 10, 2000, 150, 500

Nicholas Parker, 200, 100, 6000, 300, 1400

John Parker, 150, 12, 7000, 150, 400

James Parker, 140, 40, 5000, 100, 1000

Joshua H. Parker, 70, 20, 5000, 100, 300

James Miles, 85, 15, 5000, 100, 475

James Shepperd, 80, 20, 2000, 200, 400

Benjamin Powell, 125, 75, 6000, 300, 600

Levi Merryman, 100, 73, 6000, 350, 750

Timothy Stephens, 40, 14, 4000, 100, 400

Israel Wilson, 125, 40, 5500, 150, 500

Philip Fitzsimmons, 40, 40, 1800, 100, 200

Lorenzo Patterson, 78, 15, 9000, 600, 3900

Alexander Morrison, 100, 15, 9000, 600, 1000

Archabald Kerr (Kew), 100, -, 7500, 300, 800

George W. Powell, 100, 53, 6000, 100, 1000

Elisha Brown, 300, 150, 16000, 600, 500

James Moore, 41, 12, 1800, 25, 75

W. B. Hamilton, 114, 40, 10000, 500, 2000

S. M. Parker, 100, 30, 8000, 500, 800

Henry Morrison, 140, 35, 7000, 300, 800

John Jones, 60, -, 3000, 100, 500

James Henderson, 160, 100, 18000, 800, 1885

Henry Jois, 52, 6, 4000, 100, 250

John Emerson, 14, 3, 1000, 50, 150

Joseph G. Dance, 100, 77, 12000, 500, 1200

John G. Sidell, 140, -, 12000, 250, 800

Dennis Matthews, 240, 100, 27200, 600, 2260

Jeremiah Gellott, 118, 192, 21500, 400, 1600

Sarah Jenkins, 190, 30, 30000, 500, 1600

Frederick Smith, 175, 22, 18800, 150, 550

William Boddie, 125, 40, 8000, 300, 400

Joshua Marsh, 323, 193, 17100, 300, 1385

Richard Fisher, 152, 22, 5000, 100, 380

Martin Chilcop, 100, 35, 5000, 150, 600

John Pocock, 150, 50, 8000, 200, 600
Henry Green, 50, 30, 4000, 100, 200
Henry Amassne, 17, 5, 1500, 60, 150
John Murphy, 60, 8, 2000, 100, 200
William Hall, 120, 40, 7000, 150, 640
William Wright, 100, 60, 3000, 150, 600
George Hogman, 100, 32, 6000, 400, 1125
Joshua Jackson, 60, 5, 3000, 200, 400
James Jackson, 60, 10, 3000, 100, 360
Thomas Jackson, 52, 8, 1800, 200, 500
Benjamin Jackson, 50, -, 1250, 100, 350
Robert Wilson, 58, 50, 20000, 100, 350
Wesley Royston, 150, 125, 13700, 500, 1000
John G. Patterson, 100, 56, 6000, 290, 835
Peter Bawls, 12, -, 1500, 25, 100
John McColgan, 40, 20, 10000, 200, 900
Michael Hergenrother, 20,18, 4000, 200, 200
Adam Scarf, 9, -, 1500, 150, 200
Joseph Foster, 14, -, 2000, 125, 400
Thomas Gorsuch, 25, 12, 200, 200, 200
John Montling, 15, -, 2500, 150, 200
Wm. Welcher, 23, -, 1500, 100, 200
Joshua Eagleston, 15, 16, 3000, 200, 250
Tobias Luts, 20, 17, 3000, 150, 150
Joseph Jones, 10, -, 250, 200, 200
John Lingerfelter, 10, -, 250, 200, 200
Alex McComack, 44, 32, 5000, 200, 300
Paul Gegner, 12, -, 1200, 50, 100

George Chester, 12, -, 2000, 150, 100
John Curts, 10, -, 1000, 100, 250
John Kaylor, 33, -, 2000, 200, 200
Fred Walker, 60, 40, 2000, 300, 300
Otto Granter, 60, 40, 6000, 200, 600
George Mason, 20, 4, 2500, 200, 500
John Mummy, 40, 25, 3000, 250, 250
William Corse, 122, 40, 32000, 300, 1200
Henry Strodman, 4, -, 5000, 100, 300
Lenard Koenig, 19, -, 2000, 150, 200
John Kraff, 10, -, 2000, 150, 200
John Siple, 10, 10, 2000, 125, 150
Joseph Hotsadder, 20, 15, 3000, 200, 300
John Kupplman, 16, 16, 3000, 100, 150
John A. Nortrup, 16, 8, 2000, 150, 200
Peter Murrey, 14, -, 2000, 120, 200
Daniel Melcher, 50, 59, 6000, 200, 160
Dearick Radicker, 20, 58, 10000, 300, 1000
Frank Holstader, 22, -, 2000, 150, 300
Fred Brinkman, 12, -, 1000, 150, 200
John Hoferkamp (Hoterlamp), 8, -, 1000, 125, 150
Obediah Sanders, 25, 23, 14000, 300, 500
Henry Sapp, 20, 45, 9000, 150, 300
John Gregory, 41, 20, 4000, 200, 400
William Morgan, 60, 20, 8000, 100, 400
Benj. Brandard, 4, -, 1000, 25, 100
John Bargan, 25, 16, 1100, 20, 250
John Biddison, 39, 6, 4000, 500, 860
Conrod Beng, 2, -, 1000, 25, 100
John Odo, 8, -, 1100, 50, 150
Daniel Bargan, 32, 8, 4000, 50, 200
Peter Knick, 25, -, 6000, 150, 250
John Guntrun, 10, -, 3000, 100, 200
Jacob Eigly, 22, -, 4000, 100, 200

Lawrence Hofstider, 2, -, 1600, 25, 25

George Gegner, 16, -, 2000, 100, 200

William Clagett, 103, 8, 22500, 500, 800

Valentine Butskey, 55, -, 4000, 400, 400

John Register, 60, 10, 20000, 650, 1000

Samuel Green, 60, 20, 20000, 800, 1200

Francis Baringer, 50, 50, 20000, 600, 400

John H. French, 50, 4, 16000, 200, 1200

Henry Wemp, 36, -, 2500, 200, 200

Henry Grilst, 7, -, 1600, 150, 300

Ada Erdman, 43, -, 12000, 200, 200

Philip Burgan, 12, -, 5000, 200, 900

Zenis Barnum, 35, -, 25000, 400, 2000

James Dolan, 32, -, 30000, 200, 300

Johns Hopkins, 300, 130, 200000, 1000, 2000

Horatio Whitridge, 60, 14, 20000, 150, 500

Fred Erdman, 12, -, 3000, 100, 300

Mathias Erdman, 16, -, 4000, 150, 400

John Lamley, 13, -, 2000, 200, 200

Thomas Clifford, 12, -, 1000, 100, -

Michael Wilt, 2, -, 3000, 150, 200

David Conthwrite, 30, -, 2000, 60, 60

Henry Booth, 50, 16, 2000, 150, 600

Jacob Hiss Jr., 90, 80, 15000, 150, 250

Andrew Talbert, 20, 30, 6000, 300, 500

William White, 21, -, 2000, 50, 300

William Hiss, 100, 100, 10000, 300, 200

James H. Keene, 115, 82, 25000, 300, 1700

Joseph Gorsuch, 85, 112, 10000, 150, 250

James Thompson, 18, -, 6000, 50, 100

Dominick Hamman, 35, 5, 3000, 150, 900

Philip Wales, 30, -, 3000, 100, 200

Hugh Simms, 220, 180, 15000, 1200, 2600

Mary Tyson, 156, 10, 8000, 150, 750

John Ramsey, 50, 20, 2000, 100, 300

James Reynolds, 100, 100, 8000, 400, 500

Mary Howard, 30, 12, 4000, 50, 200

Stephen Falls, 20, 75, 4000, 60, 250

Joseph B. Hide, 100, 80, 5000, 100, 500

James Remington, 150, 105, 20000, 100, 600

James Fletcher, 100, 35, 6000, 100, 400

Thomas B. Gorsuch, 300, 50, 8000, 300, 1500

Thomas Gorsuch, 250, 150, 10000, 100, 800

Thomas Clayton, 56, -, 3000, 100, 200

Lewis Kellen, 80, 30, 6000, 250, 600

Sylvester Ford, 20, 35, 4000, 150, 200

Jacob Mannikling, 60, 90, 2500, 100, 500

George Pearce, 60, 20, 2400, 100, 200

Mathias Dahler, 50, 5, 1000, 100, 300

William Flowers, 66, -, 1800, 150, 300

Henry Guyton, 35, 40, 3000, 50, 300

James Alexander, 60, 18, 6000, 200, 500

John Whitkings, 220, 20, 14000, 500, 1600

Nicholas Alexander, 40, 50, 4000, 100, 600

Moses Balwin, 80, 40, 6500, 300, 260

John Balwin, 80, 40, 2000, 300, 230

Samuel Guyton, 82, 40, 3300, 100, 150

Norvil Guyton, 50, -, 1500, 200, 150

John J. Balwin, 200, 45, 6000, 1000, 1200

Thomas France, 40, 26, 1000, 50, 150

John Harland, 40, 30, 2000, 50, 300

Theoplis Gentrey, 24, 15, 5000, 150, 1000

William Ford, 50, 30, 3000, 100, 300

Henry Isennoff, 40, -, 2000, 100, 400

John W. Hale, 140, 120, 10000, 500, 1400

Charles T. Hale, 200, 100, 1500, 500, 1500

Thomas Balwin, 18, -, 4000, 50, 300

Calip Larmer, 100, 50, 6000, 100, 400

James McBrogden, 200, 40, 19000, 500, 2500

William Fuller, 10, -, 2000, 100, 100

Casper Zeigenham, 18, -, 1200, 150, 200

Henry Bishop, 30, 25, 2000, 100, 100

Jacob Gost, 30, 5, 2000, 100, 400

Asey K. Tylor, 130, 24, 11000, 100, 800

James Carroll Jr., 600, 500, 40000, 1000, 2500

Alethen Ristean, 100, 50, 8000, 100, 250

John G. Booth, 250, 150, 20000, , 500, 1600

George Morgan, 40, 2, 8000, 150, 500

Joshua Jessop, 150, 50, 9000, 300, 1250

John Yellot (Gellot), 120, 30, 15000, 400, 200

William Pierce, 300, 200, 30000, 500, 4200

Abraham Thiertson, 200, 100, 18000, 208, 800

William Trimble, 140, 60, 15000, 200, 1000

Jacob Hartsell, 50, 50, 6000, 150, 400

Daniel Nofsinger, 100, 10, 9000, 600, 100

Peter Nufsinger, 120, 50, 2000, 1000, 1000

George Hunt, 19, 2, 3000, 25, 225

James Curtain, 125, 25, 12000, 150, 600

Robert Ford, 125, 25, 10000, 150, 1000

Richard Karr, 12, 12, 3000, 25, 125

John W. Ford, 60, 60, 3000, 50, 150

Walter Hale, 20, 100, 4000, 100, 300

Benj. Wilson, 200, 90, 9000, 400, 800

William Parker, 90, -, 1000, 200, 500

Moses Miller, 100, 35, 14000, 400, 600

Anthoney Miller, 80, 28, 9000, 300, 200

Charles Wilson, 220, 50, 18000, 1000, 1300

Samuel Rankin, 160, 20, 20000, 1000, 1000

James Duff, 42, 3, 2000, 25, 150

Seth Smith, 60, 7, 2400, 100, 200

James France, 32, 8, 2500, 100, 200

Alexander France, 50, 11, 2500, 100, 300

David Warfield, 80, 34, 7000, 250, 650

William Corns, 300, 100, 10000, 200, 500

George King, 50, -, 2000, 150, 500

James Johnson, 120, 35, 8000, 100, 200

William Knight, 60, 80, 3000, 300, 600

John Baxter, 30, 23, 1800, 100, 125

Joshua Disney, 30, 23, 2200, 100, 150

Owen Donovan, 25, 31, 2700, 70, 300

George Gilliland, 100, 60, 8000, 200, 1310

Levi Hipsley, 140, 60, 12000, 500, 750

Daniel Jenifer, 360, 200, 28000, 600, 1250

Levi Fergerson, 100, 30, 6000, 150, 400

John Fox, 120, 30, 4500, 150, 600

Thomas Smyok, 40, 10, 5000, 500, 500

Philip Philips, 50, 50, 5000, 200, 300

Isaah Baker, 70, 20, 4500, 100, 1100

Robert Halbert (Holbert), 15, -, 1600, 50, 100

Robert Proctor, 150, 120, 6000, 100, 900

Joseph Brooks, 185, 215, 16000, 300, 1100

Henry Snyder, 12, 13, 1000, 50, 150

Eli Gamble, 50, 17, 4000, 500, 400

John Bydey, 400, 500, 20000, 1000, 2400

Mary Wilson, 175, 200, 6000, 300, 520

Lenard Quinlin, 220, 100, 12000, 500, 1100

Richard Bell, 105, 51, 7000, 150, 500

Sarah E. Bell, 100, 40, 10000, 300, 400

John Kennard, 100, 30, 8000, 200, 500

John Demass, 36, -, 1800, 50, 100

John R. Burns, 20, -, 1000, 30, 150

Peter B. Haley, 18, 3, 1800, 50, 200

Benj. Flurry, 50, 30, 3000, 100, 400

John W. Orion (Onion), 40,48, 3000, 50, 150

William Hawkens, 60, 17, 3000, 100, 750

William Knight, 100, -, 2000, -, 300

James Hawkens, 100, 30, 7000, 500, 1000

Benjamin Chew, 125, 155, 6000, 100, 500

Gospel Evel, 30, 8, 1000, 50, 300

Saml. Pinkerton, 40, -, 2000, 50, 250

William Burton, 25, 25, 1000, 50, 200

Gibbin Moore, 90, 110, 3000, 75, 400

John Holland, 50, 50, 5000, 100, 400

Samuel Paterson, 150, 190, 8000, 200, 1200

John S. Hayes, 200, 900, 40000, 100, 300

Stephen Rushel (Ruphel), 100, 125, 7000, 250, 800

William Y. Day, 180, 100, 8000, 250, 1000

James Hayes, 120, 350, 8000, 100, 400

D. S. Gittings, 140, 80, 1000, 300, 1000

James Shaw, 30, 10, 2000, 50, 400

Ann Geddis, 50, 45, 5000, 300, 600

John Kennel, 33, -, 7000, 200, 500

William Edlin, 200, 73, 20000, 1000, 1500

John Mast, 100, 65, 11000, 100, 520

William Brenter, 100, 73, 8000, 200, 900

Jacob Mast, 80, 15, 7000, 500, 700

Samuel Mast, 90, 40, 9000, 500, 1000

Mark Jenkens, 200, 10, 15000, 600, 1000

George Jenkens, 60, 6, 6000, 60, 500

Richard France, 90, 5, 9000, 100, 300

Edward Jenkens, 200, 8, 16000, 600, 2800

Dennis Carter, 32, -, 2500, 50, 300

Charles Riddle, 40, 32, 5000, 200, 400

William Gorsuch, 50, 5, 4000, 150, 500

Lucretia Carman, 65, -, 4000, 100, 300

Daniel Mast, 80, 21, 9000, 300, 1000

Elijah Clayton, 65, 25, 4000, 100, 400

Thomas Gettings, 250, 100, 28000, 1000, 1600

Griffith Benner, 80, 110, 10000, 500, 550

James Barton, 125, 59, 10000, 300, 1200

John W. Barton, 80, 20, 9000, 300, 600

Jacob Seddon, 70, 50, 6000, 100, 600

Walter Dager, 34, -, 1500, 50, 400

John Bennett, 80, 129, 500, 60, 400

Benj. Freeman, 80, 60, 2500, 100, 500

Walter Allender, 200, 200, 8000, 500, 720

James Wolf, 70, 80, 6000, 200, 400

James Beard, 11, -, 900, 50, 75

George McCubbin, 12, 9, 900, 50, 150

Alfred McCubbins, 15, 5, 1000, 80, 100

John Y. Day, 70, 33, 6000, 200, 400

Ishmael Day, 40, 10, 5000, 50, 300

Jesse Garrett, 180, 60, 8000, 200, 975

Thomas Ford, 150, 50, 8000, 500, 800

Margaret Cadden (Codden), 60, 42, 5000, 200, 400

Horace Burton, 60, 30, 7000, 200, 520

Joseph Burton, 50, 18, 5000, 150, 300

Owen Burton, 60, 32, 6000, 160, 500

Jonathan Burns, 110, 30, 8000, 300, 900

Thos. Christopher, 30, 70, 1200, 30, 100

Joshua Bevans, 40, 30, 1000, 200, 120

John Rollins, 65, 35, 1500, 20, 80

Andrew Jackson, 50, 60, 1000, 20, 50

John Yourist, 4, 11, 450, -, 35

Henry Pratt, 80, 82, 2700, 100, 150

James Salsbury, 80, 100, 4000, 25, 100

John C. Hughes, 32, 18, 700, 30, 180

Wm. R. Asher, 210, 130, 5000, 125, 650

William Meaks, 100, 50, 5000, 125, 125

John W. Williams, 140, 30, 6000, 100, 600

John W. Palmer, 175, 100, 4000, 40, 140

Stephen Knowlton, 350, 140, 25000, 1200, 3000

Joshua Hitch, 150, 80, 20000, 200, 900

Maria Wilson, 40, 250, 400, -, -

Wm. Galloway, 200, 100, 20000, 250, 6000

Wm. C. Galloway, 150, 125, 6000, 125, 400

Richard Johnson, 120, 100, 6000, 500, 300

John Morrow, 120, 140, 3000, 50, -

Joseph Howard, 20, 20, 800, -, 100

Robert Lemmon, 75, 125, 2500, 20, 200

William Bedford, 25, 25, 1000, 75, 150

John Vanness, 30, 70, 1500, 25, 78

Elizabeth Thomson, 40, 30, 600, 70, -

Wesly Wilson, 40, 30, 1200, 60, 225

Horris Steele, 250, 50, 5000, 300, 150

Gen. Cadwaleder, 100, 200, 5000, -, -

Josiah League, 45, 190, 2400, 100, 150

Joseph Wilson, 8, -, 500, 50, 200

Thomas Wilkinson, 150, 120, 3000, 32, 400

Phermelia Bevans, 100, 48, 5000, 500, 1100

William Slater, 400, 800, 5000, 500, 3000

William League, 50, 100, 1500, 60, 400

Edward Armstrong, 60, 340, 4000, -, -

Jarret Hopkins, 80, 10, 4000, 500, 300

Thomas Biddison, 200, 150, 6000, 150, 6000

William Gould, 20, 20, 800, 30, 150

John Carback, 60, 100, 2500, 150, 250

Eliza Pierce, 30, 120, 3000, 50, 500

Joshiah Earl, 25, 220, 3000, 100, 150

Bernard Warts, 75, 225, 3000, 50, 100

Edward Townson, 40, 50, 1000, 100, 100

Benj. H. York, 50, 60, 2000, 100, 200

Ellen Millenday, 25, 155, 2000, 25, 100

Bradford Sickles, 200, 50, 12000, 800, 600

William Clummy, 50, 250, 3000, -, -

John Kimmel, 9, 11, 800, 25, 20

Wm. Anthoney, 250, 1400, 20000, 200, 800

George Walters, 14, 21, 5000, 200, 400

John Neiser, 80, 20, 2000, 100, 75

Abraham Smith, 60, 250, 3000, -, 50

Curtis Humer, 80, 128, 4000, 75, 150

Hezekial Wilham, 100, 250, 5000, 200, 250

Benj. Brde, 100, 150, 6000, 200, 200

Robert Evern, 180, 220, 8000, 50, 250

Robt. Harryman, 100, 186, 5000, 50, 200

Wm. H. Bruce, 90, 60, 10000, 400, 700

Owen Emerson, 60, 46, 10000, 400, 350

Joseph Pearce, 20, 260, 400, 25, 225

John Broadbent, 150, 160, 40000, 100, 300

William Gillespie, 250, 120, 10000, 400, 1000

Mathias Townsend, 210, 70, 10000, 200, 1400

James Gillespie, 30, 80, 3000, 100, 300

Edward Miller, 150, 30, 10000, 200, 2000

John Wesly, 90, 300, 5000, 50, 250

John Shaffer, 100, 30, 1600, 30, 150

John C. Brown, 150, 300, 1100, 100, 600

James Parks, 32, 6, 800, 100, 500

William Pine, 150, 200, 7000, 300, 800

William Strong, 50, 50, 1000, 100, 200

Joseph Pughe, 45, 46, 800, 50, 150

George Kinghorn, 70, 45, 3000, 70, 300

Michael Welch, 70, 30, 2000, 50, 150

Frred Lindenburger, 55, 60, 2000, 60, 150

John W. Parker, 70, 50, 2000, 100, 200

James Cowley, 100, 26, 2000, 200, 250

Francis Steever, 10, 22, 6000, 30, 90

Charles Wachsmarth, 100, 450, 30000, 500, 1250

John Reese, 11, -, 2000, 100, 50

Washington Sapp, 14, -, 1400, 25, 50

Robt. Mooney, 24, -, 15000, 200, 800

George H. Whetter, 138, 6, 35000, 500, 700

Andrew Parlett, 120, 30, 20000, 200, 1200

Rachel Smith, 125, -, 3000, 200, 500

William Bond, 75, 70, 10000, 400, 500

Simon Cummens, 10, -, 1200, 200, 100

Moretz Kauf, 30, 4, 2800, 150, 400

Francis Liehman (Lishman), 25, 5, 1800, 150, 110

Henry Reaver (Keaver), 27, 5, 2000, 200, 200

Martin Braner, 60, 100, 4000, 200, 200

Morant Frasee, 70, 30, 3000, 100, 100

Felix Vonanth, 170, 166, 1600, 1000, 800

Carvil Stansbury, 300, 240, 26000, 500, 2000

Philip Kaylor, 4, -, 600, 75, 100

Herbert Kennedy, 100, 40, 7000, 300, 500

Josph Laroy, 90, 50, 2000, 200, 400

Philip Muller, 80, 20, 4000, 250, 700

Fred Seling, 100, 100, 6000, 200, 500

Joshua League, 50, 100, 3000, 150, 175

John Hoffman, 50, 100, 3000, 160, 150

James McLone, 280, 20, 8000, 500, 1000

Henry Moore, 25, -, 1000, 75, 150

Nicholas Maxall, 25, -, 1000, 80, 150

George Barger, 11, -, 1000, 50, 100

Daniel Henry, 60, 30, 5000, 100, 300

Wm. B. Norman, 208, 122, 25000, 850, 2000

John Segmore, 200, 500, 35000, 200, 1400

William Ferby, 75, 75, 6000, 700, 400

Jacob Deiter, 50, 130, 2500, 100, 200

Joseph Kristell, 35, 90, 1100, 50, 150

John Stewart, 40, 15, 5000, 50, 250

William Black, 20, 8, 1000, 75, 250

William Clark, 31, 2, 4000, 200, 240

Chester Koon, 10, -, 600, 100, 100

David Parlett, 50, 14, 3000, 100, 300

John Fitch, 60, 66, 1200, 50, 300

Philip Clark, 25, 15, 2500, 100, 200

John A. Price, 50, 67, 15000, 200, 1000

Wilcot Tucker, 40, 100, 6000, 100, 250

William Mace, 45, 45, 6000, 150, 500

Robert Howard, 140, 3000, 60000, 3000, 4000

John R. Foulk, 50, -, 1500, 200, 200

Samuel Hall, 20, -, 1000, 100, 200

Charles Trump, 100, 68, 16000, 300, 800

Samuel Kirk, 200, 200, 30000, 300, 800

George Counselman, 40, 67, 2000, 100, 200

Thomas Fitch, 100, 145, 7000, 100, 600

Frank Shanybrook, 90, 120, 15000, 200, 500

Alexander Fuller, 22, 23, 4000, 100, 300

Charles Grissel, 25, 25, 3000, 200, 800

Casper Long, 30, 50, 3000, 200, 200

Casper Whitemore, 12, -, 1000, 150, 200

Casper Frikey, 13, -, 1700, 100, 150

Jacob McComas, 70, 40, 8000, 200, 450

Nicholas Gatch, 90, 10, 10000, 300, 350

Benjamin Buck, 70, 10, 6000, 150, 360

Philip Johnson, 60, 40, 6000, 200, 500

John Rosengarth, 25, 25, 1500, 75, 125

Lewis Steer, 22, 29, 1500, 75, 150

Martin Gitman, 25, 22, 2500, 300, 300

Peter Zane, 20, 12, 2000, 300, 300

Jacob Kaylor, 18, 20, 2000, 300, 200

Abraham Biddison, 30, 20, 5000, 200, 250

George Aheart, 25, 8, 2500, 125, 130

Martha Frankenburger, 22, 10, 3000, 200, 150

Rodrick Arnold, 20, -, 4000, 75, 200

Rudolph Evans, 100, 250, 50000, 300, 800

Loucious Caldwell, 50, 40, 20000, 300, 1000

Robert Fowler, 100, 30, 25000, 300, 1200

Edward Whittemore, 50, 38, 7000, 200, 450

Christian Mason, 5, -, 1000, 50, 150

Jacob Schriver, 6, -, 1000, 50, 150

Philip Nine, 12, -, 2500, 50, 150

Charles Rittenhouse, 50, 50, 5000, 200, 300

William Hall, 150, 100, 10000, 100, 600

Henry Smith, 40, -, 3000, 100, 200

Casper Igleman, 30, 10, 5000, 100, 150

James Hall, 80, 70, 7000, 100, 300

Christian Brand, 40, 13, 2000, 100, 400

Mathias Linthicum, 100, -, 6000, 200, 300

Wesly Corsey, 100, -, 7000, 200,720

Owen Cosney, 20, -, 2000, 200, 300

Nicholas Schrib (Scheib), 13, -, 1500, 100, 150

Christopher Benner, 30, 20, 4000, 100, 150

Geo. W. Wade, 100, 50, 4000, 200, 600

William Randell, 15, 11, 5000, 100, 150

Thomas Randell, 40, 12, 6000, 200, 800

Fred Elterman, 17, -, 5000, 100, 100

Emanuel Wade, 80, 30, 10000, 200, 600

Elias Smerden, 200, 40, 12000, 500, 600

James Forbes, 130, 10, 8000, 100, 200

James Finn, 90, 10, 8000, 100, 200

Edward Levering, 33, -, 3000, 100, 100

Michael Wartsman, 80, 25, 20000, 800, 1000

Samuel Sutton, 160, 110, 15000, 300, 1200

Dennis King, 50, 70, 6000, 100, 200

Adam Gibes, 40, 60, 8000, 150, 500

John Jenkens, 40, 5, 4000, 200, 500

Fred Dawson, 100, 150, 15000, 50, 700

James Linthicum, 30, 10, 6000, 100, 120

William McDorel(McDowel), 30, 20, 16000, 100, 200

William Henkle, 7, -, 2000, 100, 150

Charles Kuhlert, 10, -, 3000, 150, 100

Jacob Pilsch, 13, -, 3000, 100, 400

John Knon, 15, 7, 4000, 150, 600

Francis Forsett, 62, -, 30000, 250, 700

Reynold Wright, 8, 2, 2500, 100, 350

Jane Dungan, 150, 65, 25000, 150, 600

Abraham Fuller, 30, -, 10000, 200, 600

Gustavus Felts, 136, -, 12000, 150, 200

William Lusby, 29, -, 9000, 200, 300

Mary A. Kell, 50, 50, 15000, 100, 200

George Whittemore, 106, 5, 20000, 200, 250

Philip Cashmire, 200, 80, 35000, 600, 4000

Wm. Ring (King) rents for Messrs Pattersons, 100, -, 20000, 200, 630

Bankard Baer rents from Lanaham & Rusk, 4, -, 8000, 250, 175

Geo. W. Keen rents for Cantor C of Balt., 115, -, 15000, 200, 485

Josias Bard rents from Mrs. Winnegar, 19, -, 5700, 150, 300

Geo. Adell, rents for Jacob Peples (Pepler), 10, -, 5000, 200, 270

D. J. Cummings, 25, -, 18000, 100, 320. This farm is used for pasturing Cattle and Yields Four Hundred Dollars for year.

John Cummings, 15, -, 3000, 125, 605. This is a small garden lot, and the produce of the same after the family's use pays about Three Hundred and fifty Dollars.

Wm. Baker, 20, -, 16000, 150, 500

John Whittington, 13, 2, 15000, 100, 150

Wm. Hitch, 45, 5, 100000, 100, 300

Geo. W. Riggs, 20, 10, 3000, 150, 300

John McCormick (tenant), 24, -, 40000, 100, 300

John Kirby, 57, 3, 48000, 100, 400

Samuel Feast and Son (tenants), 23, -, 56000, 400, 450

Anthony Cook 1, -, 5000, 60, 50

John Feast, 2 lots, -, 42000, 60, 100

Calvert County Maryland
1860 Agricultural Census

The University of North Carolina Library under a grant from the National Science Foundation microfilmed agricultural Census records. Records were filmed at the University of North Carolina from original records at the Maryland State Library.

Columns 1, 2, 3, 4, 5, and 13 represent the following information on the census:
1. Name of Owner, Agent or Manager of Farm
2. Acres of Improved Land
3. Acres of Unimproved Land
4. Cash Value of the Farm
5. Value of Farming Implements and Machinery
13. Value of Livestock

James Humphrey, 150, 150, 6000, 100, 960
James J. Bourne, 230, 30, 10000, 114, 1010
John E. Lathion, 300, 1200, 10500, 200, 1100
Richard Craig, 45, 60, 2200, 100, 150
Thomas H. Simmons, 20, 10, 800, 100, 226
John Bedders, -, -, -, 15, 300
Ezekiel Donnels, -, 630, 5000, -, 250
Isaac Brown, 75, 30, 1200, -, 65
Thomas Tydings, 60, 40, 1200, -, 138
John Marass, 80, 20, 1000, 141, 126
James L. Tucker, 50, 30, 600, 15, 219
Mary Thomas, 20, -, 200, -, 90
James Thomas, 10, -, 100, -, 111
William Shanton, 100, 65, 1500, -, 309
George Buckhannon, 30, 40, 300, -, 179
John W. Buckler, 55, 55, 330, 10, 155
James Griffis, 100, 125, 2250, 40, 290
Joseph Smith, 75, 125, 1200, -, -
Samuel Tydings, 50, 50, -, -, 136
Ann Allen, 45, -, 150, -, 20

William Allen, 46, -, 150, -, -
Richard Ward, 46, -, 150, -, 144
William Ward, 19, 40, 95, -, 40
Lewin Bailey, 7, -, 70, -, 144
Benjamin F. Hale, -, 40, 200, -, 200
Cephus Ward, 195, 100, 2950, -, 330
Alexander Latherline, 25, 49, 370, -, 211
James Buckler, 75, 75, 300, -, 28
Jawem Tucker, 100, 50, 2000, 100, 485
Fuller S. Stevens, 50, 600, 1010, 50, 222
Thomas Chase, 5, -, 150, -, 115
Thomas Sprigs, -, 300, 600, -, 80
Charles Taylor, 50, 50, 250, -, -
William Wilkerson, 202, -, 2000, 10, 260
James Hungerford, 10, -, 80, -, 36
David Caster, 11, -, 88, -, 22
Jessie Caster (Carter), 12, -, 96, -, 40
Benjamin Hungerford, 53, -, 100, -, 28
Peter Brooks, 75, 45, 1020, 30, 255
James G. Ireland, 75, 100, 1750, 24, 321
Columbus W. Ireland, -, 60, 300, -, -
Mary L. Day, -, 120, 600, -, -
Chas. S. Sommerville, 400, 100, 10000, 500, 1279
James Hellen, 50, 63, 2000, 20, 303

James Buckhannan, 8, -, 80, 3, 33
Henry Johnson, 19, -, 90, -, 6
Ann Hutchins, 14, -, 14, 4, 97
Charles Hil, 12, -, 96, -, 12
John Ward, 11, -, 88, -, -
Robert Allinson, 23, -, 187, -, 205
James Ireland, 46, -, 460, -, -
William Galer, 19, -, 190, -, 62
Alexander Sommerville, 500, 800, 35000, 500, 2080
Thomas R. Tongue, 200, 250, 6000, 200, 1500
Gideon D. Tongue, 125, 65, 3000, 25, 588
William Johnson, 20, 75, 475, 20, 58
James Beam, 2, -, 25, -, 40
Frederick Bishop, 10, -, 100, -, -
James E. Blunt, 65, 65, 520, 30, 298
Basil Evans, 10, 60, 600, 6, 1556
Henry Thomas, 50, 100, 1000, 18, 126
Wilson B. Crook, 75, 460, 3000, 30, 236
Benjamin R. Smith, 20, 130, 450, 25, 406
Thomas R. Grove, 50, 120, 1080, 25, 307
William A. Lusby, 40, 65, 1200, 20, 210
Richard A. Breeding, 40, 160, 600, 8, 75
Wm. H. Stallings, 100, 150, 1200, 30, 245
Edmon Christian, 100, 80, 3600, -, 141
James Fraizer, 300, 100, 5000, 75, 554
Harriett Dixon, 30, 20, 400, 25, 84
Harriett Wood, 60, 30, 500, 28, 391
Brooklin Adlnumber, 45, 20, 755, 50, 327
George Roberson, 75, 75, 600, 12, 141
Joshua Y. Bowen, 100, 60, 3200, 100, 152
Walter Helen, 20, 10, 600, 10, 125

William P. Dorsey, 40, 5, 2000, 50, 500
Moriah Dawkins, 125, 75, 5000, 40, 410
Thomas Hardister, 100, 130, 1840, 30, 347
Geo. W. Hardister, 80, 100, 1800, 20, 300
Geo. L. Weems, 100, 50, 1500, 30, 351
Nathaniel T. Weems, 10, 5, 150, 10, 125
Owen W. Bowen, 75, 89, 1656, 30, 301
Robert Wood, 75, 75, 1350, 20, 285
Jane Walice, 50, 100, 1200, 10, 106
Thomas Boon, 25, 50, 600, 12, 76
Jesse Sprigs, 50, 50, 1000, 15, 45
Isaac Boon, 50, 25, 800, 15, 108
Geo. W. Dorsey, 250, 132, 9550, 400, 1600
James A. Chesley, 250, 120, 6000, 150, 890
Elizabeth S. Hance, 400, 339, 4426, 30, 572
John Chambers, 100, 150, 2400, 25, 321
James Talbot, 135, 135, 6750, 100, 670
Robert L. Dorsey, 40, 40, 800, 20, 120
Benjamin J. Bowen, 75, 125, 1200, 50, 327
David A. Ward, 100, 100, 4000, 60, 392
Jane Grover, 50, 75, 1250, 20, 111
Robert G. Dixon, 80, 170, 1500, 80, 310
Sarah J. Bowen, 150, 150, 2500, 40, 601
Wm. H. Hungerford, 100, 200, 3000, 100, 200
Nathaniel Dare, 200, 100, 3000, 50, 634
John T. Gray, 150, 300, 2000, 60, 461

Uriah Wilkerson, 100, 88, 1000, 10, 125

John J. Dowell, 80, 70, 2600, 33, 347

William E. Bowen, 100, 300, 2400, 100, 116

John T. Wilson, 100, 232, 2324, 25, 204

Richard Younger, 40, 60, 500, 10, 39

James Younger, 50, -, 500, 20, 359

Dr. Basil Dixon, 140, 140, 3000, 23, 632

Elizabeth Allen, 60, 20, 480, 20, 148

Richd. M. Johnson, 30, 110, 980, 5,75

James Lovell, 164, 60, 1712, 20, 538

Jacob Breeden, 100, 18, 944, 30, 386

Elizabeth Lewis, 150, 150, 1200, 20, 120

John T. True, 50, -, 200, 10, 108

James Green, 120, 30, 2000, 70, 325

William P. Dowell, 60, 34, 570, 20, 47

John E. Garrett (Gantt), 33, 156, 1500, 13, 117

Charles F. Garrett, 33, 16, 1500, 13, 100

Thomas J. Wilson, 287, 160, 13000, 400, 1000

Nathaniel D. Wilson, 287, 160, 13000, 400, 1000

John J. Sollers, 159, 87, 10000, 505, 1413

Anthony Tall, 125, 255, 3040, 50, 300

Richd. P. Breeden, 225, 225, 3300, 50, 770

James B. Breeden, 45, 230, 2200, 20, 127

Henry Dawkins, 20, 40, 500, 10, 180

James E. Caster (Carter), 75, 75, 720, 10, 104

Joseph A. Johnson, 80, 120, 1000, 25, 271

Lewis R. Johnson, 75, 25, 900, 20, 264

Sewell Waters, 10, 40, 1400, 25, 275

Jane L. Williams, 300, 125, 8500, 100, 1084

Thomas Bayford 60, 140, 2000, 15, 125

Benjamin Gott, 40, 60, 1000, 30, 652

Joseph Gott, 50, 50, 1000, 25, 291

Gabriel Pitcher, 35, 35, 560, 25, 148

Benj. Blackburn, 25, 30, 5000, 25, 272

Richard Barnett, 66, 34, 1500, 25, 296

John A. Magride, 200, 200, 10000, 40, 492

Benjamin Parren, 200, 200, 10000, 40, 492

Thomas Parren, 100, 200, 2000, 30, 500

Sophia Broome, 300, 150, 8000, 100, 1200

Eliza Parren, 25, 25, 400, 20, 100

John Gott, 50, 88, 1980, 15, 370

Jesse Stineth, 50, -, 250, 25, 188

Joseph Griffis, 300, 200, 12000, 270, 1117

John Robertson, 200, 200, 5000, 70, 500

John Turner, 155, 150, 6100, 100, 565

Thomas B. H. Turner, 150, 171, 5900, 100, 1068

Jesse Hungerford, 100, 25, 5000, 40, 535

Alethy E. Hellen, 200, 37, 3500, 75, 433

Chas. E. Hellen, 150, 80, 2000, 20, 275

William D. Heller, 130, 100, 1900, 20, 570

Richd. Stoneforth, 250, 150, 12000, 100, 1700

Charlot Stanforth, 75, 75, 2500, 50, 190

Elizabeth Mackall, 125, 52, 4862, 100, 628

George Peterson, 200, 63, 10000, 250, 1215

James T. Wall, 200, 65, 10000, 350, 776

John A. Sedinak, 160, 123, 9000, 100, 985

John Parran Jr., 200, 150, 10000, 200, 1310

John Parran Sr., 50, 50, 3500, 50, 100

Sarah A. Case, 150, 50, 1500, 20, 250

Walter B. Williams, 100, 100, 2300, 30, 563

Elliott Beverly, 90, 95, 2650, 35, 282

George W. Brown, 200, 150, 1750, 75, 503

John Bedders, 25, 50, 750, 20, 150

Sarah Dawkins, 100, 50, 1500, 20, 310

Rebecca Denton, 150, 100, 5000, 50, 377

Mary Morgan, 75, 75, 1800, 35, 511

George Denton, 100, 100, 4000, 30, 727

John C. Parker, 400, 152, 20000, 300, 2000

Thomas D. Yoe, 250, 100, 8000, 50, 500

John H. Broome, 200, 112, 9000, 35, 478

Nathaniel W. Broome, 400, 55, 18000, 235, 1130

Basil A. Duke, 100, 50, 4000, 50, 500

James Duke, 250, 150, 13000, 55, 1000

George Stineth, 100, 200, 3000, 25, 250

George Ross, 175, 153, 8140, 25, 862

Basis D. Bond, 300, 160, 16000, 400, 1500

Joseph Denton, 25, 15, 800, 20, 180

Alexander Dorsey, 90, 183, 3385, 80, 500

Benj. B. Gray, 133, 67, 1600, 12, 172

Samuel Gray, 224, 200, 6315, 200, 889

Martha Kershaw, 150, 100, 3500, 80, 150

William Pitcher, 224, 120, 4140, 40, 659

Thomas Gray, 10, -, 300, 8, 144

James E. Blackburn, 200, 300, 5000, 35, 350

Samuel Varden, 500, 264, 24000, 200, 2204

Sutton Fowler, 100, 80, 4000, 100, 332

James A. Bond, 284, 100, 10000, 400, 1200

William Pitcher, 340, 66, 17000, 400, 1200

Thomas Basford, 50, 140, 2000, 15, 124

Lock L. Weems, 150, 22, 1720, 200, 440

James Turner, 200, 140, 3400, 30, 350

Nathaniel Sollers, 150, 80, 4600, 40, 385

Elizabeth Sollers, 100, 50, 4500, 200, 900

Elijah Thayer, 200, 800, 8000, -, 268

Alexander Freeman, 70, 30, 2000, 25, 368

George Ireland, 130, 100, 4000, 50, 500

Mary E. Kershaw, 20, 50, 560, 25, 150

Elizabeth B. Hawkins, 20, 50, 285, 60, 160

Mary Saveille, 200, 200, 5000, 25, 880

Enoch M. Mills, 250, 250, 5000, 75, 1000

Benjamin Locks, 20, -, 800, 10, 240

Rebecca Denton, 50, 13, 430, 12, 218

William Gray, 200, 150, 5000, 30, 793

Richard Blunt, 80, 70, 1200, 50, 498

John W. Yoe (Roe, Goe), 250, 350, 6000, 55, 1280

Elenore A. Monett, 30, -, 300, 20, 83

Augustus Bowen, 30, -, 300, 35, 161

Benjamin Fowler, 300, 100, 1800, 400, 1090

John Bond, 300, 100, 10000, 400, 1000

Samuel B. Calleston, 55, 100, -, 10, 56

James F. King, 33, -, 2000, -, 260

R. J. Grierson, 66, 34, 1200, 100, 300

J. W. Hall, 250, 150, 5000, 30, 807

Cornelius Bowen (Bower), 150, 50, 4000, 75, 467

John A. Basford, 175, 40, 8000, 500, 765

Wm. Billingsley, 180, 10, 4000, 200, 200

S. R. Bird, 230, 78, 9010, 400, 1100

J. A. Morsell, 240, 50, 6000, 200, 1500

James L. Hutchins, 230, -, 8000, 75, 800

Geo. W. Dowell, 100, 40, 2800, 100, 400

Joseph Hall, 8, 8, 300, -, 43

William Jones, 375, 248, 6500, 150, 971

J. . Ogden, 20, 17, 2000, 50, 245

Robert Buckler, 50, 95, 800, 50, 730

John A. Mills, 100, 50, 1000, 40, 315

Rebecka Boyd, 100, 60, 800, 30, 150

John Kreighton, 50, 84, 1500, 500, 305

A. Harris, 130, 40, 2800, 15, 284

W. W. Bowen, 100, 150, 2500, 40, 630

Ira Young, 20, -, 1000, 25, 200

Thomas Scrivener, 125, 125, 5000, 100, 534

Sandy Stineth, 50, 50, 1000, 15, 300

Frederick Graison, 50, 50, 1000, 20, 250

Ann Johnson, 100, 100, 2000, 10, 275

Gilbert Fowler, 75, 91, 1280, 30, 500

Richard Fowler, 40, 60, 800, 10, 280

George L. Buckmaster, 25, 75, 1500, 15, 250

Frederick Roberts, 10, 6, 500, 10, 135

A. G. Bowen, 80, 80, 1500, 50, 200

William Hooper, 30, 88, 700, 5, 100

William Mead, 40, 80, 1200, 20, 300

Henry Hutchins, 60, 73, 2500, 38, 239

J. Hutchins, 50, 50, 1000, -, 169

Elijah Bowen, 135, 66, 3000, 60, 340

Christian Bowen, 100, 50, 3750, 40, 180

Sarah J. Bowen, 75, 125, 2000, 10, 340

Dennis Monett, 74, 73, 1500, 50, 300

Joshua Stinett, 45, 55, 800, 25, 185

James _. Roberson, 60, 20, 800, 25, 150

John H. Hammett, 155, 70, 2500, 40, 466

John Scrivener, unk, unk, unk, -, 574

Fielder Rawlings, 125, 75, 2400, 50, 218

Benj. Rawlings, 70, 45, 150, 50, 305

Elizabeth Hutchins, 100, 49, 2000, 50, 330

Geo. W. Boyd, 100, 60, 1000, 30, 245

Walter Dorsey, 50, 30, 800, 30, 304

Washington Dorsey, 60, 130, 1000, 10, 90

William A. Parran, 150, 150, 5000, 50, 510

Robert W. Yoe (Goe), 100, 50, 10000, 1500, 100

Walter W. Dorsey, 300, 200, 8000, 250, 1150

Z. Hungerford (agt), 265, 75, 10000, 250, 950

D. B. M. Dixon (agt) 100, 100, 3500, 50, 725

Thos. J. Hutchins, 60, 40, 2000, 50, 230

M. L. Hawley, 4, unk, 3000, -, 125

Thos. J. Hutchins, 4, unk, 400, -, -

Mary Dorsey, 10, 40, 300, 6, 220

Rebecca Dalrymple, 80, 40, 2200, 10, 100

R. E. King, 40, 20, 600, -, 100

Elizabeth King, 50, 25, 7125, 30, 380

Susan R. King, 64, -, 1000, 35, 290

Sarah W. Howard, 50, 24, 700, 10, 170

Samuel Bowen, 30, 110, 1500, 35, 130

George Hagan, 50, 20, 800, 12, 102

Daniel Bowen, 80, 30, 1500, 150, 399

Thomas Bowen, 100, 42, 1000, -, 220

James E. Smith, 53, 72, 1185, 65, 117

Juliet Jones, 1, -, 100, -, 2

B. H. Williams, 225, 225, 8000, 50, 930

William Simmons, 50, 40, 900, -, 136

Kinsey Hance, 25, 89, 1170, 10, 125

Jesse Dalrymple, 200, 146, 4844, 70, 440

L. C. N. Weems, 45, 3, 2000, 15, 230

Thomas Hutchins, 14, 14, 800, 20, 343

G. H. Bowen, 175, 130, 6100, 400, 794

Z. Bowen, 100, 15, 2100, 25, 303

Priscilla Bowen, 40, 20, 800, 25, 153

J. W. Simmons, 150, 150, 3000, 100, 1880

E. D. Simmons, 175, 16, 3820, 75, 625

Augustus Crane, 70, 70, 1400, 40, 241

Samuel Skinner, 50, 50, 100, 20, 266

Elizabeth Skinner, 100, 100, 800, 15, 345

John Dare, 60, 115, 4375, 35, 886

Wm. H. Bowen, 90, 100, 1900, 30, 265

Alexander Fowler, 100, 100, 4000, 35, 424

James Ross, 150, 65, 3500, 50, 500

James J. Crane, 35, 100, 810, 10, 197

G. D. Allnutt, 125, 50, 4375, 30, 775

Richard Dare, 130, 148, 7500, 60, 853

Thomas Herrod, 20, -, 200, 10, 161

Richard Dare, 60, 277, 1750, 10, 250

John Dare, 60, 277, 1750, 10, 250

Young Chase, 20, 100, 960, 6, 62

Sy Dorsey, 200, 200, 1600, 200, 1400

Emanuel Green, 60, 300, 1500, 10, 130

John Hardesty, 65, 33, 1960, 60, 299

Susan N. Burkett, 250, 186, 16500, 345, 2089

Henry F. Talbott, 200, 70, 7000, 100, 660

Richard Roberts, 300, 120, 15000, 100, 1640

B. O. Hance, 180, 58, 6000, 50, 500

J. G. Allnutt, 100, 93, 3860, 50, 1298

E. Alnut, 150, 150, 4500, -, 70

G. R. Crawford, 75, 75, 1500, 60, 800

J. W. Crawford, 150, 50, 6000, 60, 800

James N. Crawford, 150, 50, 7000, 60, 400

E. R. Crawford, 125, 25, 3000, 20, 600

James S. Morsell, Jr., 300, 100, 30000, 600, 2500

James N. Morsell, 200, 57, 7000, 100, 600

Richard Hance, 300, 100, 16000, 600, 1700

E. G. Hance, 150, 100, 7500, 200, 890

Jane Stuart, 50, 80, 1300, 40, 100

Levin Skinner, 350, 250, 20000, 300, 1550

Nathaniel Dare Jr., 300, 87, 12000, 300, 1288

Octavius Y. Hooper, 100, 43, 2400, 30, 160

Priscilla Rhodes, 15, 23, 304, 15, 195

Alfred Reed, 50, 50, 1860, 10, 156

George Chambers, 125, 63, 2745, 40, 800

Joshua Paddy, 125, 75, 3000, 10, 190

Benjamin Hance, 300, 88, 2000, 300, 1400

N. C. B. Hance, 200, 80, 7000, 390, 1270

Nathaniel Dukes, 400, 150, 15000, 500,2 500

James S. Morsell, 700, 700, 50000, 500, 2741

John Hutchins, 66, 34, 1500, 35, 267

William Wood, 50, 24, 750, 10, 250

Catherine Morton, 700, 400, 27500, 500, 2810

John Robinson, 100, 40, 2000, 100, 569

Collin M. Williams, 175, 125, 3000, 100, 700

W. W. Bowen, 50, 150, 1000, 10, 160

Thomas F. Moore, 25, 13, 1000, 8, 375

Virgil Gantt, 300, 300, 12000, 412, 890

Thomas L. Anderson, 120, 80, 4000, 30, 265

Hiliary Bowen, 99, 51, 1200, 22, 311

Horatio Bowen, 155, 30, 2000, 75, 634

James Leach, 200, 20, 5500, 200, 650

Richard H. Hagan, 300, 90, 12000, 400, 1550

P. Bowen, 150, -, 2500, 50, 380

Wm. Macdaniel, 250, 155, 20000, 500, 1500

William Dorsey, 100, 80, 9000, 200, 300

Enoch G. Mills, 200,78, 7340, 200, 800

John Armiger, 220, 130, 7000, 600, 1400

E. A. Norfolk, 300, 150, 9000, 100, 1460

Ellen U. Belt, 300, 50, 8000, 200, 1358

Geo. P. Wilkerson, 130, 120, 10000, 200, 1500

Levin Stanforth, 400, 200, 25000, 500, 1900

John T. Wilkerson, 120, 130, 6000, 50, 1160

Robert B. Wilkerson, 60, 20, 1000, 40, 510

Hezekiah Norfolk, 30, 15, 300, -, -

John Stanforth, 300, 100, 20000, 100, 1000

Thomas Holland, 300, 50, 15000, 300, 1250

John A. C. Hance, 300, 75, 15000, 150, 1000

John A. Blake, 130, 100, 2000, 142, 644

Thos. Hobbs, 100, 60, 1600, 40, 180

P. Stillings, 100, 50, 1500, 25, 525

G. F. Freeland, 100, 75, 3500, 50, 384

Chas. A. W. Weems 250, 150, 12000, 200, 1000

B. A. Owings, 20, 46, 2000, 50, 320

James Calleston, 30, 70, 1000, 20, 300

James J. Hance, 100, 140, 5000, 100, 100

James P. Gates, 150, 100, 5000, 100, 570

Geo. W. Fowler (Towler), 175, 75, 7000, 50, 940

Martha E. Dixon, 50, 30, 800, 15, 192

Columbus C. Dorsey, 75, 55, 1950, 25, 370

Asbury Dorsey, 125, 75, 1950, 10, 242

Robert Crawford, 75, 23, 1500, 30, 485

Samuel Phelen, 50, 34, 1200, 25, 325

John Dorsey, 150, 50, 4000, 97, 618

Edward H. Ireland, 65, 15, 1050, 40, 490

George H. Wilson, 140, 44, 5490, 590, 800

Samuel Essex, 350, 200, 11000, 800, 770

Mary A. Ryon, 50, 50, 1500, 15, 300

Virgil Wilburn, 100, 50, 3000, 15, 335

E. J. Jones, 50, 25, 2000, 20, 200

Richard Isaacs, 60, 15, 600, 30, 514

Edward Talbutt, 160, 100, 6500, 50, 757

Sy Dorsey Jr., 100, 60, 2400, 50, 600

William Hutchins, 240, 120, 13400, 300, 1195

Jno. O. Cox, 350, 96, 15000, 200, 2200

John Harrison, 240, 120, 7500, 65, 683

Priscilla Williams, 240, 120, 10500, 65, 683

Joseph Bowen, 66, 34, 2500, 30, 275

James J. Alnutt, 200, 100, 4000, 20, 300

Wm. Harrison of Jacob, 164, 100, 5280, 20, 240

Elizabeth Harrison, 11, -, 900, -, -

Lebanan Birckhead, 200, 115, 6300, 30, 500

Geo. W. Harrison, 142, 40, 2184, 35, 471

Martha Harrison, 350, 350, 12000, 150, 885

Samuel Harrison, 100, 36, 2000, 40, 250

James J. Norfolk, 160, 80, 8000, 100, 1000

M. A. V. Beckett, 500, 200, 35000, 1000, 1553

Augustus Buckmaster, 25, 25, 800, 25, 175

Henry Dowell, 160, 100, 1300, 20, 109

W. A. Mitchell, 50, 50, 2500, -, 250

John R. Quinan, 50, 50, 5000, 100, 500

S. S. Parran, 120, 75, 5500, 150, 830

Aug. R. Sollers, 300, 200, 15000, 150, 1060

John Hopwood, 50, 10, 1500, 50, 420

James Fraiger, 50, 75, 2000, 40, 325

John H. Hopkins, 250, 170, 15000, 600, 1000

Sophia Tucker, 100, -, 2500, 50, 725

James T. King, 100, 16, 2500, 50, 300

Wm. S. King, 120, 40, 4500, 100, 600

James Cox, 100, 49, 3380, 70, 236

Walter Cox, 2, 20, 2500, -, -

Thornton Dorsey, 15, 60, 2500, -, 175

Milley Essex, 120, 40, 6000, 40, 650

Robert Essex, 100, 60, 5000, 25, 450

William Harrison, 300, 100, 12000, 800, 1200

John Turner, 75, 25, 3000, 150, 475

John Somerville, 252, 121, 15000, 250, 1420

Rachel Quill, 58, 10, 2150, 35, 200

James J. Allnut, 322, 161, 16000, 300, 1200

Daniel Hardesty, 50, 24, 3000, 40, 175

Samuel Birkhead, 35, 35, 700, 20, 250

James King, 100, 8, 1800, 50, 400

Samson Chase, 100, 150, 2500, 10, 300

Benjamin Harrison, 200, 100, 6000, 300, 781

Benj. Harrison Jr., 150, 150, 6000, 20, 536

Joseph Harrison, 80, 20, 3000, 100, 200

William Leach, 100, 80, 3600, 40, 960

Samuel Trott, 120, 120, 7000, 200, 1000

Benjamin Childs, 200, 39, 8000, 260, 958

Reason Sherbert, 40, 39, 1987, 20, 410

Thomas Sherbert, 85, 84, 3390, 15, 541

Michain Cotterton, 100, 50, 3750, 80, 552

Sophia Tucker, 60, 40, 2500, 30, 855

George King, 125, 25, 3000, 30, 370

Rebecca Dixon, 50, 30, 800, 26, 400

Samuel Watson, 33, 30, 1300, 20, 290

Collin Chambers, 40, 10, 2800, 20, 200

Robert F. Wood, 80, 40, 5000, 25, 200

Francis Stevens, 50, 56, 730, 15, 200

James Poore, 12, 8, 307, 10, 150

William Birkhead, 100, 75, 2400, 36, 430

Clement Dorsey, 60, 40, 1500, 50, 300

Samuel Birkhead, 100, 100, 6000, 40, 250

Sutton Watson, 40, 21, 1200, 15, 200

William Margues, 100, 50, 6000, 50, 650

John Trott, 150, 50, 8000, 150, 580

Daniel Kent, 325, 200, 21200, 325, 1425

Abecor Wood, 75, 40, 2000, 15, 250

William Mouldin, 150, 75, 4000, 8, 350

George W. Scaggs, 85, 30, 1000, 10, 300

Joseph King, 500, 500, 40000, 150, 1500

John T. Peavises, 90, 40, 3000, 50, 890

John B. Morsell, 2, -, 800, 20, 175

Martha W. Speakneall, 17, -, 2500, 25, 140

O. C. Harriss, 400, 100, 20000, 300, 1900

William Corcorn, 200, 69, 13450, 257, 1270

Jane Harrison, 200, 150, 7750, 60, 1000

Jno. R. Mackall, 200, 135, 13400, 60, 960

James T. Brisco, 350, 170, 30000, 450, 2000

John F. Ireland, 200, 52, 10000, 100, 725

Elijah W. Martin, 120, 80, 7000, 100, 725

John D. Ward, 110, 55, 5000, 30, 485

Thomas J. Graham, 300, 75, 16000, 400, 2600

Alexander Wilkerson, 150, 115, 10600, 100, 667

Thomas B. Gibbons, 350, 140, 15000, 500, 1000

Priscilla Fowler, 101, -, 4000, 250, 600

Elizabeth Suderland, 200, 40, 12350, 60, 1419

William Reynolds, 100, 30, 5200, 25, 375

Richard Roberts, 400, 100, 30000, 498, 2156

Elizabeth Garner, 169, 100, 10000, 200, 800

Richard Gantt, 150, 100, 7000, 200, 1000

George Younger, 200, 100, 9000, 50, 700

John J. Brown, 500, 200, 28000, 370, 2000

Benjamin Elliott, 250, 83, 13320, 300, 1885

P. Norfolk, 67, 33, 4000, 40, 350

William C. Fowler, 70, 22, 6000, 150, 700

Richard Ward, 136, 100, 5000, 50, 606

Thomas Lane, 100, 20, 6000, 50, 674

Poone Sanderland, 75, 25, 5050, 60, 655

David S. Smith, 45, 47, 4600, 40, 582

F. A. Smith, 42, -, 400, 20, 400

Ann W. Smith, 160, 36, 9900, 70, 1613

John W. Ward, 40, 22, 1200, 30, 200

Elizabeth Ward, 50, 10, 3000, 50, 400

Richard Ward, 100, 30, 6000, 50, 1000

William D. Smith, 200, 50, 8000, 60, 875

Fielder Smith, 170, 70, 11500, 80, 700

John T. Dove, 95, 40, 4050, 100, 550

John T. Dove, 45, 38, 2000, -, 325

Elisha Howes, 150, 125, 6875, 30, 900

John W. Howes, 80, 175, 3750, -, 475

Thomas H. Jones, 200, 77, 11080, 105, 1142

William Hardesty, 66, 34, 3000, 15, 800

Joseph Deal, 200, 135, 13400, 300, 1150

Peter Thomas, 100, 73, 1920, 100, 620

J. S. Smith, 110, 20, 7000, 200, 550

John Bosley, 150, 150, 9000, 25, 425

Thornton Parker, 100, 70, 3400, 20, 300

William Stuart, 60, 60, 6000, 20, 150

Thomas Childs, 160, 40, 8000, 50, 648

Richard Ireland, 150, 50, 8000, 30, 746

Dr. H. Stuart, 130, 15, 6000, 150, 730

Thomas H. Smith, 100, 19, 6000, 200, 600

John Petherage, 300, 50, 16000, 400, 1155

Mary E. Taneyhill, 120, 50, 8800, 100, 565

Wm. H. Dowell, 224, 40, 15000, 100, 1500

Wm. Whittington, 75, 37, 1800, 20, 475

Wm. H. Speaknall, 100, 60, 8000, 800, 1169

Wm. H. Speaknall, 80, 20, 5000, -, -

Richard Lambertt, 30, 7, 2380, 20, 350

Thomas Plummer, 150, 150, 10000, 300, 1250

James T. Chaney, 100, 70, 8000, 300, 900

John Carr, 170, 120, 12000, 150, 1000

Cosmo Sanderland, 75, 26, 6000, -, -

L. M. K. Griffith, 130, 100, 9000, 50, 850

C L. Griffith, 80, 121, 12500, 500, 1500

William H. Walker, 85, 45, 4000, 10, 125

Lewis Cheney, 150, 50, 8000, 400, 1318

Chas. Johnson, 25, 5, 1500, 100, 230

James Higgins, 140, 100, 16000, 200, 1000

David Careand, 170, 70, 12000, 200, 1502

Jacob Ward, 100, 100, 13000, 900, 770

Mary Plummer, 100, 26, 5040, 50, 600

Thomas Chew, 100, 45, 6000, 20, 350

George D. Lyles, 110, 15, 7000, 150, 700

Alvaris C. Wilson, 320, 170, 20000, 490, 1550

Philomena Chew, 110, 30, 6000, 130, 700

Thomas P. Whittington, 130, 74, 4080, 10, 250

John W. Spicknall, 100, 80, 9000, 100, 700

Thomas Mackall, 250, 50, 12000, 500, 1500

Basil Brady, 525, 150, 26260, 1000, 2250

Joseph Prout, 125, 75, 10000, 40, 750

William Ireland, 70, 54, 1500, 40, 375

Richard Griffith, 100, 20, 6000, 40, 1200

Wilson Crosby, 100, 26, 6700, 40, 800

William Newell, 75, 25, 3000, 50, 400

Benjamin Dowell, 86, 10, 4500, 50, 450

John Scrivener, 200, 90, 11600, 40, 800

Susan Tinge, 55, 15, 1200, 60, 215

Sarah Welsh, 100, 80, 3000, 50, 675

Augustine Wood, 51, -, 2000, 75, 345

Sutton J. Watson, 45, 20, 1000, -, 250

Samuel Lyles, 150, 50, 6000, 60, 1260

Samuel Gibson, 100, 37, 11000, 50, 400

Wm. P. Hardesty, 115, 17, 5200, 200, 950

Singleton Childs, 100, 33, 4000, 50, 600

Wm. Harrison, 100, 40, 6000, 50, 650

Jno. T. Osborn, 225, 75, 7500, 200, 900

Joseph Turner, 115, 112, 11000, 35, 800

John C. Phibbon, 90, 100, 4750, 15, 500

William Hardesty, 120, 15, 3555, 75, 850

John S. Dowell, 150, 50,4000, 80, 700

Margaret Sanderland, 180, 20, 5200, 100, 550

John Magee, 200, 100, 10000, 30, 800

Richard Graham, 300, 100, 12000, 100, 1400

Campbell Graham, 180, 120, 12000, 100, 640

Mary A. Norfolk, 100, 50, 4000, 100, 650

James Lyons, 100, 60, 5000, 30, 850

Henry F. Gibson, 100, 30, 3900, 125, 400

Samuel Gibson, 100, 20, 2400, 60, 530

Harrison Turner, 120, 50, 8000, 150, 540

Thomas W. Kerson, 375, 200, 22800, 233, 1779

Augustus Gibson, 100, 379, 4795, 75, 550

John T. Turner, 250, 106, 12460, 105, 1346

Joseph Norfolk, 40, 80, 3600, 25, 260

Elizabeth Ogden, 40, 20, 1200, -, -

Richd. E. Thomas, 120, 60, 3620, 50, 700

Richd. Stamford, 40, 145, 3000, 50, 110

R. H. T. Norfolk, 200, 10, 6000, 100, 450

James Sopha, 197, 100, 5740, 30, 630

Thos. B. Robertson, 300, 150, 8000, 150, 2000

Binson Robertson, 150, 150, 6000, 25, 700

George Robertson, 120, 21, 5000, 200, 600

Alexander Banker, 100, 53, 3000, 50, 425

William Ogden, 100, 50, 4500, 20, 550

John P. Wailes, 175, 125, 12000, 500, 1930

Sophia Dorsey, 16, -, 640, 50, 400

Joseph Blake, 600, 200, 48000, 420, 2790

Thomas J. Chew, 400, 200, 30000, 150, 2800

Julius Wilson, 275, 125, 18000, 250, 1550

David Earl, 80, 60, 1800, 40, 350

Samuel Boothe, 75, 80, 3600, 20, 400

Benjamin Ireland, 70, 70, 1200, 10, 450

Kinsey Jones, 100, 60, 2700, 25, 650

Stephen Boothe, 60, 60, 1000, 10, 375

Samuel Fowler, 100, 50, 4500, 150, 500

Richd. Quill, 25, 25, 1500, 20, 250

Robert Gibbons, 200, 137, 12000, 250, 1197

John Lyons, 200, 60, 7800, 50, 800

Thomas Blake, 300, 150, 18000, 250, 1700

Benjamin Talbott, 200, 60, 8000, 200, 650

Richard H. Dowell, 130, 20, 3000, 150, 450

Joseph B. Wailes, 116, 19, 10000, 200, 1025

Uriah Johnson, 175, 39, 6420, 80, 800

Henry Sanderlin, 100, 100, 5000, 60, 250

William Stevens, 91, 15, 4240, 30, 431

Samuel Griffith, 60, 30, 1500, 60, 350

Joseph Stevens, 6, -, 180, -, 45

John B. Howard, 400, 333, 26990, 50, 850

Elijah Boothe, 40, 20, 1200, 15, 100

James Jones, 100, 160, 4800, 25, 200

J. Y. Kent, 200, 67, 13370, 500, 3145

J. Y. Kent, 250, 86 16800, -, -

J. Y. Kent, 300, 200, 15000, 350, 1270

James Taylor, 20, -, 600, 20, 175

Jonah Y. Barker, 275, 225, 20000, 300, 2500

Thomas V. Blake, 140, 10, 10000, 400, 450

Richard H. Ward, 10, -, 700, 12, 100

Louis Griffith, 348, 139, 20000, 700, 3000

Henry Owings, 284, 100, 16000, 400, 1000

Samuel Owings, 350, 320, 24000, 505, 2286

Robert Birckhead, 38, 39, 2500, 25, 400

Virgil Slattings, 100, 100, 10000, 100, 575

Robert Ward, 81, 25, 4000, 4, 469

William Childs, 200, 50, 10000, 300, 1300

Wm. W. Childs, 50, 55, 1000, 30, 700

William Welsh, 220, 85, 9150, 165, 1137

William A Webb, 85, 50, 3000, -, -

Charles Ward, 60, 24, 2100, 20, 425

Jacob Chaney, 80, 41, 4000, 25, 425

Isaac Read, 60, 40, 2000, 15, 175

Jesse Trott, 70, 50, 3000, 60, 300

# Caroline County Maryland
## 1860 Agricultural Census

The University of North Carolina Library under a grant from the National Science Foundation microfilmed agricultural Census records. Records were filmed at the University of North Carolina from original records at the Maryland State Library.

Columns 1, 2, 3, 4, 5, and 13 represent the following information on the census:
1. Name of Owner, Agent or Manager of Farm
2. Acres of Improved Land
3. Acres of Unimproved Land
4. Cash Value of the Farm
5. Value of Farming Implements and Machinery
13. Value of Livestock

Elijah Cade, 250, 190, 4000, 30, 500
Edward Saulsbury, 250, 50, 6000, 100, 500
Robert K. Collison, 150, 52, 3000, 50, 250
Henry Wilson, 170, 20, 5000, 100, 500
John Rich, 80, 8, 1500, 15, 80
Levi Dukes, 245, 15, 6000, 150, 800
Ary Raughley, 140, 90, 6000, 75, 450
David Raughley, 140, 35, 5500, 60, 250
Benjamin Ayres, 180, 42, 6000, 50, 300
William P. Richardson, 200, 80, 6000, 40, 700
William G. Hensey, 237, 137, 8000, 250, 500
Joshua Logan (Legare), 225, 75, 7000, 30, 300
William A. Williams, 150, 38, 4000, 30, 200
William Williams, 240, 50, 7500, 80, 750
Everett Godwin, 125, 50, 3500, 100, -
Alexander Cook, 40, 15, 1500, 10, -
George W. Taylor, 150, 50, 5000, 600, 400
John Wells, 300, 20, 3500, 40, 300

Isaac Gibson, 220, 40, 5000, 50, 400
Alexander Matthews, 190, 30, 5000, 150, 650
George W. Wilson, 350, 110, 10000, 300, 800
Robert Wilson, 164, 20, 4000, 40, 300
William A. Ford, 270, 50, 12000, 300, 1000
William Cade, 100, 45, 1500, 30, 200
James Roe, 350, 150, 5000, 40, 350
Wesly Roe, 70, 13, 1000, 30, 300
Ezra Hitch, 75, 38, 2500, 150, 300
Richard C. Hicks, 80, 40, 2000, 40, 300
James Pierce, 180, 68, 300, 70, 380
Thomas Eaton, 244, 80, 5000, 100, 600
Ezra Parriss, 150, 50, 1800, 40, 200
William Knotts, 100, 100, 3000, 30, 180
Julius Smith, 200, 100, 4000, 20, 200
Robert Havely, 275, 222, 15000, 250, 1000
Aaron Shepherd, 10, 5, 300, 25, 150
William Wright, 26, 5, 400, 25, 120
William B. Taylor, 80, 20, 1000, 15, 100
Luther W. Jewell, 120, 33, 3000, 150, 200

John W. Clarke, 130, 20, 4000, 300, 400

William B. Stevens, 80, 66, 2500, 40, 200

Samuel H. Anderson, 150, 50, 6000, 350, 1000

Charles Waynan, 150, 50, 3000, 50, 300

William Jackson, 80, 55, 1200, 30, 150

James Shields, 220, 100, 4000, 250, 650

George Spooney, 140, 60, 4000, 100, 1000

Jonathan Tyler, 70, 63, 4000, 150, 400

Henry Byres, 40, 51, 1200, 40, 100

Charles Cottingham, 80, 20, 3500, 35, 250

Thomas Towers, 70, 10, 2500, 40, 300

Mary E. Fountain, 250, 50, 10000, 90, 700

Frisby Elliott, 70, 30, 1500, 25, 200

John Ringgold, 65, 30, 2000, 30, 200

Robert R. Emerson, 150, 50, 7000, 125, 600

Thomas Atkinson, 140, 90, 4000, 100, 550

Richard C. Carter, 200, 65, 8000, 500, 900

Thomas Lister, 160, 50, 4000, 200, 600

Benjamin D. Pratt, 250, 100, 4000, 100, 700

Elizabeth Carmoll, 100, 10, 1500, 30, 200

Isaac Ireland, 15, 55, 6000, 80, 220

Thomas Sparklin, 170, 50, 4000, 100, 680

Robert Wayman, 100, 100, 2000, 35, 220

Charles E. Janell, 450, 125, 12000, 300, 1000

John C. Janell, 120, 60, 2500, 40, 400

Thomas Robinson, 120, 30, 2000, 30, 380

Joseph Holden, 150, 30, 2000, 30, 250

James Goodring, 100, 100, 4000, 60, 270

William T. Ringgold, 180, 60, 3000, 40, 150

Parden Keetes, 130, 40, 2500, 30, 250

John Carroll, 350, 100, 5000, 100, 600

Martin Robinson, 100, 60, 2000, 30, 350

Thomas Hackett, 40, 36, 3500, 50, 600

William A. Barton, 50, 50, 5000, 150, 400

William Wilkinson, 20, 30, 3000, 180, 350

John H. Holt, 60, 450, 4000, 50, 350

Denwood Long, 150, 30, 4000, 70, 400

George Fisher, 100, 30, 3000, 70, 250

John R. Fountain, 30, -, 1000, 30, 150

Robert J. Jessop, 100, 30, 3000, 700, 400

William Long, 100, 40, 2000, 30, 200

Sylvester Smith, 220, 80, 3500, 80, 600

Elisha Ringgold, 150, 80, 2500, 30, 150

James A. Barwick, 200, 50, 6000, 200, 400

Charles A. Roe, 200, 40, 2500, 40, 300

Robert Roe, 150, 30, 2000, 40, 350

Andrew S. Green, 130, 20, 2000, 50, 450

Wilson Dukes, 100, 50, 1800, 40, 200

Robert J. Grell, 200, 100, 3000, 30, 300

Phillip Mullikin, 200, 71, 2500, 40, 700

Christopher King, 100, 120, 1000, 30, 90

Foster Green, 200, 120, 3000, 30, 500

William Connolly, 150, 133, 6000, 200, 9000

William Connolly Jr., 150, 75, 2500, -, 400

Richard Byers, 20, -, 400, 20, 250

Elizabeth Flynn, 200, 50, 2500, 40, 270

D. Knotts, 130, 70, 6000, 50, 200

John Knotts, 250, 103, 10000, 500, 1000

Wm. Tatman, 142, 120, 2000, 30, 300

Joshua Seward, 200, 57, 2000, 400, 200

John Williams, 350, 64, 4000, 150, 800

William Slaughter, -, -, -, -, 220

Henry Wilson, 150, 50, 1000, 30, 200

James Swann, 200, 50, 2500, 100, 700

Francis Shields, 150, 25, 3000, 75, 280

Joseph Slaughter, 120, 30, 2000, 30, 200

William H. Martin, 200, 10, 2500, -, 200

John H. Emerson, 40, 20, 1500, 50, 200

Edmond Hamitt, 200, 75, 3000, 100, 300

Robert Smith, 80, 15, 1200, 40, 200

Jacob Slaughter, 150, 50, 1000, 40, 100

William Wilson, 75, 20, 800, 35, 200

Walter Massey, 325, 90, 6000, 200, 600

Boon Lockerman, 130, 20, 1000, 30, 220

William Wilson, 200, 30, 3000, 70, 500

Ezekiel Dean, 180, 50, 3000, 40, 500

John Young, 130, 50, 3000, 50, 220

John Price, 60, 30, 1500, 30, 240

James Jones, 150, 50, 1500, 25, 100

Wm. B. Massey, 22, -, 2000, 50, 280

Harriett Turner, 130, 70, 7000, 100, 700

Joshua McGenigal, 33, 20, 700, 50, 100

Esma Lowe, 200, 175, 4000, 150, 450

Bayard Hevalen, 200, 95, 5000, 60, 400

Wm. C. Satterfield, 100, 100, 2500, 80, 700

John Allen, 100, 150, 3000, 50, 800

John Simpson, 200, 75, 4500, 50, 700

Hambleton Bennett, 50, 25, 1500, 30, 150

Samuel Demming, 140, 60, 2500, 70, 500

Robert Dunning, 50, 20, 1000, 30, 120

William Draper, 300, 100, 2500, 35, 120

Isaac Thomas, 40, -, 700, 30, 200

Allen Lockerman, 16, -, 500, 30, 150

Isaac Lewis, 90, 80, 1000, 25, 120

Benjamin Atwell, 45, 35, 1500, 25, 100

John Bishop, 200, 160, 3000, 50, 550

Owen Morgan, 125, 75, 2500, 70, 300

Isaac Bailey, 115, 60, 2500, 60, 480

Nehemiah Draper, 20, 1, 400, 20, 120

William Driggist, 120, 70, 2000, 25, 250

James Merrick, 100, 70, 1000, 30, 150

William Lowe, 60, 40, 1000, 50, 200

Alexander Saulsbury, 20, -, 700, 30, 125

James H. Tuff, 20, -, 700, 30, 80

Peter S. Morris, 30, -, 700, 20, 90

Whittington Ricketts (Pickett), 30, 30, 800, 20, 100

John W. Christopher, 200, 70, 2000, 30, 220

John A. Blanch, 40, 30, 700, 20, 180

Peter Phillips, 50, 30, 700, 25, 100

Joshua Wilson, 90, 110, 1000, 20, 100

Thomas Duland, 100, 40, 1200, 35, 300

William Lorden, 100, 96, 2000, 40, 290

John Feney, 200, 60, 2500, 100, 300

Thomas Janell, 139, 30, 6500, 100, 450

William Delahay, 160, 50, 6000, 50, 350

John D. Pippin, 100, 30, 2500, 60, 200

William Downes, 150, 50, 2000, 30, 60

Jesse W. Reed, 100, 70, 2000, 80, 400

George Brown, 100, 70, 2000, 40, 60

James Langrell, 250, 50, 2000, 50, 300

George Hutchins, 150, 180, 3000, 30, 100

James M. Whitley, 180, 40, 3000, 60, 400

Robert E. Cahall, 225, 100, 3000, 70, 500

Henry Straughn, 193, 50, 5000, 90, 600

Richard Comegys, 90, 10, 2000, 35, 250

Samuel Coursey, 350, 150, 8000, 90, 680

Henry Onell, 200, 130, 3000, 40, 400

Thomas Plummer, 200, 70, 3500, 70, 600

John Dawson, 150, 150, 3500, 60, 1000

Perry Canaday, 100, 50, 1500, 30, 250

Thomas A. Roe, 210, 105, 5000, 100, 600

Thomas Potts, 500, 400, 12000, 300, 1600

Noah Seward, 227, 100, 4000, 200, 350

James J. Clark, 400, 200, 10000, 100, 700

Joseph Anthony, 150, 100, 4000, 100, 400

John Reid, 200, 125, 3000, 200,700

Hester Eaton, 350, 310, 5000, 55, 400

Peter Draper, 200, 233, 5500, 90, 450

John Shewbrook, 60, 43, 500, 35, 80

David Canaday, 80, 20, 1000, 35, 400

John E. Starkey, 250, 40, 4500, 100, 700

James Wood, 100, 90, 1000, 30, 150

Josiah Jessop, 160, 90, 4000, 60, 350

William Slaughter, 30, 14, 400, 20, 120

Wallace M. Bradly, 100, 50, 800, 30, 150

William Roe, 280, 40, 6000, 150, 700

Aaron Henry, 75, -, 1000, 30, 400

Henry C. Veasey, 160, 40, 6000, 150, 350

Greensberry W. Ridgley, 800, 300, 11000, 70, 800

Edward Wood, 60, 40, 600, 35, 100

Charles H. Wood, 120, 30, 1000, 40, 250

John Wood, 80, 20, 600, 30, 220

John S. Bell, 178, 40, 3500, 40, 300

Thomas Bell, 400, 68, 5000, 100, 500

Marcellin Plummer, 170, 30, 4000, 60, 450

Denard Willis, 80, 20, 800, 30, 120

James Pippin, 170, 30, 4000, 70, 500

Tristan Pippin, 120, 40, 1600, 50, 300

John H. Dukes, 120, 40, 2000, 70, 400

Thomas Butler, 70, 30, 1500, 35, 300

Thomas Butler Jr., 120, 50, 1500, 20, 150

Robert Anthony, 200, 105, 4000, 85, 400

Ambrose Dill, 250, 50, 2000, 30, 200

George Bell, 70, 20, 1000, 40, 500

Perry Outten, 40, 10, 400, 20, 70

Margarett A. Saulsbury, 120, 125, 1200, 40, 200

William F. Harvey, 60, 20, 500, 20, 200

Francis Wilson, 150, 62, 2000, 40, 320

Solomon Melvin, 75, 15, 1500, 50, 300

Robert E. Graham, 60, 12, 1500, 30, 220

William Cohee, 120, 50, 1200, 30, 200

Julie Carter, 200, 200, 6000, 150, 1000

Samuel Miller, 100, -, 1500, 30, 300

James Melvin, 200, 50, 3700, 80, 380

Jesse Mitchell, 150, 10, 300, 50, 200

John Vance, 60, 22, 700, 40, 350

Thomas Sewell, 230, 50, 3000, 35, 300

Thomas Saulsbury, 200, 100, 3800, 80, 650

Thomas Hardesty, 122, 122, 3000, 40, 350

John Wright, 100, 88, 4000, 60, 400

Benjamin Long, 270, 80, 4000, 70, 300

William S. Levitt, 150, 75, 6000, 70, 500

James T. Richardson, 100, 35, 1500, 40, 250

Margarett Porter, 20, 20, 1000, 30, 120

James S. Rickards, 200, 100, 4500, 70, 400

Samuel Emerson, 100, 30, 1800, 40, 350

James P. Selby, 180, 55, 5000, 75, 600

Elijah Porter, 70, 150, 1200, 40, 150

Erastus Anderson, 100, 40, 1800, 35, 250

Henry Brown, 130, 110, 1800, 40, 240

Norriss Wilson, 150, 65, 3750, 80, 600

Robert Jarnell, 175, 50, 10000, 300, 1000

William Gellett, 150, 25, 4000, 75, 600

William R. Jarnell, 125, 21, 2500, 40, 250

Richard Legg, 165, 60, 2500, 90, 450

James Carlisle, 150, 70, 2500, 65, 450

John Thawley, 65, 35, 1500, 40, 400

Benedict Wyatt, 150, 50, 3000, 900, 450

Frank Cosden, 100, 25, 1200, 30, 290

Lewis Cook, 120, 40, 1600, 30, 320

Robert Dean, 130, 30, 1500, 35, 700

Elizabeth Macknahan, 120, 60, 1000, 40, 300

Robert Hignutt, 150, 150, 2000, 50, 420

Cornelius Wright, 70, 45, 700, 40, 250

Thomas Andrew, 90, 15, 600, 30, 160

Jeremiah Reed, 100, 40, 700, 25, 220

John W. Bullock, 140, 20, 1500, 40, 320

William H. Robinson, 180, 50, 1700, 30, 175

William Alford, 150, 50, 1000, 30, 190

Elijah Williamson, 15, 7, 500, 30, 170

Andrew Corey (Covey), 84, 20, 800, 30, 260

Lena Nicholas, 135, 60, 1200, 40, 400

Peter Dean, 100, 70, 1500, 30, 250

Samuel Nicolas, 70, 40, 1500, 40, 400

Uriah Corey, 90, 20, 150, 30, 170

Jacob Towers, 160, 80, 920, 30, 250

Sarah Davis, 40, 60, 500, 30, 150

Andrew Towner, 40, 65, 500, 30, 85

Thomas Arnett, 35, 10, 350, 25, 200

Daniel Nicols, 60, 35, 600, 30, 130

Robert Edgell, 60, 40, 700, 30, 145

Peter Jester, 18, 13, 300, 20, 150

William Raseum, 70, 80, 500, 35, 180

James Nicols, 140, 55, 2800, 50, 400

John Dean, 30, -, 600, 20, 60

Robert Nicols, 55, 50, 1000, 30, 350

Wesly J. Nicols, 50, 80, 1000, 25, 200

Daniel Connolly, 75, 90, 1500, 25, 180

James Bodass, 60, 130, 1000, 25, 130

Nancy Harris, 100, 250, 1000, 25, 60

Andrew Harriss, 40, 40, 250, 20, 55

Sampson Satterfield, 22, 20, 200, 5, 45

Richard Bullock, 100, 150, 2800, 80, 470

William Paine, 90, 60, 800, 40, 150

Ezra Collins, 75, 35, 900, 50, 200

William James, 22, 10, 300, 20, 130

Daniel Eaton, 10, 20, 280, 20, 90

Shadrick Dean, 10, 40, 400, 20, 95

Martha Stevens, 90, 30, 1000, 20, 45

William Buley, 10, 20, 300, 20, 55

Peter Sullivane, 140, 33, 2000, 50, 390

Andrew Griffith, 90, 90, 1000, 40, 100

William Griffith, 50, 10, 1000, 45, 300

Asbury Stamford, 70, 30, 500, 20, 40

Gabriel Birley (Buley), 120, 240, 1800, 30, 300

Elijah Hignutt, 80, 12, 500, 20, 180

Elijah Towers, 75, 100, 1700, 35, 140

John Collins, 75, 75, 1000, 30, 120

Samuel Wilkins, 30, 70, 1000, 40, 150

Jacob Whitaker, 80, 20, 1500, 60, 220

Matthew Johns, 30,77, 800, 25, 45

Benjamin F. Flemming, 250, 50, 3000, 60, 400

Thomas F. Griffith, 150, 50, 2000, 40, 190

Charles H. Todd, 150, 68, 3000, 60, 400

John Cheezum, 70, 30, 1200, 30, 150

George Andrew, 100, 30, 300, 30, 275

John Smith, 20, -, 250, 15, 50

Henry Harriss, 130, 50, 2000, 30, 200

William Willoughby, 100, 20, 2000, 40, 300

Luke Andrew, 75, 50, 1000, 25, 60

Ruben Williamson, 40, 80, 500, 20, 70

Peter Harris, 100, 100, 1000, 40, 190

Benjamin Lewis, 50, 50, 900, 25, 268

William Phillips, 60, 40, 800, 25, 200

Charles McRush, 8, -, 600, 20, 129

James Butler, 79, 20, 1200, 30, 300

Peter Willis Sr., 9, -, 800, 20, 150

Peter Willis Jr., 140, 50, 1200, 40, 160

Thomas Dukes, 60, 7, 1200, 50, 350

Martha Franton, 10, -, 200, 15, 40

Theodore Handy, 75, 40, 3000, 60, 250

William Andrew, 100, 20, 3000, 40, 250

Thomas F. Corkrin, 200, 200, 4000, 50, 450

James Ross, 150, 50, 2500, 40, 220
James H. Covey, 50, 25, 800, 25, 200
John Nicols, 120, 30, 7000, 100, 400
Peter Collins, 170, 100, 700, 30, 250
Thomas Waddle, 60, 40, 700, 30, 70
Edward Frampton, 70, 75, 1000, 25, 100
John Z. Williamson, 15, 5, 250, 15, 150
John Dear, 100, 60, 1200, 35, 360
James L. Pane, 90, 60, 700, 30, 140
Henry Andrew, 50, 20, 500, 20, 40
William A Tellir (Tellis), 175, 40, 1500, 35, 400
William Foster, 60, 40, 500, 25, 60
James Pane, 70, 51, 1500, 30, 200
Elias P. Hopkins, 50, 20, 1000, 30, 220
Eben Jackson, 100, 55, 1000, 35, 320
Mary Foster, 100, 50, 800, 25, 150
Richard Willoughby, 60, 60, 300, 20, 60
William Nance (Nanine), 77, 70, 500, 20, 40
Levin Robinson, 100, 60, 1000, 30, 300
William Trice, 80, 60, 450, 25, 100
Martin Trice, 90, 16, 600, 25, 150
William Sisk, 90, 125, 1200, 35, 200
William Hand, 100, 50, 1500, 30, 240
Joseph Mitchell, 100, 50, 1500, 40, 200
James Todd, 10, -, 200, 15, 150
James Dowdel, 80, 20, 700, 35, 200
William Blades, 30, 20, 500, 20, 135
John B. Todd, 25, 15, 700, 35, 150
George Breeding, 75, 40, 800, 20, 100
Elijah Friend, 35, 59, 375, 15, 70
Robert Sena, 30, 30, 400, 15, 50
John Dillen, 100, 30, 1000, 35, 400
Perry Friend, 50, 50, 400, 15, 100
John Webb, 130, 150, 2000, 45, 450
Asbury Friend, 20, 34, 500, 25, 300

Joshua Adams, 200, 300, 5000, 40, 100
William N. Smith, 300, 500, 10000, 250, 1010
John Cheezum, 110, 50, 1000, 40, 220
Levin Cheezum, 35, 15, 1200, 30, 220
Linden Molock, 40, 60, 800, 25, 46
David Waddle, 50, 25, 1200, 50, 300
Lena Covey, 30, 30, 750, 30, 100
Elijah Thomas, 125, 250, 3750, -, 450
Adam Blay, 10, 5, 250, 20, 60
Robert Sena, 20, -, 300, 15, 50
Thomas Wright, 25, 12, 400, 25, 200
Stephen Blay, 30, -, 300, 20, 55
William Berkett, 50, 20, 350, 20,100
Frederick Berkett, 50, 20, 350, 20, 70
Henry Holmes, 100, 50, 3500 150, 850
Garrison Blades, 25, 75, 2500, 40, 190
James Adams, 100, 70, 2000, 30, 250
Henry Friend, 20, 10, 250, 15, 60
Elisha Adams, 75, 60, 1500, 35, 200
Bennett Patten, 30, 78, 1800, 45, 300
Henry Nicols, 80, 20, 1200, 40, 240
Benjamin Lecompt, 80, 20, 1200, 40, 65
William H. Stevens, 150, 67, 3000, 50, 420
Henry Chase, 15, 20, 350, 20, 40
Thomas Cheezum, 120, 70, 1800, 40, 240
John D. Farquhanson, 250, 58, 7000, 150, 200
William E. Perry, 100, 250, 1500, 40, 200
Perry Covey, 150, 70, 7200, 45, 350
Charles Perry, 50, 10, 600, 20, 100
Thomas Dillen, 50, 10, 600, 20, 100
Benjamin Nicols, 100, 69, 1600, 50, 275

William Perry, 100, 50, 1200, 40, 100

Nathaniel Perry, 30, 10, 500, 20, 150

James E. Duglass, 200, 200, 8000, 175, 1300

William Jenkins, 40, 80, 1500, 30, 220

Joseph Harrison, 150, 50, 2000, 40, 300

Willoughby Sharp, 30, 30, 600, 25, 130

Henry Irvin, 30, 30, 800, 25, 100

Foster Hubbard, 15, -, 250, 15,18

William H. Stafford, 100, 50, 2000, 70, 300

John O. Stafford, 75, 15, 1500, 50, 100

George Andrew, 100, 75, 1800, 40, 30

Henry Shields, 300, 65, 7000, 75, 800

Allenson Corkrin, 20, 78, 700, 25, 95

Caleb Eaton, 20, 5, 1000, 35, 120

Henry Corkrin, 125, 25, 2500, 45, 300

Lowder Hubbard, 150, 20, 2500, 40,70

Elias Hubbard, 30, 20, 1000, 40, 220

Richard Willis, 55, 20, 1200, 40, 60

Matthew Patton, 100, 100, 2000, 40, 200

William Ross, 30, -, 300, 25, 75

Wright Hubbard, 22, 25, 600, 25, 150

William Wherdleton, 160, 30, 2500, 45, 260

Henry Dear, 35, 20, 950, 35, 190

William Gooter, 15, 2, 300, 25, 100

Alexander Whitely, 140, 3, 2500, 50, 370

George F. Wherdtey, 60, -, 1200, 40, 250

William Williamson, 140, 20, 3000, 70, 400

Saulsbury Carroll, 40, -, 500, 30, 200

William H. Austin, 100, -, 1200, 40, 190

Lorenzo Jones, 180, 36, 3000, 45, 400

William T. Kelly, 70, 30, 2000, 45, 400

George Pinkett, 80, 45, 1500, 40, 235

James Lynch, 100, 20, 3500, 45, 250

Joseph Harniss (Harriss), 100, 60, 1600, 40, 250

Elisha Nicols, 68, 100, 1000, 30, 110

Solomon Holmes, 20, -, 300, 25, 60

John Kesler, 50, 50, 750, 85, 100

Thomas Hubbard, 60, 35, 800, 24, 330

Henry Kinnaman, 140, 60, 1400, 30, 300

John D. Williams, 150,70, 3500, 60, 430

Caleb Connolly, 100, 60, 600, 30, 100

Levin Williams, 50, 20, 1200, 40, 170

Maddison Williams, 170, 60, 3000, 70, 750

James Jones, 60, 40, 1200, 45, 250

Elias Cox, 180, 60, 3500, 75, 370

Collison Waddle, 100, 40, 3000, 60, 600

Samuel Webb, 40, -, 400, 20, 60

Charle Edgell, 60, 50, 700, 30, 140

Daniel B. Cannon, 100, 20, 4000, 65, 275

Absolem C. Patchell, 12, 23, 1500, 25, 270

Henry _. Patchell, 200, 65, 3500, 45, 400

Harrison Simpson, 80, 15, 600, 25, 225

James Chase, 60, 40, 600, 27, 75

John F. Sullivan, 15, 5, 200, 15, 40

Sylvester Andrew, 30, 70, 1000, 25, 120

Phillip H. Hopkins, 50, 70, 1000, 20, 15

Edward Andrew, 80, 40, 700, 20, 90
John Todd, 150, 50, 1500, 50, 400
Christopher Nicols, 40, 30, 500, 15, 30
Thomas Valient, 100, 58, 2500, 50, 245
Henry Valient, 70, 35, 200, 40, 160
White Barrick, 200, 46, 915, 30, 130
Short A. Willis, 100, 480, 6000, 400, 130
Edward Hardcastle, 200, 260, 8000, 100, 550
Tazisey Towers, 100, 100, 700, 25, 75
Thomas Roe, 140, 130, 2500, 50, 225
John Phillips, 50, 70, 700, 30, 200
Hanett Ross, 100, 75, 1200, 40, 270
James Friend, 120, 50, 1200, 40, 370
Margarett Freexon, 30, 21, 300, 25, 70
Solomon S. Chaffinch (Schaffinch), 130, 50, 3000, 45, 300
Abraham Collins, 150, 203, 3000, 70, 350
Rennols Collins, 80, 75, 1000, 45, 200
Walter Warren, 125, 27, 1500, 70, 350
Jesse Hubbard, 100, 52, 1500, 40, 150
Henry Friend, 20, -, 450, 20, 40
Daniel Sparklin, 100, 30, 1500, 45, 300
Poulson Hubbard, 75, 35, 1200, 40, 350
Margarett Rumbold, 50, 50, 600, 40, 140
Daniel Thomas, 50, 30, 800, 25, 25
Aaron Towers, 75, 35, 750, 25, 65
Thomas Carroll, 100, 100, 1000, 40, 260
Jehew Nicols, 100, 100, 1000, 35, 200
John Blades, 110, 50, 1200, 40, 210

James H. Windsor, 140, 50, 1500, 35, 200
Thomas Wheeler, 100, 78, 1900, 30, 110
John Coursey, 20, -, 400, 15, 15
Samuel Butler, 50, 30, 450, 15, 25
Joseph Patchell, 100, 47, 1000, 45, 250
Willis Wright, 50, 23, 900, 35, 200
Uriah Sephens, 60, 39, 500, 25, 50
James Parris, 150, 250, 3500, 75, 450
David Todd, 150, 40, 4000, 70, 450
James Herryman, 8, -, 275, 15, 40
Custis Wright, 150, 76, 2000, 45, 200
Elijah Todd, 200, 62, 3000, 60, 300
Solomon Kenton, 40, 10, 500, 25, 150
Daniel Willoughby, 150, 70, 2000, 45, 350
Edward Nicols, 55, 70, 1200, 35, 110
Ann Connolly, 140, 50, 2000, 45, 200
William W. Lowe, 40, 75, 500, 30, 50
Henry Davis, 200, 100, 3000, 40, 350
William Davis, 75, 6, 2000, 40, 120
John A. Wallace, 25, -, 2000, 50, 140
Twiford Noble, 150, 90, 4000, 150, 490
William Noble, 100, 50, 3000, 150, 500
Jesse Hubbard, 150, 50, 5000, 150, 480
Robert Dickinson, 60, 15, 2500, 15, 100
Beauchamp Hubbard, 18, -, 350, 20, 50
John R. Slack, 130, 170, 1500, 50, 300
Edward Davis, 64, 30, 2000, 40, 275
Daniel J. Noble, 45, 75, 1000, 30, 60
Byard Wiliams 100, 75, 1820, 55, 400

Zebdiah Turner, 80, 40, 1200, 35, 250

Sarah Turner, 60, 60, 1000, 35, 150

Daniel P. Bewdel (Bendel), 40, 10, 300, 15, 100

Susan H. Green, 90, 30, 3000, 60, 250

Amos Flaherty, 20, 20, 400, 15, 75

Isaac Flaharty, 40, 49, 800, 20, 75

Job Flaharty, 50, 76, 950, 25, 140

John Pool, 700, 100, 3000, 95, 560

Elizabeth Hutchinson, 45, 15, 400, 20, 60

Richard Frampton, 90, 50, 1000, 80, 225

James Webb, 30, 10, 400, 20, 110

Richard Lawson, 45, 100, 600, 15, 24

John H. Nicols, 50, 52, 600, 25, 70

Curtis Towers, 60, 42, 600, 25, 100

Lemuel Hubbard, 100, 90, 3000, 100, 550

Henry Hubbard, 75, 80, 1000, 40, 275

Isaiah Blades, 189, 90, 4000, 100, 400

Eli Blades, 100, 25, 1500, 25, 210

Peter Jester, 80, -, 900, 20, 130

John Oneal, 90, 40, 700, 20, 75

Isaac Warren, 50, 77, 1000, 40, 140

Henry Covey, 70, 70, 1000, 40, 100

John Willoughby, 30, 70, 500, 30, 80

James Bryan, 20, 40, 350, 20, 60

Manless Hutchinson, 20, 40, 400, 30, 70

Ann Williamson, 70, 61, 600, 25, 130

Phillip H. Rawlins, 150, 150, 3000, 40, 200

Lewis Duhaday, 70, 30, 600, 20, 170

Thomas Nicols, 150, 50, 1000, 30, 250

Jacob R. Andrew, 40, 20, 500, 20, 130

Daniel Coubbon, 130, 70, 890, 25, 200

Becholeman Adams, 40, 10, 1000, 30, 100

James Morgan, 200, 87, 3500, 90, 450

William A. R. Griffith, 100, 61, 3000, 100, 490

James Saulsbury, 50, -, 2000, 70, 220

Thomas H. Draper, 80, 50, 1500, 70, 250

Matthew Garey, 100, 50, 1200, 50, 220

Benton H. Hobbs, 40, 20, 500, 25, 75

William Norriss, 200, 250, 5000, 100, 500

George W. Collison, 200, 53, 4000, 100, 650

Ephriem Moore, 200, 175, 3000, 90, 450

James W. Saulsbury, 100, 60, 1500, 50, 450

Edward Dalzel, 150, 50, 2000, 40, 290

Jackson Samons, 75, 25, 600, 25, 70

Thomas F. Garey, 250, 70, 4000, 175, 1000

Charles Carnish, 120, 80, 2500, 40, 250

Benjamin A. Richardson, 100, 100, 2500, 40, 200

# Carroll County Maryland
## 1860 Agricultural Census

The University of North Carolina Library under a grant from the National Science Foundation microfilmed agricultural Census records. Records were filmed at the University of North Carolina from original records at the Maryland State Library.

Columns 1, 2, 3, 4, 5, and 13 represent the following information on the census:
1. Name of Owner, Agent or Manager of Farm
2. Acres of Improved Land
3. Acres of Unimproved Land
4. Cash Value of the Farm
5. Value of Farming Implements and Machinery
13. Value of Livestock

Pages for this county are out of sequence.

Wm. Hooper, 41, 10, 1500, 50, 500
G. Warfield, 40, 50, 10000, 150, 850
A. Long, 21, 5, 1000, 25, 200
J. Nausbaum, 100, 30, 2000, 200, 400
C. Franklin, 66, 10, 600, 10, 140
A. P. Franklin, 28, -, 800, 10, 140
N. Franklin, 100, 20, 5000, 50, 300
W. H. Franklin, 90, 15, 1500, 50, 200
J. T. Franklin, 40, 10, 1000, 10, 250
R. H. Franklin, 50, -, 1200, 10, 150
T. Smith, 20, 17, 1500, 70, 250
D. Zele, 200, 50, 5000, 200, 700
F. J. Crawford, 220, 30, 6000, 150, 1000
E. Henderson, 60, 20, 1000, 10, 150
S. Choate, 172, 40, 4500, 100, 400
A. Greenwood, 82, 10, 300, 50, 300
A. Albaugh, 90, 30, 3000, 100, 300
T. J. Kelley, 140, 10, 3000, 150, 400
G. Barr, 130, 300, 3000, 50, 400
E. Lindsay, 180, 130, 1000, 400, 1200
R. D. Gorsuch, 160, 20, 6600, 350, 400
J. Selman, 300, 200, 20000, 400, 2400
S. Bowers, 47, 8, 100, 50, 250

W. Farvel, 35, 20, 10000, 75, 225
R. Powel, 48, 12, 900, 10, 300
L. Barnes, 50, 5, 1000, 25, 200
J. Zele, 33, 2, 700, 10, 80
Wm. Yohn, 160, 60, 70000, 100, 700
C. Bowers, 40, 20, 1500, 50, 200
J. Cook, 34, 3, 1000, 25, 200
N. Nuton, 33, 2, 1200, 10, 150
S. Smith, 56, -, 10000, 100, 300
R. Criswell, 96, 40, 400, 200, 800
R. Zile (Zele), 90, 28, 3000, 100, 500
J. Frizzle, 110, 56, 4000, 300, 2000
L. Frizzle, 147, 100, 5000, 250, 1000
J. Selman, 172, 50, 6000, 100, 700
W. Barnes, 25, -, 500, 10, 180
T. A. Barnes, 58, 20, 800, 30, 200
George E. Smith, 40, 10, 800, 20, 175
P. Zepp (Zopp), 40, 30, 800, 50, 150
W. Zepp, 54, 30, 1000, 100, 300
John Zepp, 100, 20, 10000, 75, 315
C. Waiters (Warters), 200, 60, 15000, 200, 600
T. Linthicum, 78, 20, 1500, 100, 300
S. Davis, 150, 50, 60000, 150, 800
Robt. Dade, 350, 150, 15000, 300, 1000
S. R. Warters, 108, 65, 10000, 100, 700

John Pool, 75, 25, 1500, 50, 75
N. Warters, 70, 40, 40000, -, -
W. Gosnell, 70, 62, 1800, 200, 600
W. P. Gosnell, 100, 40, 1800, 150, 350
Lee G. Grimes, 10, 80, 2500, 125, 400
W. S. H. Dorsey, 130, 25, 3000, 40, 400
L. Gosnell, 70, 30, 800, 75, 300
W. Gosnell, 30, 10, 700, 10, 90
B. Penn, 100, 50, 2500, 30, 270
John Harrison, 90, 30, 2000, 40, 260
B. S. Poole, 200, 48, 1800, 50, 250
Thos. Fleming, 65, 80, 1800, 100, 70
C. Fleming, 65, 80, 1800, 100, 400
L. Selman, 75, 5, 1000, 75, 225
S. Clagett, 80, 20, 800, 100, 500
J. Gillias, 60, 60, 3000, 75, 400
J. McQuay, 38, 36, 2500, 100, 300
J. Cover, 164, 45, 6000, 200, 800
R. Condon, 167, 70, 2000, 200, 300
W. Harkley, 50, 20, 3500, 50, 400
N. S. Engle, 145, 25, 4000, 100, 1000
Roger Franklin, 225, 25, 400, 100, 400
John S. Young, 55, 5, 500, 10, 60
W. H. Barnes, 160, 40, 3000, 150, 500
H. Helery, 17, 3, 1000, 20, 100
David Buckingham, 32, 8, 600, 30, 270
C. Franklin, 45, 20, 1500, 100, 200
W. R. Steele, 100, 125, 5000, 400, 600
S. J. Grimes, 47, 47, 1000, 100, 100
W. W. Picket, 120, 50, 6800, 20, 200
N. Sprinkle, 23, -, 2500, 60, 125
J. C. Gist, 200, 80, 14000, 500, 1200
W. Smith, 124, 150, 2500, 50, 400
A. Smith, 56, 50, 2000, 25, 200
A. Bloom, 80, 30, 400, 100, 120
J. Cook, 85, 10, 300, 100, 400
D. Bloom, 45, 35, 600, 20, 250

E. Paught (Piught), 135, 90, 7000, 60, 450
L. T. Bennett, 200, 250, 6000, 300, 600
Wm. Fuzzle, 140, 100, 3000, 60, 300
E. Cook 90, 70, 3000, 100, 300
C. George, 100, 30, 2300, 50, 200
J. Harp, 44, 6, 700, 10, 150
S. Gorsuch, 150, 50, 6000, 100, 600
J. P. Skidmore, 50, 30, 1500, 100, 400
A. J. Skidmore, 52, 8, 2500, 20, 300
J. Penn, 80, 132, 3000, 100, 500
E. Picket, 125, 25, 5000, 25, 200
Upton Rankles, 95, 35, 2000, 100, 300
Bel Buckingham, 180, 20, 2500, 30, 600
J. Penn, 190, 25, 2000, 100, 500
J. Penn, 98, 70, 2000, 25, 100
D. Dudderer, 200, 74, 5500, 75, 700
H. Devall, 85, 20, 2500, 20, 350
L. Franklin, 150, 30, 3000, 100, 300
W. Gunn, 61, 25, 1500, 100, 300
B. Spurner, 75, 15, 1800, 20, 150
P. Lowman, 72, 30, 1600, 150, 600
J. Lowman, 50, 30, 1000, 40, 250
Z. Berchears, 80, 20, 5000, 22, 200
R. Henry, 100, 50, 3000, 50, 300
Samuel Fogle, 80, 10, 1000, 50, 250
A. Davis, 160, 50, 3000, 100, 500
P. W. Davis, 100, 60, 200, 20, 200
U. L. Dorsey, 141, 120, 8000, 200, 1200
A. Buckingham, 70, 30, 800, 100, 200
S. Baker, 86, 30, 200, 60, 300
T. Davis, 200, 100, 6000, 200, 700
W. Harrison, 18, 18, 1000, 10, 100
H. S. Davis, 250, 150, 10000, 200, 1000
S. England, 40, 7, 3600, 20, 300
B. Clary, 88, 20, 2500, 25, 200
Henry Bessard, 150, 50, 6000, 150, 500
F. A. Switzer, 40, 6, 100, 15, 200

J. C. Pennington, 185, 5, 6000, 700, 600

T. B. Ourings, 700, 300, 20000, -, -

John Hood, 140, 40, 2000, 50, 600

J. Teaner, 130, 30, 300, 100, 350

S.Elgin, 50, 3, 700, 50, 250

L. Dorsey, 167, 100, 7000, 200, 900

U. More, 250, 150, 4000, 100, -

N. Gosnell, 90, 6, 1400, 50, 250

Charles Preket, 100, 10, 1200, 25, 200

S. Brasheur, 200, 60, 2000, 100, 300

Vact Hammond, 70, 70, 1200, 30, 200

Otho Dotey, 60, 40, 1000, 20, 150

C. Brasheur, 157, 40, 2000, 100, 300

R. Duvall, 215, 215, 8000, 100, 600

M. Growe, 62, 10, 2500, 30, 600

J. Hood, 175, 40, 3000, 150, 500

N. Harrison, 225, 50, 3000, 40, 700

W. Harrison, 24, 15, 500, -, 100

J. Spurner, 169, 50, 6000, 300, 700

G. Downey, 102, 4, 4000, 46, 300

H. Danner, 95, 5, 3500, 100, 600

J. H. Rain, 85, 15, 50, 75, 300

J. T. Smith, 90, 10, 1500, 25, 250

G. Dorsey, 130, 20, 3000, 100, 500

W. Knauf, 100, 6, 1100, 75, 350

J. W. Cockran, 10, 250, 2000, 200, 500

J. Brown, 78, 10, 800, 60, 300

H. Brown, 135, 45, 1800, 20, 300

J. Mulinson, 55, 5, 800, 10, 100

B. Hood, 97, 25, 1200, 50, 250

E. Myers, 50, -, 500, 30, 200

B. Bennett, 90, 20, 1200, 100, 400

Julius B. Banit, 200, 300, 6000, 200, 800

Jno. M. Dorsey, 150, 50, 9000, 50, 700

Eli Hewett Sr., 110, 30, 50000, 195, 600

J. C. Wadlow, 40, 65, 2500, 200, 520

J. T. Steel, 100, 25, 6000, 200, 700

Jesse Frederick, 5, 25, 100, 50, 200

Johna Frizzle, 10, -, 200, 100, 100

William Lindsey, 2, -, 50, 10, 150

John Frizzle, 110, 50, 4000, 250, 500

William Frizzle, 27, -, 1000, 50, -

Josephine Carter, 15, -, 500, 10, 35

Robert Lee, 100, 26, 3000, 150, 400

Joshua Shipley, 160, 80, 17000, 600, 1200

Elias Brown, 220, 80, 7000, 200, 600

John G. Pearce, 218, -, 3500, 200, 500

Augustus Houbs, 5, 100, 500, 50, 250

S. T. C. Brown, 250, 100, 10000, 500, 2500

Nathan Porter, 14, 2, 400, 50, 75

Cornelius Shipley, 150, 50, 7000, 400, 900

Robert _. Shipley, 150, 60, 6000, 300, 1200

John M. Clemans, 60, 30, 2000, 200, 60

Lewis Scribner, 45, 25, 700, 80, 100

John W. Brown, 20, 11, 500, 20, 200

John W. Scribner, 20, 15, 500, 25, 200

Henry Sprinkel, -, -, -, 100, 700

Nicholas Dorsey, 50, 4, 1000, 20, 125

George H. C. J. Bush, 60, 32, 2000, 50, 100

Micajah Rogers, 200, 300, 10000, 100, 600

Henry Devries, 150, 110, 500, 200, 300

William T. Devries, 120, 50, 400, 150, 500

Lloyd Brown, 100, 62, 3000, 100, 300

John O'Neal, 60, 80, 2500, 50, 400

Osten Arenton, 18, -, 400, -, 100

Thomas Arenton, 40, 12, 800, 10, 100

Josiah Brown, 85, 18, 1500, 100, 200

Thomas L. Lee, 45, 5, 900, 75, 250

Wm. P. Gorsuch, 100, 36, 5000, 175, 1000

Elizabeth Shipley, 75, 30, 3000, 200, 400

Richard Barnes, 12, 2, 700, 20, 150

Susanna Warfield, 350, 50, 1400, 250, 1000

O. H. Owings, 50, 7, 4000, 50, 500

George Y. (T.) Wethered, 130, 34, 7000, 130, 600

Elizabeth Barnet, 70, 47, 2000, 10, 125

Jesse Hollingsworth, 100, 85, 5000, 50, 500

James George, 300, 200, 15000, 600, 2100

Wm. Gaiter, 250, 50, 9000, 50, 500

William C. Polk, 175, 75, 10000, 200, 1200

Charles W. Hood, 250, 150, 2000, 800, 2100

W. W. Warfield, 270, 80, 10000, 150, 900

Daniel Terme, 30, 15, 2000, 30, 120

Wm. R. Ridgley, 75, 30, 2000, 30, 400

P.W. Webb, 200, 90, 15000, 300, 600

Dr. Brown, 110, 30, 2800, 100, 400

Geo. Patterson, 1350, 450, 70000, 2000, 15000

Zadoc Water (Waiter), 220, 200, 10000, 100, 600

Margaret Brown, 75, 50, 7000, 60, 300

George Dixon, 80, 83, 4000, 100, 300

William Eden, 360, 340, 20000, 200, 1500

Richard Waters, 55, 10, 1900, 140, 300

Warren G. Little, 60, 40, 2500, 200, 300

David Prugh, 150, 100, 5000, 400, 600

Joshua Lee, 100, 3, 2000, 200, 800

Joshua Lee Jr., 100, 20, 2000, 150, 500

Gilbert Place (Peace), 70, 38, 1500, 30, 300

Gassaway Ralling, 23, 12, 1200, 70, 150

Agustus E. Dorsey, 250, 200, 10000, 200, 1000

David A. Dorsey, 280, 100, 12000, 300, 1500

Robt. M. Hewett, 70, 36, 3000, 50, 300

Augustus Stricker, 70, 30, 1500, 50, 200

Riser Jenkson, 750, 50, 4000, 150, 300

Augst. Picket, 50, 10, 1000, 100, 200

Cornelius Jenkins, 30, 38, 1000, 40, 300

Larkin Shipley, 110, 30, 3000, 100, 400

Thos. Richardson, 25, 28, 800, 10, 300

Hanson Jenkins, 30, 120, 1000, 30, 200

Maranda Barnes, 153, 100, 5000, 21, 175

Wm. Bradenbaugh, 1200, 900, 40000, 500, 2500

Jacob Bradenbaugh, -, -, -, -, -

Elias Brandenbaugh, 100, 20, 100, -, 70

William Selman, 200, 70, 3000, -, 305

Larkin Shipley, 200, 200, 5000, 50, 600

Thomas Glenan, 80, 80, 4000, 200, 900

Philoman Welsh, 40, 30, 400, 50, 150

Lewis H. Shipley, 90, 26, 1400, -, 100

Francis Harp, 56, 15, 1000, 30, 200

Henry Shipley, 100, 125, 7000, 200, 200

James Morgan, 70, 62, 4000, 100, 200

Aaron Gosnell, 100, 30, 300, 100, 200

Brice Ways, 16, 10, 1000, 20, 200

Henry T. Betts, 190, 170, 8500, 350, 1000

Wm. G. Shipley, 111, 60, 5000, 300, 612

Aquilla Day, 50, 64, 4000, 150, 200

S. R. Gore, 120, 40, 5600, 200, 800

Leesa Gore, 160, 70, 8000, 100, 500

Joseph Callaway, -, -, -, 30, 400

Robert Murray, 125, 150, 6000, 120, 500

W. H. Haines, 31, 6, 1000, 100, 400

M. G. Selby, 100, 20, 3000, 150, 500

B. M. Campbell, 244, 100, 9000, 300, 1000

Christian Devries, 45, 80, 2100, 150, 400

John Devries, 50, 82, 1000, 100, 400

John O. Brown, 120, 70, 4000, 25, 300

Peter Davis, 100, 23, 1200, 140, 300

Amos Wilson, 80, 100, 3600, 50, 150

Wm. Picket, 120, 20, 1000, 30, 200

H. Leatherwood, 200, 100, 7000, 80, 400

A. Gosnell, 46, -, 400, 60, 200

E. Buckingham, 100, 100, 2500, 100, 200

P. Gallias, 103, 103, 2000, 20, 100

John Flemen, 55, 50, 850, 50, 250

Melton Picket, 70, 28, 700, 40, 150

G. Buckingham, 32, 30, 800, 40, 200

P. Grove, 200, 700, 3000, 50, 250

Wm. Conoway, 220, 110, 7000, 100, 400

Jesse Gee (Lee), 90, 50, 1000, 100, 300

J. Parrish, 77, 60, 200, 100, 400

Jacob Heltertrielle, 12, 33, 900, 20, 150

F. Dorsey, -, -, -, 100, 400

A. Wilson, 180, 180, 7000, -, -

E. Hudson, 100, 100, 400, 150, 400

L. Wilson, 78, 25, 2000, 200, 175

M. Parrish, 100, 60, 3000, 100, 500

R. Parrish, 75, 30, 1000, 50, 150

G. Shipley, 100, 150, 4000, 200, 500

C. Shipley, 70, 70, 1500, 100, 250

A. Brown, 70, 32, 600, 20, 150

J. H. Lindsay, 80, 142, 5000, 300, 500

J. Edmondson, 100, 300, 4000, 100, 300

H. S. Buckingham, 50, 46, 1200, 50, 100

L. Gardner, 80, 79, 1200, 50, 400

J. Parrish, 120, 80, 3000, 200, 300

L. Hoff, 40, 14, 800, 50, 200

B. Shipley, 80, 76, 2000, 200, 200

C. Shipley, 125, 125, 2500, 200, 200

J. Shipley, 100, 15, 1500, 200, 200

N. Gore, 160, 11, 8000, 500, 1000

S. Warfield, 70, 46, 2500, 100, 250

J. Sundergill, 150, 50, 300, 100, 200

A. Brown, 59, 40, 1200, 100, 100

R. Lions, 100, 52, 2000, 100, 200

J. Smith, 75, 75, 1500, 100, 500

G. Shipley, 60, 45, 3000, 100, 900

A. Shipley, 125, 155, 2000, 150, 350

R. Pool, 90, 90, 1500, 200, 300

Dr. R. Thompson, 125, 28, 5000, 150, 800

O. Buckingham, 125, 95, 3000, 100, 400

L. Gorsuch, 80, 21, 2000, 100, 100

F. Gardner, 66, 30, 2500, 100, 300

O. Buckingham, 65, 30, 2000, 100, 400

W. H. Harden, 145, 30, 1100, 100, 700

J. Gore, 161, 30, 10000, 300, 130

J. Selby, 131, 40, 10000, 100, 900

J. Baseman, 250, 200, 1000, 300, 1000

C. Bennett, 150, 150, 5000, 150, 600

O. Pennington, 35, 20, 600, 10, 90

J. Frizzle, 70, 30, 2000, 25, 275

J. Ritter, 110, 31, 3000, 100, 400

L. W. Manroe, 150, 25, 10000, 300, 100

N. D. Norris, 125, 50, 7000, 300, 1200

W. Roberts, 102, 8, 1500, 50, 250

S. Harden, 80, 17, 4000, 300, 700

S. Patterson, 64, 64, 3000, 60, 400

H. Gore, 40, 10, 1000, 40, 260

Z. Conoway, 200, 200, 5000, 200, 800

E. Hancock, 200, 20, 4000, 200, 300

W. Beasman (Baseman), 80, 66, 2000, 150, 400

P. Bennett, 328, 120, 12000, 400, 750

S. Bennett, 170, 90, 6000, 500, 750

E. Cassell, 50, 25, 900, 10, 90

W. Whalen, 195, 50, 7000, 200, 300

T. Conoway, 188, 30, 2000, 120, 9

J. Becraft, 45, 31, 2000, 50, 150

F. Oliver, 70, 38, 3000, 200, 300

L. Jacob, 26, 5, 500, 20, 80

S. Parker, 25, 15, 600, 50, 50

A. Bennett, 180, 45, 5000, 150, 500

J. Phillips, 50, 30, 2000, 50, 450

W. Phillips, 50, 30, 2000, 50, 350

J. Scribner, 220, 40, 5000, 200, 500

N. Lusby, 100, 30, 2000, 50, 300

J. Slack, 100, 47, 4000, 200, 300

L. Johnson, 160, 40, 4000, 100, 400

W. Beam, 170, 13, 4000, 350, 1400

W. Osler, 96, 10, 1000, 200, 300

W. Bennett, 350, 50, 600, 200, 700

J. Carter, 90, 50, 3000, 150, 250

W. Allen, 110, 110, 10000, 200, 600

J. Fite, 100, 70, 4000, 150, 650

E. Ireland, 280, 100, 15000, 300, 1000

S. Bentz, 528, 400, 40000, 1400, 2000

T. Mercer, 135, 40, 3500, 200, 400

M. U. Shipley, 64, 20, 1700, 100, 400

S. Mercer, 116, 60, 7000, 100, 500

W. Penn, 82, 31, 1200, 100, 400

J. Stucker, 75, 25, 2000, 25, 175

J. Reckord, 71, 40, 2500, 100, 400

Perry Harp, 67, 50, 3000, 25, 275

J. Shipley, 100, 82, 2500, 50, 400

A. Shipley, 100, 100, 300, 100, 500

Jacob Frinter (Frinzer), 100, 70, 4000, 150, 350

Phillip Lowdenlager, 100, 4, 1800, 30, 800

David Buckler, 25, 15, 800, 30, 250

George Foster, 55, 25, 3000, 150, 400

Aaron Stocksdale, 200, 25, 5000, 300, 600

Josiah Stocksdale, 150, 25, 3000, 60, 500

David Richard, 65, 40, 2000, 50, 300

John Rill, 35, 12, 1500, 150, 350

Phillip Sipe, 75, 12, 2800, 100, 200

Jeremiah Ebaugh(Elaugh), 150, 80, 3500, 200, 600

David Houck, 60, 30, 2000, 200, 500

Melchos Algen, 120, 30, 4000, 200, 500

Joshua Algen, 100, 100, 4000, 200, 500

Erasmus Fowble, 70, 30, 2000, 100, 200

John B. Chesnutt, 120, 47, 5000, 400, 500

Emanuel Fowble, 54, 20, 1600, 50, 250

Jackson Belt, 60, 6, 1000, 100, 350

Jonas Deal, 120, 20, 5000, 500, 600

Jacob Miller, 120, 50, 4000, 15, 500

Wm. Shipgager, 60, 17, 1500, 100, 300

Catherine Murray, 100, 110, 6500, 250, 700

Elisha Wheeler, 90, 10, 3500, 200, 450

Elias Houck, 188, 32, 7000, 200, 1000

Richard Richard, 250, 150, 15000, 100, 400

Danul Null, 80, 20, 3000, 80, 200

Elizabeth Cox, 80, 25, 4000, 50, 250
John Lamott, 70, 30, 2500, 55, 15
John Murray, 30, 2, 4000, 30, 120
Andrew Shaffer, 25, 6, 2000, 200, 400
Henry Buckle, 29, 8, 4000, 300, 350
David Grogg, 77, 5, 3500, 125, 325
Henry Stansbury, 120, 8, 3500, 200, 500
David Richard, 200, 30, 2000, 50, 130
Charles Richard, 200, 30, 2000, 50, 130
John H. A. Frank, 40, 16, 1400, 30, 180
John H. Frank, 150, 50, 2000, 100, 300
Amos Harris, 21, 10, 1000, 50, 80
John Paine, 130, 90, 5000, 130, 650
Noah A. Brown, 60, 70, 3000, 150, 400
Leonard Rill, 50, 5, 1200, 50, 200
Melchor Harris, 31, 7, 800, 10, 75
Joseph Armacost, 80, 80, 3100, 150, 400
Conrad Snell, 47, 5, 1000, 50, 150
John Doffler, 47, 15, 1000, 100, 70
Elias Harris, 130, 83, 3500, 350, 1000
Joshua Harris, 10, 10, 1000, 25, 40
Isaac Hoffman, 100, 70, 2500, 50, 600
George Richards, 150, 150, 3000, 150, 250
George Houck, 60 80, 3500, 100, 400
Adam Shaffer, 60, 20, 1000, 50, 200
George Houck Jr., 320, 155, 4000, 200, 1000
Jesse Brown, 60, 50, 2500, 70, 150
George Ruby, 80, 31, 2400, 200, 400
William Corbin, 50, 13, 1000, 50, 150
Lewis Green, 90, 40, 3500, 100, 600
John Lasswell (Lasnell), 50, 7, 1000, 75, 150

John Minchy, 87, 20, 1900, 100, 200
William Abbott, 67, 7, 1200, 100, 300
John Trine, 79, 15, 2000, 50, 250
John Green, 62, -, 1200, 15, 100
Adam Zimmerman, 50, 3, 1300, 95, 300
Ephraim Stansbury, 130, 12, 3500, 100, 400
George Derr, 100, 52, 2500, 10, 40
Henry B. Houck, 135, 45, 3900, 300, 400
Jeremiah Mathews, 100, 50, 4000, 200, 350
Joseph Lippy, 180, 39, 3900, 300, 500
Michael Snider, 45, 15, 1200, 50, 400
William Albanph, 80, 25, 1500, 50, 300
Joshua Bosley, 140, 70, 6000, 100, 500
Christian Finch, 120, 43, 7500, 200, 600
George Gross, 30, 20, 1500, 50, 250
Peter P. Smith, 110, 50, 2900, 100, 350
Henry Fowble, 36, 10, 1100, 50, 250
David Fowble, 75, 28, 1700, 75, 200
Jacob Tingling, 100, 50, 3000, 150, 400
Joseph Burnell, 100, 13, 4000, 75, 300
David Dunn (Bunn), 80, 32, 3000, 200, 300
James Bosley, 85, 25, 2100, 50, 300
David B. Houck, 40, 10, 2000, 200, 200
Daniel Cox, 65, 13, 3000, 30, 150
William Reniman, 100, 76, 4000, 150, 700
George Rinehart, 50, 10, 1500, 150, 400
Samanth Whelhelm, 60, 20, 1000, 100, 300

Ephraim Murray, 30, 20, 1000, 60, 150

Jeziah Murray, 75, 20, 1100, 40, 300

Jacob Rineman, 35, 17, 900, 50, 150

David Shaffer, 25, 5, 1000, 40, 150

Joshua Caltriton, 98, 40, 4000, 160, 220

John J. Boblitz, 60, 31,1000, 80, 400

John Killaugh, 65, 37, 1100, 50, 175

Joshua Stricklin, 30, 32, 1000, 75, 150

Henry Alban, 50, 50, 2500, 90, 300

John Shaffer, 22, 13, 700, 40, 250

William Fritz, 26, 3, 800, 30, 300

John Shafer, 22, 20, 1500, 75, 150

Jacob Shafer, 26, 40, 4000, 100, 200

John P. Shafer, 20, 25, 2800, 120, 250

Jacob Ely, 80, 6, 500, 60, 150

Leonard Belt, 50, 10, 1000, 75, 250

John Ebaugh (Elaugh), 20, 13, 700, 50, 250

John Wink, 24, 30, 1000, 100, 186

George Alban, 15, 25, 1000, 80, 200

Henry Hare, 80, 10, 2000, 150, 800

Thomas Fisher, 45, 17, 1250, 120, 150

John Shaner (Shawer, 15, 5, 600, 50, 100

William Hare, 35, 5, 800, 40, 150

Peter Shaner, 40, 15, 1500, 100, 150

Nicholas Gross, 65, 18, 1200, 120, 200

Daniel Shaner, 20, 6, 500, 30, 50

John Wareham, 80, 20, 1500, 75, 300

George Coltriser, 70, 32, 3000, 150, 500

Jacob Coltriser, 100, 40, 4000, 200, 800

George Utz, 90, 25, 2500, 140, 400

John Dayhoff, 150, 50, 4000, 160, 500

George Simper, 24, 3, 1000, 80, 150

Wm. Boblitz, 100, 100, 3000, 200, 150

Richd. Harriss, 81, 13, 1000, 75, 300

Andrew Snider, 30, 5, 1000, 80, 100

Susan Armacost, 72, 40, 3000, 150, 400

Alexander Fowble, 90, 15, 4000, 160, 210

James Cattentar, 90, 10, 3000, 160, 180

Abijah Collison, 16, -, 1600, 50, 100

Jeremiah Cottritor, 55, 20, 1600, 20, 100

Henry Kellar, 85, 15, 2000, 75, 600

William Houk, 40, 5, 1500, 120, 400

Conrad Boall, 17, 30, 900, 30, 60

Joshua Algier, 29, 14, 800, 50, 800

Nicholas Algier, 105, 104, 3000, 100, 400

Henry Leister, 140, 65, 3200, 100, 400

Hezekiah Jorden, 90, 7, 1400, 50, 100

John Abbott, 90, 12, 1600, 150, 50

Henry Kellar, 80, 9, 2000, 150, 100

Jacob Sellers, 110, 50, 1900, 230, 500

Samuel Zopp, 50, 19, 900, 70, 100

Henry Ruby, 90, 17, 2800, 150, 800

Henry Zimmerman, 120, 80, 2000, 200, 1000

Ctherine Zimmerman, 90,15, 2400, 30, 250

John Barnes, 85, 11, 2000, 50, 220

Ephraim Nachime, 90, 29, 1300, 200, 500

John Wink, 80, 30, 3000, 40, 500

Jacob Shaffer, 100, 25, 1200, 50, 400

Henry Shaffer, 60, 30, 900, 40, 200

Jesse Warner, 60, 18, 2800, 80, 250

David Shaffer, 52, 10, 4500, 150, 450

Godfrey Kriclor, 65, 15, 2100, 60, 300

John Stansbury, 200,1 40, 3000, 100, 300

Thomas Bolley (Rolley), 100, 19, 3000, 150, 250

Henry Tate, 80, 12, 1900, 250, 230
Michael Baker, 70, 13, 2000, 75, 250
Ephraim Baker, 70, 13, 1900, 50, 200
Daniel Nemerger, 41, 4, 3000, 50, 300
Daniel Groves, 100, 74, 3900, 100, 400
Michael Ritter, 200, 45, 8000, 450, 600
Henry Snider, 50, 70, 1700, 100, 300
John Stephon, 60, 10, 1400, 71, 320
Samuel Witmer, 40, 15, 2500, 100, 120
Jacob Gimmell, 100, 15, 1500, 75, 450
Jacob Shanon, 30, 13, 1200, 60, 210
Jacob Shanon, 60, 40, 2500, 200, 500
Daniel Snider, 30, 12, 900, 50, 100
Josiah Leister, 100, 47, 3000, 50, 250
Simon Shaffer, 130, 20, 600, 500, 800
John A. Brown, 100, 26, 4000, 100, 400
David K. Brown, 52, 20, 1080, 50, 200
Noah L. Bixler, 140, 45, 5000, 200, 500
Christopher Rogs, 90, 32, 1700, 225, 250
Christian Rogers, 150, 72, 5000, 220, 600
Frederick Ritter, 88, 20, 3400, 100, 200
Daniel Reese, 100, 20, 6000, 125, 500
Ely Hall, 66, 10, 3400, 100, 500
Christian Roger, 195, 5, 8000, 200, 700
John Rogers, 150, 45, 6000, 100, 500
Edward Mathias, 130, 36, 4000, 300, 350
John Tingling, 60, 20, 3900, 200, 400

John J. Haines, 80, 80, 4900, 90, 125
George Utz, 56, 36, 2900, 150, 500
Joseph Wiemer, 34, 16, 2800, 70, 200
George Wentz, 100, 91, 6000, 50, 400
Henry Whierhiem, 100, 100, 4500, 100, 700
John Whierhiem, 100, 15, 4300, 100, 500
Peter Myers, 170, 80, 10000, 500, 800
Jacob Shaffer, 70, 40, 8000, 600, 600
George Bixler, 100, 50, 6500, 30, 500
Sarah Stephoon, 60, 6, 1100, 50, 150
Christian Stormer, 100, 10, 1100, 70, 450
Benjamin Lippy, 200, 87, 9000, 400, 600
Lewis Leters, 56, 81, 2300, 130, 200
John Sellers, 200, 70, 6700, 300, 850
John Shaffer, 150, 60, 9000, Burnt Up, 2000
Joseph Chalfont, 700, 60, 10000, 200, 450
John Vance, 80, 25, 6500, 250, 200
Cornelius Wantz, 93, 18, 5000, 200, 700
John Hinkle, 30, 40, 3300, 150, 250
Ely Slager, 140, 60, 8000, 400, 1000
Joseph Hoover, 220, 60, 16000, 200, 900
Oliver Whearheim, 100, 130, 9500, 200, 900
Perry Wine, 70, 7, 2000, 25, 200
Henry Whearhiem, 10, 40, 9000, 200, 700
Dana Panebaker, 100, 50, 5000, 200, 700
George Frank, 100, 30, 2000, 175, 600
David Hofacker, 130, 70, 4500, 500, 700
Lemuel Lawson, 100, 13, 1000, 100, 300

Frederick Slagler, 100, 9, 3000, 150, 250

Josiah Kimel, 30, 10, 2000, 250, 400

Ely Warner, 70, 20, 2500, 200, 400

Jonas Warner, 100, 15, 2900, 100, 400

Peter Lockerbaugh (Lockerlaugh), 100, 85, 3900, 150, 500

Zachariah Abaught, 68, 32 4000, 120, 350

Henry Shaffer, 60, 13, 2400, 130, 3500

Frederick Miller, 50, 10, 2900, 100, 200

Emanuel Hurdl, 105, 45, 5500, 250, 600

Lewis Snyuder, 100, 23, 3900, 150, 220

John L. Boyer, 92, 10, 2400, 100, 200

Daniel Walker, 90, 10, 2000, 75, 230

Ephraim Shue, 49, 10, 1800, 90, 150

Noah Gamics, 54, 14, 1400, 75, 180

Peter Bailey, 80, 4, 1500, 90, 200

Jonah Gauz (Ganz), 68, 14, 2400, 85, 300

Chas. Shaffer, 68, 17, 3000, 100, 300

Mary Zimmerman, 100, 10, 900, 50, 215

Jacob Zimmerman, 64, 16, 900, 50, 300

John K. Zimmerman, 150, 150, 7000, 300, 500

Joel Bollinger, 65, 80, 8000, 400, 800

Ephraim Williams, 24, 13, 1500, 75, 130

William Barley (Borley), 100, 25, 2500, 175, 300

Benjamin Bowers, 100, 23, 3000, 165, 300

Jacob F. Shaffer, 100, 15, 3400, 200, 700

John Fair, 100, 27, 6000, 190, 650

Henry Fair, 100, 12, 2500, 100, 350

Henry Lucabaugh, 100, 40, 1700, 90, 450

Jesse Watts, 100, 85, 4000, 120, 500

William Kagle, 100, 9, 3000, 100, 300

Peter Warner, 94, 19, 2000, 75, 350

Willaim Fair, 60 10, 1000, 50, 100

Daniel Bollinger, 40, 10, 1700, 50, 200

Jacob Hozocker, 200, 90, 6000, 500, 760

Adam Luchabaugh, 100, 34, 3000, 75, 350

Samuel Hoffacker, 180, 120, 7000, 300, 800

Joshua Stransbury, 50, 10, 2000, 72, 400

Jacob Deitzer, 90, 10, 7500, 200, 550

Daniel Bixler, 140, 50, 8000, 200, 900

Jno. W. Suatbury, 40, 30, 1500, 20, 300

Ephraim Tracy, 200, 70, 7000, 500, 600

Samuel Miller, 90, 30, 2100, 300, 400

George Brown, 104, 50, 3000, 150, 800

William Walter, 100, 80, 5000, 150, 300

Adam Bucher, 100, 20, 5000, 20, 600

Phillip Yost, 55, 15, 2100, 75, 350

John Lockerbaugh, 60, 15, 5500, 80, 190

Peter Peterman, 140, 30, 6500, 130, 300

Jacob Mandray, 200, 44, 4000, 175, 500

Christian Fulmer, 62, 13, 1200, 75, 130

Aaron Courmine, 100, 230, 800, 140, 450

Henry Strevey, 100, 14, 3000, 120, 350

Jacob Miller, 65, 35, 6000, 200, 500

Adams Rohrbaugh, 140, 60, 7500, 200, 700
Nathan Couts, 30, 15, 1200, 75, 250
John Krah, 170, 50, 9000, 300, 900
Amos Wolfgang, 100, 34, 6000, 130, 300
George Grove, 90, 30, 9000, 100, 700
Michael Learher (Leacher), 61, 10, 1800, 40, 300
Jacob Bowman, 100, 20, 6500, 100, 600
Henry Warner, 100, 15, 3000, 100, 300
Henry S. Warner, 100, 5, 4900, 180, 450
John Bince, 120, 7, 3000, 50, 320
Frederick Yost, 67, 14, 1150, 60, 160
Sarah Rohrbaugh, 100, 50, 6000, 140, 350
Reuben Kuhns, 97, 1, 3000, 150, 400
Jacob Bomager, 60, 10, 1500, 70, 250
Godfrey Bomager, 50, 3, 1300, 60, 125
William Kade, 100, 17, 2000, 80, 350
William Zepp, 100, 4, 1300, 75, 300
Jesse Shannon, 97, 5, 2000, 100, 400
Samuel Lilley, 39, 14, 1200, 75, 150
Christian Hunt, 120, 40, 7500, 175, 380
Samuel Stansbury, 75, 10, 900, 75, 360
John Bowman, 36, 15, 2400, 30, 200
John Reading, 34, 10, 1400, 60, 150
Henry Jones, 110, 60, 4000, 300, 500
David Bachman, 295, 160, 25000, 350, 1200
Jacob L. Yost, 94, 24, 1500, 25, 120
John Weaver, 180, 67, 1140, 70, 450
Benj. Bixler, 350, 150, 15000, 200, 1000
Benj. Zepp, 100, 60, 3000, 150, 350
Andrew Smerch, 140, 40, 8000, 200, 750

Adam Frickriger (Fuckrige), 75, 25, 2400, 200, 250
Herny Winn, 200, 50, 5000, 250, 600
Phillip Yost, 50, 20, 1000, 60, 250
Henry Creame, 80, 20, 3000, 100, 200
Halmos Tasto, 55, 20, 1000, 30, 100
Emanuel Zepp, 38, 40, 1300, 40, 200
Michael Deet, 61, 13, 1000, 25, 200
William Wample, 100, 30, 1300, 100, 350
Jacob Stoner, 50, 25, 1000, 50, 150
Henry Zepp, 45, 12, 1300, 50, 160
Henry Houck, 90, 19, 3500, 100, 400
Edward Spangler, 100, 60, 5000, 140, 600
William Steagner, 100, 90, 3900, 150, 500
Henry Ganett (Garrett), 50, 56, 3000, 180, 480
Peter Steagner, 80, 20, 100, 50, 230
Joseph Stonecife, 66, 33, 2000, 50, 200
Peter Trindler, 55, 14, 1900, 60, 180
John Foreman, 90, 45, 2000, 70, 200
David Mathias, 100, 39, 1500, 170, 380
Henry Wilson, 100, 25, 3000, 60, 380
Jacob Eglehart, 34, 14, 900, 70, 500
David Garrett (Ganett), 120, 17, 3400, 150, 900
Joseph Rinehart, 120, 15, 3000, 150, 650
Andrew Lace, 100, 50, 3900, 40, 500
George Faultestine, 90, 10, 3410, 130, 300
William Coumine, 50, 50, 3900, 30, 300
Josiah Boose, 50, 13, 2900, 75, 100
Phanuel Wents, 100, 25, 6000, 250, 600
Peter Bixler, 90, 30, 2900, 100, 800
John Fultz, 100, 85, 3900, 300, 600
Uriah B. Sulivan, 150, 30, 5000, 200, 500

James Rhodes, 166, 44, 9000, 250, 900

John Geiman, 150, 100, 8500, 250, 900

Fredrick Bowage, 90, 10, 1900, 75, 250

Henry Miller, 100, 10, 2500, 100, 600

Henry Minter, 100, 20, 2400, 90, 235

Jacob Miller of M, 100, 90, 2500, 100, 450

Jesse Hare, 92, 10, 2400, 80, 230

Aaron Miller, 100, 100, 4900, 130, 140

Geroge Wheonhiem, 300, 100, 3900, 100, 600

John Boring, 90, 30, 3000, 50, 250

Mordecai Boring, 40, 30, 1900, 60, 190

Daniel Coottriton, 140, 60, 3500, 95, 300

Henry Falkstine, 140, 40, 2900, 80, 1200

David Buthart, 100, 40, 4300, 150, 400

Adam Hiseley, 24, 10, 2000, 75, 300

Jacob Showers, 130, 100, 9000, 200, 1000

Henry E. Beltz, 29, 8, 2225, 175, 565

Henry Brinkman, 20, 20, 1800, 50, 155

George Crouse, 17, 16, 400, 50, 213

Adam Showers, 120, 110, 7000, 200, 900

Daniel Bowman, 32, 10, 1600, 175, 200

E. Lambert, 80, 3, 3000, 50, 400

J. Roop, 150, 40, 10000, 500, 1200

H. Townsend, 55, 8, 4500, 30, 300

E. Words (Worels), 40, 10, 2000, 30, 300

J. Smith, 145, 35, 7000, 400, 800

J. Yerger (Geiger), 122, 25, 9000, 500, 1000

Isaac Lambert, 15, -, 2000, 20, 300

L. Buckingham, 30, 14, 1200, 30, 150

P. Hanes, 100, 40, 4000, 50, 400

Z. Zepp, 32, 3, 1200, 15, 200

S. Wilt, 36, 7, 1200, 15, 200

J. Deihl, 130, 30, 7000, 200, 800

J. Snider, 50, 13, 3500, 75, 200

A. G. Barnes, 42, 3, 1500, 25, 300

J. Bail, 115, 25, 6250, 100, 500

D. Gilbert, 170, 30, 10000, 100, 650

W. Repp, 97, 14, 10000, 500, 1000

D. Harman, 100, 19, 7000, 100, 600

H. _. Marker, 150, 3, 11000, 100, 600

J. W. Engles, 158, 50, 14000, 100, 1000

D. Cassell, 113, 4, 7000, 400, 700

J. Shewery, 191, 12, 20000, 1000, 2000

S. Lambert, 104, 1, 8400, 100, 700

W. Sauble, 114, 18, 9000, 100, 700

R. Haines, 110, 7, 10000, 100, 800

J. H. Richardson, 70, -, 7000, 100, 450

D. Nichodemus, 60, 30, 7000, 100, 500

John Engler (Engles), 155, 5, 14000, 200, 1500

J. H. Lovell, 160, -, 12000, 200, 800

A. Poole, 125, 4, 9000, 200, 700

N. Shipley, 90, 10, 8000, 100, 600

N. Forester, 120, 30, 6000, 75, 700

L. N. Baile, 118, 30, 8000, 100, 800

S. Baile, 25, 25, 1000, -, 75

N. Nichodemus, 190, 4, 14000, -, 300

H. Cassell, 80, 5, 6800, 50, 300

H. Hines, 30, 3, 2500, 30, 300

J. Baile, 128, 12, 1000, 50, 700

D. Baile, 90, 10, 3500, 25, 300

S. Diehl, 132, 5, 10000, 150, 1200

D. Engle, 160, 2, 16000, 200, 1500

Ben Bowers, 116, 10, 4000, 50, 500

A. Cere (Carr), 72, 8, 1000, 30, 200

P. Cook, 90, 10, 200, 50, 600

R. Smith, 54, -, 5000, 100, 600

Jacob Reed, 36, -, 1400, 10, 150
William Bail, 100, 43, 300, 300, 500
Alexander McAllister, 78, 22, 2400, 10, 250
Americus Shoemaker, 70, 11, 2000, 150, 250
John T. Shriver, 70, -, 2000, 80, 200
George Avilt (Avitt), 20, 6, 1200, 20, 75
George Warehan, 100, 58, 1600, -, 100
John Little, 50, 5, 1400, 10, 200
John Kregelo, 65, 13, 2000, 40, 300
John Kuekey, 120, 25, 300, 150, 500
Philip Six, 100, 33, 3000, 200, 600
William Shaw, 300, 75, 20000, 100, 1000
Elizabeth Beckum, 4, -, 250, -, 30
Thomas Shaw, 300, -, 10000, 500, 1200
John Fogle, 60, 5, 1250, 50, 300
Joshua Null, 100, 25, 2500, 300, 400
George W. Kegle, 100, 23, 2500, 40, 400
Abraham Sheets, 100, 34, 4000, 50, 200
John Stackslager, 81, 35, 4000, -, 120
Jesse Heck, 30, 2, 2000, 50, 200
Jacob Valentine, 48, 13, 2000, 50, 175
James Short, 80, 25, 2500, 200, 1000
Washington Short, 80, 25, 2500, 200, 1000
John Ridinger, 80, 18, 1500, 150, 200
Jacob Newemer, 90, 25, 1100, 75, 300
Washington Galt, 140, 83, 7000, 300, 700
Contee Raitt, 120, 35, 7000, 300, 600
William Blubaker, 120, 50, 3900, 50, 250
William H. Crowe, 120, 40, 5000, 200, 650

William Keefer, 80, 40, 3000, 100, 250
Thomas Reidisul, 190, 30, 12000, 400, 1075
Thomas Jones, 120, 80, 6000, 50, 250
William Keyser, 150, 45, 4000, 150, 355
Henry Bishop, 45, 12, 1000, 40, 250
Sarah Bishop, 13, -, 850, -, 45
Philip Smith, 60, 10, 1400, 60, 300
David Baumgartner, 100, 30, 3000, 300, 400
Stephen Smith, 110, 50, 3500, 75, 750
Augustus Arnold, 180, 25, 6000, 150, 700
Samuel E. Haw, 80, 25, 2000, 75, 250
Henry Baumgartner, 190, 70, 6000, 150, 775
David Fancy, 100, 55, 2000, 50, 200
Peter Baumgartner, 130, 25, 8350, 150, 70
Frederick Ohler, 75, 25, 1000, 50, 150
James A. Clabaugh, 80, 23, 2600, 150, 2000
John Clinin, 95, 25, 2000, 200, 500
Charles Fair, 106, 38, 1950, 170, 500
John Bishop, 175, 45, 5000, 400, 500
Lemuel Fair, 140, 40, 3000, 150, 300
Gideon Hiteshue, 95, 21, 2000, 25, 278
John Hiteshue, 90, 35, 2700, 200, 500
John Clabaugh, 158, 55, 2000, -, 100
James Clabaugh, 158, 55, 2000, -, 100
William Vaughn, 65, 8, 2000, 50, 400
Samuel Hess, 150, 58, 5000, 200, 575
R. J. Jamison, 50, 25, 7000, 100, 250
John Bishop, 110, 40, 2000, 41, 300

George Young, 95, 40, 2000, 150, 400

Jane Null, 33, 7, 800, 25, 150

Richard Hill, 50, 30, 2000, 150, 174

William Frame, 67, 1, 1000, 100, 300

William H. Hess, 125, 15, 2500, 150, 300

William Shoemaker, 70, 27, 4000, 250, 823

David Kephold, 80, 20, 3000, 200, 520

John Curnes, 35, 8, 1500, 20, 100

John Shriver, 80, 20, 3000, 100, 600

A. N. Hess, 150, 25, 8000, 200, 525

John Fair, 100, 50, 5008, 2500, 550

Gabriel Stever, 95, 20, 5000, 300, 550

John A. Ruidollar, 115, 20, 8000, 325, 575

John Shoemaker, 190, 100, 6000, 200, 600

Samuel C. Bower, 18, -, 3000, -, 150

James Thompson, 115, 45, 2625, 150, 520

James Rowe, 47, 15, 1600, 40, 150

Benjamin Showk, 21, 9, 2500, 250, 150

Samuel Swope, 268, 40, 10000, 500, 1070

Emanuel Lambert, 31, 12, 1770, 75, 370

James Burke, 31, 12, 1770, 75, 370

James Eckinrode, 32, 8, 1900, 60, 200

Oley Reaves, 80, 29, 4000, 150, 250

L. A. Hiser, 87, 5, 1500, 50, 200

S. Slaugenhaupt, 75, 15, 1500, 80, 300

John W. Jones, 100, 20, 3000, 100, 550

Samuel Bowers, 80, 32, 2500, 200, 300

John Reaves, 100, 30, 4000, 100, 775

Jesse Lafar, 125, 10, 5600, 50, 525

E. T. Spalding, 62, 150, 2000, 50, 150

John T. Six, 100, 30, 5000,75, 525

John Goode, 100, 32, 4000, 200, 600

James Airing, 85, 14, 1600, 50, 200

Daniel Hawn, 35, 26, 1200, 60, 200

John N. Davis, 152, 25, 5400, 200, 900

John Hess, 45, 20, 1600, 10, 220

Eliza Connover, 100, 45, 4000, 200, 450

Fredeick Reaves, 40, 10, 1600, 50, 200

Frederick Harmer, 140, 50, 5000, 30, 150

James Haner, 90, 20, 1600, 75, 250

Washington Reeves, 62, 18, 2000, 100, 250

Samuel Null, 180, 25, 7500, 350, 650

Andrew Kechus, 42, -, 1500, 50, 200

James McAllister, 130, 36, 10680, 500, -

Mary Eline, 49, 4, 2020, 600, 575

John Harner, 30, 20, 2700, 150, 400

Daniel Mehring, 85, 20, 18000, 400, 700

Levi Mehring, 90, 30, 300, -, 200

James Feiser, 100, 26, 5440, 600, 620

George Halterbrick, 140, 28, 4450, 300, 440

Sebastian Obold, 120, 8, 5000, 50, 425

Levi Murrin, 100, 70, 4000, 200, 475

George Reindollar, 180, 65, 9000, 50, 300

James Coons, 80, 22, 4000, 120, 450

Joseph Weltz, 130, 30, 5000, 500, 1008

Cunard Share, 200, 45, 10000, 500, 1000

H. D. Mehring, 125, 35, 4500, 350, 500

James Knox, 130, 39, 6000, 200, 600

Emanuel Stary (Story), 200, 70, 10000, 250, 1350

David Harmer, 45, 35, 7800, 40, 178

Nicholas Fringer, 60, 30, 2000, 45, 225

John Roger, 85, 15, 2500, 100, 425

Francis Larbar (Larban), 53, 15, 3000, 50, 225

W. States, 15, -, 1000, 50, 150

S. Clingan, 100, 30, 3000, 200, 500

J. Reindollar of J., 30, 10, 2000, 75, 200

M. Fringer, 78, 12, 5000, 100, 350

James Reindollar, 70, 18, 250, 95, 350

Simon Harman, 48, 3, 2000, 50, 200

Jacob Null, 120, 30, 4500, 200, 430

John Thompson, 130, 40, 8000, 500, 1200

John Rinner, 80, 28, 3500, 200, 450

John H. Mootrick, 155, 25, 15000, 500, 1280

John Hilterbrick, 47, 26, 7000, 225, 100

Samuel Gath, 95, 12, 5000, 200, 600

John R. Knox, 75, 55, 4000, 200, 475

Jcob Snyder, 115, 30, 3000, 150, 460

Samuel Harman, 75, 30, 3000, 100, 300

Joseph Harnish, 90, 22, 3000, 120, 150

Amos Flickinger, 94, 12, 3500, 50, 250

John Keseling, 50, 10, 3000, 100, 240

Paul Kacs, 67, 30, 2500, 30, 250

Thomas Langley, 150, 26, 6000, 400, 640

Samuel Galt Sr., 195, 80, 130, 75, 1000, 1300

Sterling Galt, 125, 25, 7000, 300, 860

Samuel Kesling, 90, 20, 2500, 150, 525

John Kuhns, 94, 16, 3000, 140, 575

Mrs. C. Hawk, 16, -, 2000, 25, 140

H. Hahn, 50, 20, 3000, 135, 430

James Nivele (Nivell, Wivell), 100, 22, 2500, 200, 600

Dr. M. Reindollar, 85, 21, 3000, 250, 350

Jerome L. Baumgartner, 75, 38, 4500, 200, 505

James Adlesperger, 120, 51, 4800, 100, 325

George Keseling, 120, 10, 3000, 75, 250

Samuel Hahn, 130, 75, 4000, 150, 500

George Smith, 82, 12, 1500, 50, 375

Barny Snyder, 76, 28, 3500, 200, 350

S. Baumgartner, 100, 36, 4000, 125, 500

W. Reaves, 700, 40, 300, 65, 507

J. Baldwin, 9, 12, 800, 30, 200

W. Hallerberry, 45, 6, 1000, 100, 475

H. Shriver, 100, 34, 7000, 200, 500

J. Hallerberry, 200, 40, 10000, 550, 825

P. Markers, 45, 10, 1800, 120, 275

S. Babylon, 73, 2, 3000, 550, 575

Jacob Sheets, 50, 20, 1500, 150, 275

Jonas Spangler, 59, 8, 1800, 150, 300

Benjamin Flegle, 40, 10, 3000, 100, 500

D. Frock, 45, 19, 2500, 50, 150

J. Frock, 60, 15, 2000, 150, 300

J. Clutts, 130, 20, 4500, 150, 425

J. L. Peper, 350, 100, 4000, 2500, 1550

W. H. Gance, 45, 6, 1800, 50, 225

W. Gaulden, 85, 20, 3500, 200, 250

S. Newcomer, 85, 15, 3000, 150, 300

James Crouse, 120, 25, 4590, 250, 700

A. Garner 80, 15, 4000, 200, 620

J. Shoemaker, 37, 5, 1000, 25, 250

Jacob Shriver, 70, 6, 5000, 275, 525

William Fogle, 50, 8, 2200, 125, 520
George Snyder, 130, 30, 4000, 150, 375
Ann Stultz, 114, 10, 4000, -, 300
William Hiner (Hines), 150, 71, 5000, 250, 80
Abraham Shriver, 110, 31, 3800, 150, 450
William Crabbs, 80, 20, 3000, 25, 375
John Crabbs, 70, 30, 2580, 200, 400
D. Martin, 180, 40, 11000, 380, 1340
William Furney, 120, 45, 4000, 200, 690
S. Shriver, 130, 25, 3000, 75, 450
W. Southgate, 120, 25, 3000, 300, 600
G. Benner, 53, 6, 2000,75, 300
William Shriver, 110, 24, 3000, 75, 650
H. Reindollar, 59, 3, 2000, 50, 210
J. Muring, 90, 10, 3500, 175, 425
J. Snyder, 160, 50, 5000, 600, 800
Wm. Reindollar, 110, 28, 3500, 100, 475
G. Crapster, 65, 10, 3000, 400, 300
F. Crabbs, 300, 240, 6000, 300, 965
Maria Morelock (Mordock), 49, 62, 4000, 300, 350
Catharine Petey, 100, 100, 19000, 400, 7501
Rebecca Warner, 47, 15, 1500, 100, 200
William Bunst (Banst, Baust), 90, 8, 7000, 600, 450
Amos Callow, 125, 26, 8000, 300, 400
Daniel Cottitor, 110, 6, 15000, 300, 350
Christian Veaners, 96, 15, 1400, 150, 200
Jacob Myers, 10, 100, 2000, 250, 350
Edward Harvy, 100, 68, 5000, 150, 600
Jesse Hesson, 95, 15, 2000, 150, 400

William Deckenchart, 100, 30, 2900, 150, 450
John Myers, 120, 15, 5000, 200, 600
Charles Desvilligs, 85, 15, 8000, 600, 400
John Hesson, 35, 15, 2000, 150, 200
William Clousher, 100, 5, 3400, 100, 350
Abbot Bankert, 65, 14, 2000, 100, 450
Ephraim Bankert, 100, 6, 2000, 150, 350
John Fleagle, 130, 20, 4000, 300, 400
Peter Erb, 60, 70, 1000, 100, 200
Frederick Halurik, 110, 57, 3000, 150, 400
George W. Hall, 125, 20, 8000, 200, 800
Henry Crowel, 100, 20, 2000, 180, 300
Josiah Babylon, 240, 29, 15000, 300, 700
Samuel Whearhan, 160, 17, 14000, 400, 1200
John Whearhan, 120, 40, 400, 300, 500
John Lampert, 39, 30, 1000, 120, 300
Josiah Crawford, 28, 28, 1400, 90, 200
Samuel Fitz, 20, 20, 1500, 80, 180
John Gottz (Gotty), 37, 15, 2500, 90, 200
William Luster, 114, 20, 4000, 190, 500
Jeremiah Rinehart, 200, 40, 12000, 500, 900
Eli Hesson, 75, 14, 2500, 50, 400
William Lampert, 60, 16, 2200, 150, 200
John Humbert, 99, 10, 4500, 150, 600
David Myerly, 75, 19, 2000, 130, 300
John Wantz, 104, 45, 7000, 300, 400

David Hilbrand, 69, 30, 3000, 150, 00

Jacob Hanfed, 100, 40, 4000, 200, 500

Jesse Heagle (Feagle), 94, 10, 3000, 250, 300

Abraham Hann, 35, 210, 13000, 80, 200

Jacob Fogeslong, 100, 29, 3000, 150, 600

Elinor Hydes, 33, 14, 1100, 50, 200

William Duleplan, 100, 15, 2500, 60, 300

Jacob Moukley, 150, 73, 3100, 400, 900

James Hesson, 95, 35, 7000, 300, 500

Andrew Arther, 100, 25, 1700, 2500, 300

Henry Lell (Sell), 100, 14, 4000, 300, 500

William Worley, 140, 26, 6000, 200, 600

John Myers, 100, 15, 6000, 300, 450

Jame Reaber, 90, 18, 1700, 150, 150

George Worley, 100, 14, 2000, 125, 260

Jacob Bradly, 83, 16, 1500, 130, 300

Elizabeth Brady, 150, 40, 3000, 160, 450

Joseph Young, 100, 30, 4000, 300, 550

Noah Fermake, 90, 18, 3000, 150, 250

Michael Ludwic, 90, 30, 1600, 75, 250

Jacob Erb, 30, 40, 9000, 400, 900

Benjamin Reaber, 100, 29, 3000, 200, 422

Adam Fraity, 60, 22, 3600, 150, 300

Joseph Fraity (Faity), 107, 49, 9000, 900, 1500

Noah Plowman, 80, 30, 3900, 75, 200

John Eckard, 138, 26, 7000, 150, 200

Andrew Haines, 90, 10, 3000, 70, 300

Nathan Crumback, 100, 18, 1500, 90, 200

Barbary Boston, 140, 60, 5000, 300, 600

Jacob Koons, 80, 25, 3000, 300, 700

David Stoner, 150, 60, 6000, 450, 700

William W. Dallas, 300,75, 50000, 2200, 1000

John N. Star, 100, 23, 10000, 600, 900

John Arther, 30, 10, 2000, 100, 300

Hanson T. Webb, 87, 20, 4500, 250, 600

David Bare, 160, 41, 4000, 2000, 800

Jacob German, 35, 35, 4000, 50, 200

Jacob Yon (You), 90, 10, 3600, 300, 400

William Lance, 26, 19, 3000, 150, 200

Sarah Warner, 45, 36, 1500, 80, 180

Rogers Burnic, 200, 100, 6488, 290, 700

Jacob Murdock, 130, 34, 5400, 350, 500

Jacob Babylon, 47, 30, 4000, 130, 200

Alexander Little, 160, 56, 10000, 500, 800

Andrew Myers, 120, 90, 8000, 200, 600

Jesse Babylon, 100, 60, 5000, 300, 500

Michael Babylon, 30, 31, 2000, 75, 200

John Siles (Siler), 120, 14, 8000, 300, 600

George Warner, 119, 25, 6000, 400, 650

Abraham Nurbaugh, 100, 89, 6000, 500, 650

William Formett (Formatt), 100, 50, 4000, 500, 750

Joseph Bowers, 55, 60, 2000, 300, 380

Jonas Myers, 150, 40, 4000, 200, 800

William Babylon, 160, 12, 4000, 300, 650

Jacob Beaks (Burks), 100, 18, 4000, 300, 400

Joseph Myers, 45, 15, 9000, 150, 475

Josiah Bankert, 105, 14, 5000, 165, 575

Jacob Eckert, 100, 50, 10000, 400, 900

Jacob Brown, 70, 13, 1500, 100, 200

William Nausbum, 52, 15, 3000, 100, 200

Samuel Unger, 100, 30, 3000, 150, 400

Joseph Winters, 100, 22, 2500, 300, 450

Solomon Stonner, 100, 40, 9000, 520, 900

Nathan Stonner, 130, 16, 4000, 400, 300

Lersam Curts, 90, 10, 3000, 150, 250

Tobias Cobar (Coban), 150, 26, 12000, 20,700

Nathan Meredith, 25, 30, 1000, 100, 300

John Gramer, 90, 15, 2000, 175, 350

Ephraim Bankert, 130, 20, 100, 400, 750

Joal Myers, 140, 20, 7500, 300, 500

Mary Banks, 50, 10, 3000, 75, 300

Abraham Appler, 130, 100, 11000, 400, 950

David Hildeband, 80, 10, 8000, 300, 800

Hiram Davis, 75, 10, 4500, 100, 600

Emanuel Stoner, 100, 14, 5000, 300, 650

William Stulz (Steely, Stuly), 200, -, 10000, 150, 400

Jonathan Switzer, 16, 7, 3000, 150, 300

Thomas Sepad (Lepad), 180, 20, 14000, 400, 800

Joal Furquaharson, 96, 30, 6000, 400, 500

William Hughes, 70, 10, 2000, 150, 450

Joseph Moore, 80, 4, 8000, 400, 600

Abraham Wolf, 180, 10, 10000, 300, 600

Ulrick Misters, 117, 10, 6000, 150, 750

Ezra Shroer, 70, 4, 500, 100, 400

William Shepard, 100, 28, 2800, 150, 550

Ephraim Bopp (Boss), 150, 26, 9000, 300, 700

George Blizzard, 100, 65, 12600, 300, 750

John M. Torny, 46, 40, 2400, 20, 350

Ephraim Stoner, 100, 20, 10000, 400, 750

Granville Harris, 160, 25, 18000, 800, 1200

Davis Leightner, 90, 10, 1000, 300, 600

Nathan Haines of Th, 200, 60, 10000, 500, 700

Edmund Yingling, 104, 18, 5000, 200, 550

Robert Haines, 200, 200, 15000, 200, 600

Joel Harris, 100, 60, 15000, 400,700

David Crumbaker, 27, 45, 1500, 100, 250

Willam Engler, 54, 13, 4500, 175, 450

Clumbus Engler (Engles), 110, 27, 6000, 300, 450

Ruben Winters, 18, 19, 1000, 100, 200

William Hitchen, 100, 29, 8000, 300, 500

Levi Caylor, 16, 14, 1000, 70, 150

Levi Engler, 150, 50, 10000, 500, 500

Samuel Winters, 200, 14, 7000, 400, 700

Mary Benedict, 59, 30, 4000, 300, 390

A. H. Sonsarg (Sonsarq, Sonsany), 200, 40, 1700, 100, 400

Jacob Sauble, 100, 40, 2000, 400, 900

George H. Brown, 53, 10, 3800, 150, 600

John J. Jordan, 130, 23, 9000, 400, 600

Henry Drake, 55, 40, 8000, 150, 450

John Boop (Roop), 100, 17, 6000, 50, 300

Levi Cookson, 114, 54, 9000, 475, 700

Denis Cookson, 79, 19, 8000, 400, 500

Edward Zolecker, 170, 40, 18800, 400, 1500

Thomas Sheperd, 130, 15, 10000, 500, 900

Willing Yingling, 100, 54, 974, 200, 900

Mordecai Harris, 50, 1, 4000, 100, 400

Joseph Weaber, 100, 25, 6000, 250, 700

Henry Foss, 160, 25, 9000, 400, 600

William Wagner, 92, 14, 8000, 300, 600

Michael Wagner, 170, 30, 8000, 400, 475

Abraham Weaber, 40, 30, 200, 140, 300

Margaret Bromet, 200,70, 9000, 40, 700

John Koome, 95, 14, 6500, 300, 200

Thomas Bolbist, 175, 10, 1300, 200, 375

Elias Coppersmith, 40, 13,1500, 175, 300

Lewis Fortmott, 195, 14, 8000, 300, 750

Augustus Ensors, 150, 15, 9000, 500, 900

Daniel Roop, 84, 13, 7000, 100, 600

James Warner, 114, 7, 4500, 200, 475

Joab Babylon, 105, 30, 3500, 300, 350

John Babylon, 150, 30, 12000, 300, 1500

Samuel Borkum, 180, 19, 10000, 400, 1000

Charles Myers, 80, 73, 1000, 150, 300

John Roberts, 40, 10, 6000, 300, 350

John Furgerson, 60, 13, 4000, 150, 250

Hezekiah Yingling, 14, 15, 1250, 300, 300

Henry Read, 11, 10, 2500, 175, 140

Michael Roberts, 360, 40, 1500, 200, 1000

Davis J. Roop, 200, 17, 10000, 300, 700

Elias Yingling, 45, 20, 1500, 20, 100

William Yingling, 20, 3, 4000, 50, 200

W. J. Mitten, 21, -, 2000, 30, 200

J. Smith, 125, 5, 10000, 300, 800

J. Fisher, 110, -, 8800, 300, 400

J. Manning, 23, -, 3000, -, 100

L. Srumbt, 27, 10, 2500, 50, 20

Wm. Rease, 4, -, 300, -, 25

J. Beaver, 75, 75, 5500, 200, 700

C. Boyle, 144, 20, 13000, 200, 1200

J. K. Longwell, 100, 100, 1200, 100, 300

J. Yingling, 70, 30, 11000, 300, 500

F. Shipley, 100, 54, 2000, 200, 600

J. Conaway, 100, 30, 3800, 100, 400

J. Wagner, 15, 10, 1000, 20, 95

L. Eichelberger, 16, 4, 300, 5, 75

R. Fuzzell, 150, 50, 1500, 10, 100

E. Owings, 148, 50, 2000, 100, 500

David Owings, 156, 6, 1400, 50, 1000

Vacht. Buckinghm, 33, 4, 1800, 100, 350
P. Ceary, 44, -, 1200, -, 100
J. Homes, 34, 16, 1500, 30, 300
J. Bouenle, 95, -, 3000, 100, 300
S. Foutz, 120, 20, 6000, 100, 400
J. Malchorn, 77, -, 6000, 100, 400
A. Foulker, 60, -, 4000, 100, 400
J. Fringer, 50, 5, 5000, 75, 300
J. Beavers, 150, 50, 5000, 20, 20
Joshua Lockard, 70, 30, 2800, 70, 400
S. Ogg, 45, 15, 1000, 50, 100
W. Lurfel, 26, 4, 1500, 10, 200
E. Fowlker, 53, 10, 1500, -, 100
W. Fowler, 53, 3, 1500, 5, 300
T. Fowler, 75, -, 800, 20, 150
R. Fowler, 25, 3, 1000, 20, 300
W. Beaver, 30, 20, 1200, 40, 300
F. Brothers, 53, 3, 800, 30, 200
B. Davis, 94, 4, 1200, 30, 200
S. Carr, 200, 100, 4000, 100, 800
J. S. Shepley, 100, 30, 1600, 50, 300
J. Smith, 27, -, 400, 20, 200
J. Forester, 94, 5, 3000, 100, 400
Jesse Williams, 121, 50, 7000, 100, 400
J. Harris, 75, 15, 1050, 100, 150
J. Carr, 100, 51, 4000, 50, 300
J. Appler, 98, 2, 4000, 25, 200
W. Stoner, 87, 40, 4000, 50, 400
S. W. Cook, 120, 20, 5000, 50, 300
J. Robertson, 250, 150, 6000, 100, 1200
D. Goodwin Jr., 140, 15, 10000, 100, 700
M. Gist, 120, 20, 4000, 100, 700
H. Dorsey, 120, 20, 14000, 160, 1000
R. Loge, 40, -, 300, 50, 200
S. Samuel, 44, 4, 4000, 100, 400
M. Samuel, 45, 3, 2000, 50, 300
E. Lynch, 113, 20, 7000, 100, 700
J. Blizzard, 95, 10, 6000, 100, 700
J. Myerley, 15, 5, 3000, 50, 175
J. Goodwin, 80, 5, 8000, 50, 400

S. Myerley, 72, 10, 7000, 50, 300
J. Orendorff, 148, 100, 15000, 300, 1500
J. Logus, 50, 10, 5000, 100, 280
S. Stephenson, 138, 50, 10000, 300, 1500
L. Zile, 130, 17, 8000, 150, 800
W. Williams, 28, -, 950, 20, 100
W. Smith, 14, 2, 1400, 5, 100
W. Grumbine, 43, -, 1200, 10, 250
L. Manaham, 93, 16, 3500, 100, 550
N. Pennington, 100, 30, 13000, 200, 800
A. Cassell, 110, 30, 8000, 200, 700
D. Engler, 150, 150, 20000, 200, 900
J. W. Beardan, 149, 90, 15000, 500, 900
J. Shafer, 160, -, 10000, 300, 1000
D. Shriver, 115, 25, 15000, 300, 100
A. Shriver, 160, 20, 18000, 800, 1400
D. Moul, 17, -, 1200, 10, 150
S. Vanbibber, 155, 150, 20000, 60, 1000
J. Stout, 33, 1, 2000, 50, 200
W. H. Orendorff, 150, 10, 10000, 200, 1000
J. B. Snoden, 29, -, 600, 50, 300
S. Doyle, 20, -, 1900, 10, 200
W. Durbin, 35, 5, 3000, 20, 200
W. Barnes, 120, 12, 6000, 50, 600
G. Duvall, 65, 5, 3000, 50, 300
A. Koontz, 94, 10, 7000, 100, 300
John B. Stoner, 9, -, 100, 5, 175
John Young, 144, 8, 9000, 200, 1500
J. Schwergart, 104, 9, 6000, 30, 700
W. Rinehart, 110, 10, 8000, 200, 1200
W. Morelock (Murdock), 200, 8, 10000, 300, 1200
H. Morduck, 30, -, 6000, 20, 300
A. Rogs, 243, 30, 20000, 300, 1500
W. Lowery, 100, 6, 5000, 50, 300
J. Smith, 55, 6, 5000, 50, 300
J. Stoner, 94, -, 5000, 150, 700
J. Marker, 108, 18, 7000, 100, 600

D. Roop, 100, 10, 10000, 100, 700
J. Royer, 200, 68, 12000, 200, 1200
J. Wheyery, 143, 30, 10000, 200, 500
H. Brown, 108, 40, 8000, 300, 1500
J. G. Capets (Cassets), 11, -, 2300, 5, -
A. Boyer, 200, 100, 900, 100, 600
D. Reese, 152, 40, 7000, 400, 1000
J. Roop, 190, 80, 20000, 400, 1500
P. Dulzour, 70, 20, 3500, 100, 600
J. Wolf, 25, -, 2000, 20, 100
J. Deffenbaugh, 19, -, 1200, 20, 200
M. Wentz, 90, 5, 3000, 50, 400
D. Brennan, 50, 13, 3000, 30, 300
J. Zehn, 92, 20, 2000, 50, 300
J. Petry, 143, 10, 7000, 300, 700
J. Soluner, 100, -, 800, 100, 1200
G. Weeks, 35, 10, 800, 10, 100
J. Zacharias, 133, 37, 12000, 200, 880
C. Schwergart, 168, 15, 10000, 200, 1200
D. Byers, 100, -, 7000, 100, 500
J. Hasson, 90, 10, 6000, 100, 200
G. Schaeffer, 87, 10, 1000, 100, 500
W. Hook, 125, 20, 2500, 50, 200
John Boggs, 123, 10, 10000, 200, 1000
J. F. Shade, 56, 10, 4000, 100, 300
A. Shade, 68, 20, 2000, 100, 600
A. Fuhrman, 70, 42, 4000, 100, 300
P. Bankert, 85, 30, 4000, 100, 800
J. Dudrow, 75, 18, 1800, 50, 200
D. Wentz, 42, 3, 900, 25, 75
V. Wentz, 34, 2, 1200, 25, 300
G. Fite, 80, 4, 4000, 50, 400
F. Cotrighter, 75, 50, 6000, 100, 600
D. Wentz, 100, 10, 6000, 100,700
J. Sawyer, 90, 15, 3000, 100, 800
John Miller, 160, 70, 800, 100, 1100
J. D. German, 27, -, 1500, -, 75
J. Miller, 140, 60, 7000, 100, 800
H. Bael, 106, 2, 6000, 100, 600
M. Lynch, 120, 30, 6000, 25, 125
Geo. Stoner, 30, 5, 4000, 50, 300

P. German, 125, 10, 8000, 200, 600
P. Slagle, 20, 1, 200, -, 100
J. H. Happe (Hasse), 146, 50, 8000, 500, 700
J. Myerley, 90, 5, 4000, 50, 300
W. S. Stansbury, 64, 20, 2000, 50, 700
G. Leppe (Seppe), 86, 40, 500, 50, 500
A. Bish, 7, 10, 5000, 70, 350
D. Koontz, 60, 40, 4000, 50, 500
L. Myers, 55, 25, 2000, 25, 150
J. H. Frock, 25, 20, 1000, 20, 150
J. H. Mirbriddle, 30, -, 1000, 20, 250
John Fred, 44, 40, 4000, 100, 350
E. Deodbiss, 35, -, 9000, 45, 200
L. Young, 28, 2, 1000, 10, 100
John Wentz, 30, -, 1000, 10, 125
W. Bish, 175, 50, 8000, 300, 1000
M. Frock, 92, 28, 3000, 100, 600
M. Frock, 100, 48, 5000, -, -
G. Wehan, 100, 40, 5000, 200, 400
D. E. Ruggle, 122, 8, 6000, 100, 800
W. Backman, 192, 100, 8000, 200, 1200
W. Backman, 5, 30, 4000, -, -
John H. Millar, 125, 30, 4500, 50, 350
J. Stonesifer, 125, 23, 8000, 100, 700
M. Yingling, 100, 20, 6000, 100, 700
D. Crowl, 17, 2, 1600, -, 100
E. Bixley, 150, 30, 5000, 100, 800
C. J. Beggs, 140, 10, 7000, 100, 500
N. Shaeffer, 138, 8, 7000, 200, 600
L. Schaeffer, 168, 55, 1200, 200, 1500
J. Guinan, 146, 1, 10000, 200, 500
J. Myers, 138, 28, 8000, 1000, 600
J. Orendorff, 150, 50, 10000, 200, 2000
J. Little, 125, 15, 8000, 200, 500
D. Baugardner, 122, 10, 50, 50, 400
D. Eveley, 128, 2, 6500, 20, 100
J. Schaeffer, 32, -, 6000, 100, 300
M. Zacharias, 92, 10, 300, 75, 500
D. Leister, 124, 7, 5000, 100, 800

M. Oarster, 140, 5, 5000, 200, 700

M. Miller, 106, 8, 5000, 150, 500

S. Reese, 17, 3, 1500, 20, 100

J. Reese, 130, 50, 1000, 200, 900

A. Reese, 110, 20, 4000, 100, 700

D. Crowell, 36, 4, 1600, 50, 300

D. Schaeffer, 160, 20,7000, 200, 800

G. Orendorff, 100, 40, 8000, 300, 800

J. Crowell, 40, -, 3000, 50, 200

J. Miller, 180, 50, 10000, 100, 900

A. Schaeffer, 20, -, 1800, 20, 200

A. Winters, 131, 15, 10000, 100, 1000

J. Henry, 150, 10, 6000, 100, 800

J. Bankers, 60, 20, 2500, 200, 700

J. A. Megles, 72, 10, 1000, 30, 300

D. Stone, 58, 7, 1800, 15, 200

D. W. Danner, 106, 106, 5000, 50, 300

A. Weaber, 80, 125, 12000, 100, 700

J. Swalzbaugh, 60, 10, 2000, 50, 300

J. Powder, 40, 40, 3000, 20, 100

L. Wamples, 70, 5, 8000, 50, 500

J. Englar, 140, 20, 8000, 200, 1000

D. Orendorff, 100, 24, 7000, 50, 600

J. C. Moore, 441, 8, 4000, 100, 300

L. Zepp, 110, 40, 8000, 100, 600

F. Wampler, 70, 32, 5000, 50, 400

J. Powder, 170, 60, 12000, 1000, 2000

E. Buckingham, 110, 25, 4000, 50, 600

J. Buckingham, 130, 40, 18000, 75, 700

H. Milton, 96, 2, 7000, 25, 500

W. Fowler, 48, 3, 1600, 25, 300

Samuel A. Lauven, 110, 45, 6000, 100, 500

J. H. Busby, 35, 20, 5000, 100, 400

Saml. Baker, 16, 5, 1000, 500, 75

John Niven, 22, 6, 700, 100, 75

John Watinfelt, 20, 22, 800, 200, 100

Nelson Williams, Against G. Jacobs, 50, 75

William Louge (Longe), 50, 25, 2000, 500, 200

Edward Barnes, 16, 19, 1000, 5, 250

Henry Hoffman, 3, -, 350, -, 30

John Wagoner, 36, -, 2000, 300, 300

Andrew Easton, 68, -, 1200, - 165

Robert K. Capler (Caples), 50, 27, 800, 10, 150

Luke Wagers, Tenant of G. Jacobs, 100

James Edmondson, 100, 50, 500, 50, 100

John W. Barber, 40, 30, 700, 150, 200

Elizabeth Ogg, 50,78, 1000, 10, 250

George Smith, 50, 78, 1000, 10, 250

William Bloom, 10, 50, 600, -, 70

Samuel Caples, 50, 117, 1500, 30, 150

Henry Crooks, 75, 100, 1200, 50, 200

George Trumb, 100, 50, 2000, 50, 200

John Frick, 35, 18, 1000, 40,100

Andrew Bigot, 40, 10, 60, -, 95

Mary Ogg, 30, 5, 1000, -, 20

Christian Undesock, 11, 3, 1000, 5, 50

Leander Roseburg, 22, 28, 1000, 50, 250

John Baker, 10, -, 650, -, 45

Burgess Nelson, 50, 248, 5000, 100, 400

Andrew Whelham, 40, 20, 1000, 50, 180

John Mizel, 20, 10, 300, 5, -

William Bloom, 5, 65, 600, -, 25

John Davis, 100,75, 3000, 500, 500

Elisha Buckingham, 20, 19, 100, 10, 50

John Baker, 20, 80, 1000, 20, 270

Mary Cresnel, Tenant of G. Jacobs, 15

Richard Frizel (Fuzel), Tenant of G. Jacobs, -, 17

Ruben Longin, Tenant of G. Jacobs, -, 20

Ann Smith, 65, 17, 1600, 35, 200

Nimrod Gardner, 160, 50, 10000, 400, 700

Ninrod Bergotts (Bergatts), 60, 25, 3000, 50, 300

John Blizard, 5, 40, 500, 10, 40

Jacheniah Jordan, 30, 7, 280, 20, 50

George Barber, 50, 40, 1800, 50, 100

William Williams, -, 32, 450, 40, 50

William Bloom, -, 25, 250, 15, 138

Margaret Beard, 25, 28, 2000, 30, 30

Elisha Griffin, 40, 8, 2000, 50, 150

Ludwick Flank, 100, 80, 1500, 50, 250

John Shafer, 12, 4, 500, 25, 170

Nicholas Ogg, 100, 260, 1400, 100, 200

Elizabeth Glamtia, 100, 54, 6000, 100, 250

Hanson Davis, 75, 15, 1800, 100, 450

Lewis Phillips, Tenant of G. Jacobs, 150, 200

Pery Barnes, 50, 2, 900, 20, 150

Hanson Bonthlon, 150, 30, 4500, 200, 100

John Pool, 63, 20, 1500, 50, 300

Nathan Gorsuch, 40, 37, 2900, 30, 300

Nicholas Buckingham, 40, 100, 4000, 2, 130

Nathan Pool, 15, 20, 400, 10, 70

Oliver Pennington, 75, 78, 1500, 30, 150

Lloyd Shipley, 35, 5, 300, 30, 120

James Murphy, 12, 13, 550, 5, 130

Thomas Gorsuch, 60, 30, 4500, 300, 600

Amelia Miller, 60, 35, 1000, 10, 150

Michael Phillips, 50, 4, 4500, 100, 400

George Barnes, 15, 9, 500, 35, 500

John W. Gorsuch, 125, 40, 5000, 600, 800

Isaac Keller, 20, 50, 400, 10, 170

William Arnold, 13, -, 400, 15, 125

Lloyd Pool, 150, 50, 4500, 10, 100

Ashbury Phillips, 26, 70, 1000, 50, 160

Abraham Prugh, 120, 85, 5000, 400, 600

Robert Wilson, 70, 10, 2000, 100, 200

Frederick Shipley, 100, 100, 2500, 150, 370

Joseph Pool, 300, 155, 10400, 500, 500

James Lavish, 50, 20, 5000, 100, 270

George W. Shipley, 50, 18, 3500, 20, 200

Jacob Wickest, 60, 35, 400, 20, 250

Elijah Woolery, 90, 10, 300, 20, 50

Elias E. Stockdale, 190, 50, 4800, 100, 300

Edmund Stockdale, 170, 90, 6000, 200, 424

Thomas Stockdale, 100, 125, 5000, 150, 400

Lewis Stockdale, 100, 125, 5000, 200, 600

Mason Barnes, 80, 20, 1100, 70, 500

Joshua Murry, 100, 80, 1800, 50, 400

Benjamin Barnes, 65, 30, 2500, 115, 120

Benjamin Haines, 60, 20, 1000, 50, 150

John Pool, 80, 4, 1500, 100, 300

George W. Gorsuch, 180, 50, 3000, 200, 700

William Lamott (Lanott), 240, 95, 8000, 500, 1000

Sophia Jordon, 100, 50, 2500, 10, 150

George Jacobs, 1000, 1500, 20000, 500, 1136

Mariah Williams, 50, 6, 1100, 50, 200

Ann Smith, 18, 2, 900, 2, 55

J. Lowell Grove, 45, 7, 1800, 50, 150

Esra Griffen, 40, 48, 1150, 70, 100
Louisa Griffey, 80, 29, 2500, 50, 300
Louis Gorsuch, 67, 28, 2500, 175, 300
William Gardner, 60, 47, 2000, 150, 400
Joseph Haines, 40, 15, 900, 30, 100
John T. Stockdale, 70, 40, 2500, 150, 550
Greenbery Williams, 43, 10, 250, 150, 200
Mary Caples, 125, 25, 2500, -, 200
Jacob Knight, 12, 6, 1000, -, 60
William Caples, 100, 50, 4000, 700, 900
Conrad Grass, 15, 15, 500, 10, 50
Peter Wolf, 45, 6, 1700, 120, 240
Peter Sipe, 45, 5, 600, 15, 100
Charles Nimrod, 20, 5, 500, 20, 120
John Lockert, 40, 10, 900, 30, 250
Mary Roberts, 100, 36, 900, 10, 200
Edwin Holand, 40, 12, 1300, 75, 200
Michael Fry, 16, 2, 700, 10, 50
John C. Rinehart, 18, 4, 900, 10, 50
Thomas Brown, 150, 70, 7000, 100, 1200
Rebecca Tanney, 40, 2, 3500, 20, 150
Daniel Hering, 130, 36, 900, 400, 700
Christopher Woolery, 30, 50, 1000, 50, 250
Cornelius Armcost, 50, 30, 2000, 100, 300
Anthony Arnold, 15, 15, 900, -, 40
William Shilling, 40, 20, 1000, 50, 190
Elias Brothers, 200, 100, 4000, 100, 1000
Charles Arnold, 25, 20, 1000, 200, 200
Hyintha Belzard, 150, 90, 7000, 300, 400
James Williams, 100, 50, 3504, 150, 500
Daniel Sincon, 14, 2, 500, 10, 50

James W. Hook, 120, 50, 3000, 75, 220
Elizabeth Murray, 100, 50, 1300, 75, 325
Joseph Ward, 70, 30, 1500, 25, 200
William Ward, 100, 20, 4000, 50, 300
Stephen Curler, 70, 5, 1500, 10, 150
John H. Uhler, 40, -, 3500, 100, 225
Adam Hoover, 30, 10, 3000, 130, 1200
Daniel Armagest, 30, 30, 2000, 50, 100
Thomas Demos, 60, 25, 6000, 60, 350
Mordecai Cockey, 200, 100, 1000, 100, 700
M. G. Cockey, 30, 6, 5000, -, -
Daniel Frazer, 200, 200, 9000, 400, 1000
Washington Ward, 15, 8, 2000, 10, 100
Martha Ward, 50, 10, 3000, 30, 110
Abraham Leister, 90, 11, 5500, 150, 300
Jacob Stork (Storp), 14, 2, 800, 20, 50
John Burns, 5, -, 2000, 50, 250
George Leister (Lester), 130, 27, 6500, 100, 300
Daniel Hawn, 23, 8, 800, -, 70
John Armacott, 120, 10, 6000,100, 700
David Shriver, 65, 16, 3500, 100, 500
Daniel Bush, 165, 90, 7000, 300, 600
Owen Buckingham, 60, 25, 4000, 100, 200
Jacob Gamber, 35, 8, 1900, 50, 300
Jacob Constant, 45, 15, 1300, 45, 200
Elizabeth Boyer, 75, 10, 1900, 20, 400
Joshua T. Barnes, 100, 130, 3000, 150, 400

Thomas Spencer, 16, 10, 1000, 75, 225

Reuben Troyer, 75, 31, 2000, 100, 420

Jacob Troyer, 45, 95, 4000, 30, 200

George Richard, 200, 100, 8000, 200, 900

William Fowble, 70, 20, 4000, 150, 600

Henry Zepp, 25, 10, 1000, 75, 130

Cornelius Buckley, 260, 110, 12000, 300, 700

Jacob Caples, 200, 160, 3500, 200,700

Eliza Stockdale, 150,79, 3500, 50, 300

Noah W. Stockdale, 45, 5, 900, 50, 400

Noah Stockdale, 110, 60, 5000, 100, 400

Frances Butler, 210, 85, 15000, 200, 1100

Dennis Grimes, 85, 25, 300, -, 250

Alexander Hoover, 185, 65, 100, 100, 700

Yost Martin, 100, 26, 5500, 50, 400

Henry Bond, 60, 30, 2500, 350, 150

Margaret Bond, 150, 50, 5000, 150, 400

Charles Williams, 45, 8, 1500, 100, 200

Benjamin Bond, 190, 55, 7000, 300, 300

Washington Jones, 50, 30, 2500, 100, 500

Noah Brown, 60, 10, 2000, 40, 300

Benjamin Williams, 45, 12,1700, 50, 250

Henry Miller, 20, 23, 900, 100, 250

John Elserran, 120, 69, 1800, 100, 700

Samuel Martin, 40, 19, 1500, 50, 201

Daniel Stull, 200, 70, 5400, 200, 300

George Mumagh, 65, 11, 1800, 100, 200

John Hoff, 200, 75, 5000, 300, 900

John Uhler, 70, 15, 200, 300, 480

Henry Hobbrock, 30, 6, 1400, 50, 250

Alexander Shipley, 45, 32, 4000, 100, 350

Hezekiah Caples, 100, 45, 4000, 100, 300

William Caples, 100, 60, 5500, 100, 700

Phillip Flatter, 200, 80, 7500, 200, 600

Amos Algier, 150, 100, 5000, 200, 400

Peter Flatter, 20, 5, 600, 10, 200

Charles Brown, 27, 2, 1500, 20, 400

John Flatter, 150, 75, 5500, 150, 600

Kinzey Taylor, 120, 18, 4000, 200, 400

John H. Chen (Chew), 115, 40, 7000, 200, 600

Osiah Oursler, 70, 10, 3500, 100, 220

Conrad Rust, 50, 6, 900, 50, 200

Samuel Taylor, 70, 20, 1100, 50, 200

Eliza Taylor, 110, 90, 1900, 151, 450

Samuel Taylor, 60, 39, 1000, 50, 150

Jacob Lipho (Lippe, Lippo Leppo), 50, 18, 1200, 100, 300

Elijah Leppo, 40, 25, 1600, 50, 160

Isaac Green, 135, 65, 1700, 50, 300

Jesse Magee (Mager), 100, 110, 3500, 100, 200

John Stansbury, 40, 6, 1000, 25, 100

Augustus Magee (Mager), 30, 5, 1600, 100, 200

John Swerden, 35, 3, 1000, 50, 250

John Shamburg, 200, 206, 9000, 80, 700

Isaac Green, 80, 80, 1800, 100, 500

Daniel Long, 100, 50, 2800, 150, 350

Ludwic Long, 60, 75, 2800, 250, 300

George Aslough, 25, 24, 1500, 50, 200

William Homes, 40, 10, 1000, 40, 150

William Lockert, 60, 55, 1200, 50, 300

Frederick Devilbliss, 61, 43, 4000, 80, 1000

Henry Dexter, 38, 1, 1800, 50, 350

William Long, 60, 70, 4000, 150, 500

Joseph Eskerman, 75, 16, 2500, 100, 510

Joseph Stansbury, 100, 60, 123000, 400, 1000

Sophia Roades, 50, 14, 4000, 100, 200

Henry Knoder, 45, 5, 3000, 150, 200

Singleton Haines, 90, 16, 4000, 20, 500

Jeremiah Baylor, 55, 17, 5000, 200, 500

Eli Erb, 180, 129, 3625, 300, 440

Henry Warner, 20, 100, 7000, 180, 200

S. Myers, 75, 40, 2800, 300, 550

J. W. Frock, 100, 12, 4000, 400, 575

William Frock, 130, 37, 5000, 50, 210

E. Myers, 145, 40, 7000, 1000, 1000

G. King, 110, 95, 5000, 150, 850

Samuel Erb, 100, 15, 2500, 50, 375

J. Stonisfer, 90, 36, 2000, 100, 225

Jno. E. Hohn, 40, 21, 600, 150, 300

John Myers, 100, 35, 2500, 200, 550

C. Broders, 28, 18, 1800, 100, 200

L. Lister, 125, 50, 7000, 200, 525

J. Flickinger, 50, 10, 2000, 125, 275

C. Erb, 160, 45, 5000, 400, 890

S. L. Tingling, 200, 100, 6000, 250, 600

M. J. Frock, 55, 30, 3000, 250, 500

Uriah Flegle, 60, 50, 3000, 80, 375

John Erb, 150, 153, 10000, 500, 300

Samuel Erb, 80, 63, 3000, 75, 200

P. E. Myers, 85, 35, 4500, 150, 700

Perry Rumbio, 7, -, 800, 250, 600

H. P. Myers, 12, 20, 2000, 50, 300

J. A. Humbert, 95, 25, 4000, 250, 800

J. R. Bowersox, 60, 21, 2500, 200, 600

W. Stansbury, 35, 25, 1000, 100, 230

Wm. Crouse, 50, 30, 1300, 50, 280

P. Zingling, 25, 13, 1000, 25, 200

S. Stewart, 100, 20, 1500, 25, 350

Joseph Erb, 34, 30, 1000, 50, 270

S. Bankard, 60, 30, 2000, 57, 175

Nancy Hairo (Haire, Haird), 75, 6, 2000, 100, 285

S. Yingling, 47, 3, 1800, 57, 175

P. Mearinhamer, 150, 100, 4000, 225, 600

G. Lippe (Sippe), 100, 50, 2500, 50, 150

D. Lippe, 90, 62, 4000, 60, 375

D. Matthews, 55, 28, 2000, 100, 340

C. Wibling, 70, 30, 2100, 175, 300

J. Baum, 60, 60, 2400, 100, 300

Wm. H. Yingling, 65, 35, 3500, 250, 500

John Lippe, 60, 37, 2000, 75, 300

A. Yeiser, 100, 80, 1500, 50, 325

E. Yeiser, 90, 57, 3000, 175, 500

D. H. Rudolph, 75, 10, 1725, 50, 175

H. Frock, 100, 36, 2000, 100, 328

L. Bish, 30, 20, 1200, 50, 175

W. Bish, 31, 12, 1000, 30, 275

C. Wisner, 200, 100, 3800, 300, 850

S. Guting, 150, 60, 4000, 250, 700

S. Orter, 75, 24, 3000, 100, 600

J. Maris, 150, 78, 6000, 150, 790

C. Bankert, 70, 15, 2500, 200, 300

J. Bankert, 55, 50, 2800, 80, 375

W. J. Bankert, 50, 25, 1000, 100, 300

C. Erb, 70, 65, 2000, 75, 300

J. Wisner, 225, 143, 6500, 600, 4100

M. Troxell, 70, 34, 2500, 100, 400

P. Arter, 60, 40, 2000, 200, 350

D. B. Arhart (Barhart), 300, 400, 10000, 500, 1200

M. Smith, 33, 4, 1000, 50, 200

C. Duce, 50, 14, 1500, 40, 175

J. Steiner, 50, 20, 800, 25, 200

H. H. Jones, 24, 6, 800, 50, 150

J. Miksell, 80, 62, 2800, 150, 425

Wm. B. Miksell, 45, 30, 3300, 125, 350

J. Baumgardner, 130, 45, 5000, 200, 600

P. Marets, 80, 22, 5000, 300, 700

John Starner, 35, 15, 1200, 75, 225

Jacob Hess, 100, 50, 2500, 100, 250

D. Leese, 70, 5, 1200, 80, 300

U. H. Reinecker, 33, 7, 1800, 50, 200

W. H. Hal, 100, 40, 3000, 250, 375

Isaac Hesson, 160, 77, 7000, 225, 700

Wm. Burgeon, 150, 50, 2500, 150, 550

P. B. Miksell, 67, 15, 3000, 100, 210

R. Fuse (Fare, Faso), 75, 28, 800, 100, 175

D. Leppe, 75, 20, 2000, 120, 100

P. Kump, 50, 10, 3000, 60, 350

Henry Miksell, 150, 50, 3000, 300, 700

L. Leister, 60, 25, 1800, 100, 190

A. Trover, 40, 6, 1000, 100, 150

John Jones, 140, 60, 550, 200, 400

Geo. Little, 38, 12, 700, 30, 150

G. Bowman, 130, 30, 3500, 400, 455

D. Leppe, 100, 15, 2000, 150, 425

E. E. Myers, 138, 6, 4000, 154, 250

E. Legare, 100, 30, 4000, 325, 525

G. W. Weaver, 27, 5, 1500, 40, 180

W. J. Feiser, 200, 50, 11500, 700, 1000

S. Messinger, 85, 20, 3000, 250, 550

J. Blonkerd, 105, 30, 3000, 200, 600

J. Wantz, 105, 30, 3000, 20, 600

D. Crouse, 40, 10, 1500, 100, 325

P. Frownfelter, 60, 29, 1800, 250, 325

H. Black, 150, 40, 3000, 25, 75

E. G. Heagy, 60, 80, 2500, 150, 275

H. Messminch, 30, 20, 1000, 100, 360

D. Souse, 90, 20, 1800, 150, 400

D. Lerrue (Lenre, Lessre), 50, 9, 1500, 100, 200

A. Long, 65, 6, 1550, 40, 400

J. Molter (Motter), 100, 30, 2300, 200, 410

J. E. Herd, 110, 50, 4000, 130, 235

J. Kutrer, 95, 34, 2000, 125, 400

B. Hesson, 130, 25, 3300, 150, 500

Wm. Molter (Motter, Motler), 106, 60, 3000, 250, 580

P. Snyder, 90, 60, 2000, 80, 400

J. Hull, 38, 12, 1000, 40, 110

J. Troxell, 75, 20, 2500, 75, 350

J. Reindollar, 80, 35, 2600, 70, 175

Paul Eck, 70, 9, 2000, 60, 325

J. W. Troxell, 167, 25, 3000, 200, 320

J. E. Hahn, 170, 80, 10000, 275, 800

H. Myers, 48, 19, 2000, 200, 425

J. Study, 170, 40, 8500, 450, 940

S. Cover, 200, 50, 10000, 500, 1350

G. Roons, 120, 20, 8600, 520, 600

G. Dutrow, 125, 100, 7500, 500, 1000

C. Freser, 125, 25, 7000, 300, 700

S. Stoneisfur, 95, 20, 7000, 100, 325

P. Shott, 55, 12, -, 30, 150

Daniel Shull, 18, 4, 1000, 40, 150

M. Shull, 48, 32, 1800, 45, 200

D. _. Bowersox, 36, 8, 1200, 75, 200

J. Study, 120, 30, 4000, 250, 400

J. Bachtell, 125, 25, 500, 250, 503

J. Lawyer, 100, 36, 5000, 50, 700

F. Stinger, 83, 9, 1000, 150, 200

J. Rorback, 57, 11, 1800, 100, 300

G. Humbert, 180, 53, 4700, 500, 750

D. Kemp, 75, 45, 4300, 275, 425

J. Kemp, 30, 10, 1800, 75, 225

J. U. Wolf, 35, 10, 1000, 50, 200

W. Ruhrback, 170, 20, 90, 100, 350

A. Stonesifer, 72, 30, 4000 -, 130

S. Fuser, 40, 3, 1000, 40, 140

Sarah Maurd (Mainrd, Maine), 250, 50, 15000, 500, 950

Joseph Brown, 100, 60, 5000, 400, 800

T. Lynch, 55, 50, 3000, 50, 250

E. Bankert, 70, 15, 2000, 125, 375

A. K. Shriver, 75, 81, 7000, 200, 410

Wm. Shriver, 80, 60, 5000, 200, 650
J. E. Dodur, 175, 47, 12000, 500, 750
A. Koontz, 100, 45, 4800, 250, 800
B. Schlyder, 35, 15, 2500, 150, 275
S. Myers, 50, 6, 2300, 250, 325
J. Murdock, 110, 34, 5500, 350, 900
D. Routson, 90, 14, 100, 25, 175
J. Hahn, 60, 10, 2000, 75, 275
P. Frownfelter, 65, 40, 4000, 150, 425
G. Dutrow, 185, 30, 5000, 200, 925
D. Fuser, 150, 50, 4000, 400, 00
Jno. Delaplane, 140, 30, 8000, 250, 525
S. Winemiller, 130, 40, 10000, 100, 300
Wm. Frock, 150, 22, 3000, 100, 425
Jno. Davidson, 175, 25, 5500, 300, 700
J. W. White, 110, 50, 7000, 600, 780
C. Angel, 43, -, 1500, 20, 300
D. Harmon, 90, 14, 3000, 50, 200
J. Nipple, 150, 50, 6500, 300, 500
John Six, 145, 15, 5000, 200, 728
R. Stonesifer, 140, 23, 5000, 100, 1100
Jacob Clutts, 130, 49, 7000, 400, 600
Wm. Baumgarner, 120, 40, 4000, 300, 500
Adam Molter, 80, 26, 3000, 100, 500
A. S. Zentz (Zenty), 10, -, 500, 25, 200
John Weighright, 250, 68, 8000, 500, 150
Daniel Sell Jr., 200, 50, 10000, 330, 850
S. Harman, 115, 20, 2000, 250, 525
Daniel Boale, 350, 65, 10000, 550, 1854
Perry Egler (Eyler), 180, 32, 5000, 300, 750
J. Whitner, 130, 10, 3500, 125, 615
H. Fox, 200, 20, 6000, 100, 550
Wm. Valentine, 150, 34, 6000, 400, 700
J. Dudrer, 60, 15, 2800, 150, 450
Jacob Taylor (Sayler), 100, 70, 6500, 500, 1220
John Stornber (Stomber), 48, 8, 1800, 125, 425
S. Willhide, 125, 40, 8500, 550, 880
David Goff, 220, 50, 7500, 300, 920
Deo. Angell, 100, 30, 3500, 152, 450
L. F. Byerly, 133, 40, 8650, 550, 1075
H. Dern, 131, 9, 5000, 400, 660
Jno. Diffendale, 175, 50, 7500, 200, 650
R. Sanders, 150, 100, 8000, 400, 700
G. Beale, 135, 40, 5500, 450, 600
G. Misering, 110, 35, 6000, 220, 720
J. Goon (Gorn, Grose), 80, 24, 4000, 25, 400
Mrs. S. Rufsnyder, 150, 50, 6000, 150, 500
Mrs. C. Warsche, 195, 34, 9000, 45, 400
S. Angel, 47, 4, 2000, 50, 200
Wilson Scott, 225, 143, 20000, 150, 600
John Nipple, -, -, -, -, 700
L. Brown, 100, 13, 3000, 150, 260
Wm. Mering (Moring), 150, 41, 8000, 575, 640
S. Diffendale, 180, 8, 3000, 200, 360
J. Yingling, 95, 17, 2500, 20, 380
E. Warner, 176, 90, 6000, 100, 800
J. Lynn, 23, 8, 1000, 110, 205
John Delaplane, 35, -, 1500, 100, 620
J. W. McAlester, 115, 19, 5000, 300, 620
D. Lynn, 150, 30, 5000, 75, 400
J. D. Bowman, 125, 25, 3800, 120, 250
Mrs. E. Byerly, 9, 40, 3500, 75, 300
L. Harris, 85, 48, 7500, 400, 650
J. Roop, 100, 60, 600, 175, 678
J. Fuser, 23, -, 3000, 100, 300
J. W. Angel, 110, 35, 4000, 300, 700
P. Roons, 140, 60, 6000, 280, 775

J. Myers, 65, 8, 2500, 100, 375
D. Stoner, 100, 40, 5800, 40, 280
E. Johes (Jones), 80, 11, 4500, 35, 175
Wm. Cumbacher, 100, 30, 7500, 175, 700
G. Angel, 9, 15, 2600, 50, 275
J. Shanetts, 160, 44, 10000, 500, 900
D. Buffington, 180, 30, 11605, 500, 900
H. Shule, 60, 10, 1800, 50, 340
J. Davis, 81, 25, 3500, 125, 485
G. Reifsnyder, 80, 26, 3200, 500, 620
P. W. Hann, 125, 63, 5000, 450, 640
J. Harmon, 105, 25, 3000, 100, 450
J. Buffington, 150, 59, 1000, 400, 1570
J. Roons, 110, 50, 4800, 400, 575
A. E. Null, 100, 56, 6500, 420, 600
S. Johnson, 60, 6, 2800, 100, 350
A. Hiteshue, 160, 27, 7000, 50, 355
J. Hess, 75, 5, 4000, 300, 600
Simon Coppersmith, 161, 8, 6000, 2000, 550
John Garner, 98, 8, 4000, 150, 650
E. Yingling, 100, 60, 3000, 200, 600
D. Wilson, 14, 16, 1200, 50, 100
G. W. McConkey, 130, 70, 9500, 300, 550
Thomas Cook, 100, -, 2000, 125, 150

Cecil County Maryland
1860 Agricultural Census

The University of North Carolina Library under a grant from the National Science Foundation microfilmed agricultural Census records. Records were filmed at the University of North Carolina from original records at the Maryland State Library.

Columns 1, 2, 3, 4, 5, and 13 represent the following information on the census:
1. Name of Owner, Agent or Manager of Farm
2. Acres of Improved Land
3. Acres of Unimproved Land
4. Cash Value of the Farm
5. Value of Farming Implements and Machinery
13. Value of Livestock

Pages for this county are out of sequence.

David Stadts, 140, 12, 10000, 300, 12000

A. D. Caulk owner, 109, 9, 8000, 100, 800

James Gillispie owner, 25, -, 4000, 50, 200

John B. Morton, owner, 300, -, 10000, 150, 1000

John Calwell owner, 111, -, 6000, 150, 400

John F. Stephens owner, 180, 55, 10000, 200, 600

Wm. Calwell owner, 180, 55, 10000, 200, 600

Benjamin Hessey agent of Ginn, 160, 40, 8000, 100, 1000

Joseph Hutchinson agent of W. Wilson, 130, 10, 7000, 200, 1000

John B. Gonce agent of W. Wilson, 200, 50, 10000, 100, 800

F. R. Rochester agent of W. Wilson, 200, 60, 15000, 400, 1000

John S. Townsend agent of W. Wilson, 200, 60, 120000, 800, 1200

Joseph Senicker agent of W. Wilson, 200, -, 12000, 150, 600

James Hennedoy agent of Ginn, 26, -, 6000, 100, 600

John M. Flinthand owner, 180, 20, 10000, 300, 900

Chris D. Sweatman owner, 140, -, 10000, 200, 900

Dr. W. C. Perkins agent of Ginn, 150, 33, 15000, 200, 1200

Thomas Lusby owner, 140, 20, 15000, 100, 400

John W. Davis agent of Mort & Price, 450, 160, 33300, 300, 1500

George W. Price owner, 400, 50, 31500, 300, 200

John J. Lockwood agent for R. Lockwood, 200, 25, 12000, 500, 2000

John R. Halson agent for B. Lockwood, 200, 20, 11000, 175, 1500

John F. Becks agent of B. Green, 260, 120, 20000, 150, 1200

John F. Becks owner, 180, -, 12000, 400, 1000

Robert H. Hays owner, 200, 20, 10000, 150, 1500

Carry Fraiser agent of J. W. Davis, 1560, 40, 15000, 100, 500

Levy Young agent of Davis, 270, 30, 5000, 50, 400

Edward Pennington agent of George Olhon, 250, 50, 10000, 200, 800

H. Morton agent of J. Dine, 300, 50, 15000, 300, 800

M. Redgrave owner, 175, 50, 15000, 500, 1200

James T. Band (Bond) agent A. C. Nowland, 209, 20, 15000, 500, 800

Society of Jesuits agent of the Jesuits, 270, 20, 18000, 500, 500

William T. Lusby, 300, 30, 20000, 100, 1600

B. Green agent of Jesuits, 280, 25, 20000, 500, 900

Martha Richnor agent of Jesuits, 100, -, 5000, 50, 300

William Price agent John Mouton, 120, 2, 8000, 100, 900

James Ford owner, 300, -, 13000, 200, 1200

George Biddle agent of Mrs. Wert, 230, 30, 15000, 250, 1200

William Eldinson owner, 300, -, 10000, 100, 1100

Caryies Dill agent J & Linch Delhuffshy, 275, 25, 20000, 400, 1000

F. B. Crookshank, 500, 120, 20000, 200, 600

J. H. Walker owner, 340, 50, 30000, 300, 1700

James Ginn agent of G. Davis Del., 300, 20, 15000, 200, 1600

Isaac Walker agent of Ginn, 200, -, 10000, 150, 1100

Arthur Johns agent of Ginn, 300, 25, 12000, 250, 800

J. T. Biddle owner, 175, 25, 14000, 300, 1000

Stephen Lofland agent of M. Garison, 250, 25, 18000, 500, 1600

Thomas Price agent of G. Davis, 250, 250, 14000, 200, 1200

David Lyman owner, 400, -, 20000, 150, 1000

Charles Jones-George Davis Del., 240, 30, 9600, 200, 1300

Thomas T. Price agent of Davis Del., 178, 75, 8000, 500, 800

Wm. M. Knight owner, 225, 108, 15000, 300, 1600

John W. Foard owner, 420, 25, 30000, 225, 2500

James W. Clark agent Rev. Dichuey, 250, 15, 10000, 200, 1000

Edward Jones agent of E. Clark, 100, 100, 3000, 100, 300

Thomas Vandike owner, 350, -, 30000, -, 800

Newell Bougus owner, 150, 25, 8000, 200, 1300

Dr. Thomas A. Rekert agent of Davis Del., 140, 60, 10000, 150, 1000

William Myers owner, 65, -, 8000, 150, 800

J. W. Huges, 60, -, 5000, -, 50

Thomas C. Crockett agent of Mrs. Lusby, 500, 300, 60000, 500, 1500

Estate of M. Eldridge owner, 360, 100, 20000, 500, 2000

Jams J. Hall agent of B. Crosklas, 600, 100, 60000, 500, 2800

Francis King owner, 200, 170, 11000, 150, 1300

John W. Dickson owner, 90, 35, 8000, 200, 600

John C. Davis agent of F. D. Crookshank, 120, -, 5000, 400, 400

Lewis Rea agent of J. Jarvis MD, 190, 30, 10000, 150, 900

John C. Furguson owner, 130, 15, 8000, 100, 800

Dr. Derney owner, 160, -, 8300, 250, 1000

David P. Davis, 150, 40, 10000, 400, 800

James S. Morrison owner, 120, 50, 8000, 500, 1100

George B. Pearce agent for John Rebald, 175, 60, 8000, 150, 1500

Perry Pennington agent for G. Davis Del., 350, 150, 25000, 1000, 2000

Thomas A. Lowise owner, 350, 50, 250000, 500, 2000

Wm. E. Etherington agent for M. Fisher, 250, 200, 12000, 400, 2000

E. D. Ethrington agent for Stephen Gillespie, 220, 40, 10000, -, 800

Thomas P. Jones agent W. G. Etherington, 180, 20, 10000, 300, 1500

John W. Jones owner, 150, 20, 12000, 400, 1000

Richard Leomon owner, 225, 50, 20000, 1000, 1400

Nathaniel Mehrin owner and agent for W. Knight, 200, 20, 8000, 200, 1200

Alexander Laws, 180, 10, 12000, 100, 1000

John Wrath owner, 200, 25, 13500, 500, 2000

John P. Vandike agent of G. C. Crookshank, 60, -, 3000, 100, 400

W. H. Yates agent for Vandike, 250, 10, 10500, 500, 1300

Hyland Benison agent for W. Knight, 250, 50, 15500, 300, 2500

James Price agent for Mrs. Right, 63, -, 3200, -, 250

George Gonce owner, 175, 30, 2500, 100, 500

Joshua Blackney agent for Wm. Henry, 150, -, 7500, 400, 800

Joseph Biggs agent M. C. Lusby, 360, 40, 20000, 500, 2000

B. B. Price agent J. Biggs, 350, 250, 30000, 300, 2000

George W. Hessey agent M. Young Del., 600, 200, 6000, 500, 3900

John Cameron agent of Foreman, 600, 66, 8000, 500, 2700

W. E. Cook agent of Reybold, 250, 50, 1800, 200, 1000

Daniel Pennington agent of Thompson, 450, 50, 30000, 450, 1800

William Robinson agent of A. Ribald, 300, 25, 9000, 250, 900

James Moore agent Doct Cunningham, 900, 25, 6000, 100, 600

Mark Manloff agent Dr. Cunningham, 150, -, 4500, 100, 300

Dr. J. T. Cunningham agent Dr. Cunningham, 900, -, 6000, 100, 750

Samuel T. Carroll owner, 100, -, 3000, 100, 500

Wm. O. R. Knight, agent Dr. Cunningham, 180, 100, 7000, 200, 900

James Pippen agent of A. Ribald, 400, 100, 25000, 300, 1600

John Husfelt agent A. Ribald, -, -, 2000, -, 400

T. V. Ward owner, 280, 80, 20000, 1000, 2000

W. H. Price agent of T. V. Ward, 275, 25, 18000, 500, 1000

Sarah W. Veasey owner, 300, 100, 15000, 300, 500

John Meradith, 180, 20, 12000, 150, 800

Wm. Timms agent of A. C. Newland, 275, 70, 15000, 250, 1600

John V. Price owner, 200, 40, 5000, 900, 1000

Daniel Penington owner, 100, 20, 3000, 100, 900

George Read agent for Dr. Welds, 125, 50, 4500, 100, 500

James Weib agent for G. Hempsey, 160, -, 7000, 150, 500

James W. Morgan agent for T. V. Ward, 93, 35, 8000, 150, 800

John C. Armstrong owner, 900, 40, 8000, 150, 1000

Ann M. Watt agent for Lockwood, 140, 30, 5000, 150, 600

John H. Hessey agent of T. V. Henry, 175, 25, 10000, 75, 600

James Robinson owner, 150, 50, 6000, 100, 200

James Biggs agent for Mrs. Lucis, 90, 10, 2000, 50, 400

Alexander Biggs agent J. C. Groome, 200, 200, 10000, 150, 130

Jane Toodin agent of S. B. Ford, 125, 25, 8000, 100, 1000

Joshua Reed agent J. C. Groome, 80, 50, 4000, 100, 500

Hayland Rice, 150, 50, 6000, 100, 650

Richard D. Akin agent for Mr. Money, 120, -, 3000, 100, 400

Benedict Lavin owner, 230, 30, 7800, 150, 600

Estate of N. P. Loyd over John Wratt, 151, -, 5000, 100, 700

A. E. W. Lusby, 130, 40, 3400, 150, 800

Joseph Magill owner, 100, 21, 6000, 400, 800

And. Pearce owner, 180, -, 8000, 500, 1200

Henry Baan owner, 60, -, 1000, 200, 200

Wm. H. Camerson (Cameron)owner, 250, 100, 15000, 400, 1100

Julia A. W. Pearce owner, 200, 30, 14000, 300, 1500

A. Richardson owner, 40, 40, 13000, 230, 1400

R. J. Milburn agent Mrs. Pearce Rich, 500, 100, 18000, 400, 1200

John W. Morgan agent for John Ribald, 200, 50, 19000, 200, 1700

David J. Lonethrow (Sonethrow) owner, 175, 100, 7250, 150, 100

John J. Buley owner, 100, -, 3000, 25, 500

Maria S. Ford owner, 400, 250, 25000, 100, 1000

Mary W. Veazey owner, 350, 350, 28000, 400, 1700

James Hessey owner, 300, 50, 14000, 200, 2500

James Criemind agent of Mr. Wheary, 100, -, 4000, 50, 300

John F. Dodson agent of Mr. Wheary, 100, 90, 5000, 100, 700

Jacob Trusty agent of Lewis, 150, 50, 4000, 50, 150

Henry Hughs agent of J. C. Groome, 14, -, 600, 50, 500

Wm. Knight owner, 400, 150, 27500, 500, 8000

Edwin G. Charlea owner, 26, -, 2000, 100, 300

Wm. R. Biddle owner, 9, -, 600, 50, 200

James Ruly owner, 22, -, 1500, 50, 400

John T. Reed owner, 150, 20, 8500, 150, 800

Scott Price agent J. R. Price, 200, 50, 15000, 200, 1400

Andrew Jervis agent Dr. Roberts, 26, -, 3000, 50, 200

Joseph H. Manloff owner, 250, -, 12500, 300, 1100

Ebenezer Clark owner A. Newland, 240, 30, 10000, 400, 1400

Wm. R. Freeman owner, 400, 200, 30000, 200, 800

A. H. Oldham owner, 200, 300, 12000, 100, 900

William Templeton agent G. W. Oldham, 200, 50, 10000, 25, 800

A. J. Pennington agent Wm. Knight, 320, 100, 25000, 1500, 500

A. J. Pennington owner, 770, 100, 2000, -, -

Maria Davis owner, 200, 20, 8800, -, 1200

Wm. H. Newland agent J. Riggs estate, 270, 20, 10000, 200, 1100

Thomas Pretimon, agent Stephen Church, 250, 80, 15000, 500, 1800

D. S. P. Wills owner agent Mrs. Lusby, 500, 300, 25000, 800, 300

David D. Davis, 320, 50, 30000, 400, 9000

J. R. E. Price agent G. Walker, 350, -, 15000, 350, 1800

Samuel Watts owner, 300, -, 15000, 250, 2000

Tom T. Weaks agent W. Knight, not, given, 11000, 200, 1400

Jacob Howell, 200, 50, 10000, 100, 600

James Walker, 150, 60, 7500, 50, 600

David M. Taylor, 250, 35, 15000, 1000, 2200

David M. Taylor, 200, 50, 15000, -, 700

Wm. Boulden, 140, -, 12000, 250, 700

Peter Clever, 220, -, 10000, 200, 1200

Samuel Smith, 200, 60, 15000, 300, 1400

B. B. Chambers, 181, -, 9050, 300, 2000

Mrs. Fulton, 90, -, 4500, 150, 500

James Boyd, 32, -, 840, 80, 150

Charles Walbert, 200, -, 16000, 300, 1000

Mary E. Boulden, 100, 25, 30000, 500, 900

James A. Boulden, 175, -, 30000, 500, 1200

Wm. Boulden, 150, 75, 7000, 300, 1000

John B. Pearce, 100, 30, 6500, 50, 250

Noble Idle, 150, 100, 7500, 300, 1800

Jakey Warren, 120, 80, 6000, 250, 300

Ezekiel Boulden, 220, 150, 15000, 500, 1800

Isaac Filter (Fitter), 300, 100, 20000, 300, 1800

William Biddle, 300, 180, 15000, 200, 2200

D. D. Pearce, 200, 80, 14400, 400, 1500

Anson Griffith, 80, 20, 5000, 100, 350

George W. Boulden, 400, 50, 22500, 450, 1400

Lambert Boulden, 90, 5, 5250, 150, 350

Johnson Boulden, 150, 20, 6000, 150, 550

H. D. Forder (Fouler), 150, 50, 10000, 150, 800

Wm. M. Armstrong, 60, 100, 4500, 50, 1000

Philip C. Plumer, 160, 10, 8500, 100, 612

John Kibler, 250, 30, 1600, 500, 1000

Mary Hudson, 132, -, 6600, 150, 600

William T. Buse, 36, -, 2000, 100, 250

Samuel Dushane, 250, 80, 9200, 250, 1150

John P. Clayton, 200, -, 10000, 300, 800

Charles L. Boulder, 300, 100, 30000, 200, 1400

Samuel Thompson, 18, -, 3000, -, 200

B. Lake, 150, 10, 5800, 400, 400

G. R. Carpenter, 80, -, 4000, 100, 900

Wm. Tllmtham, 260, 20, 16800, 350, 1200

Mary Ral, 200, -, 12000, 250, 1500

John B. Gill, 230, 20, 17500, 1000, 2000

Jacob Caulk, 200, 20, 17000, 300, 600

P. Hendricher, 400, 200, 26000, 500, 2250

B. M. Crawford, 400, 200, 30000, 800, 3000

Joseph Merrett, 400, 200, 30000, 600, 2000

William Jones, 250, 50, 22000, 600, 1400

G. L. Plithar, 180, 30, 15000, 200, 600

C. S. Hayes, 130, 30, 10000, 200, 600

C. Merrett, 120, 40, 10000, 150, 700

Joseph Price, 100, 30, 10300, 200, 600

T. H. Murphy, 200, -, 20000, 200, 900

L. S. Roberts, 260, -, 15600, 500, 1600

D. K. Price, 200, 200, 15000, 400, 1000

A. Kibler, 300, 10, 18000, 450, 2000

Thomas Massey, 250, -, 15000, 500, 1800

John Peach, 320, 100, 15000, 300, 1800

James Pearce, 500, -, 25000, 600, 3000

James E. Lewis, 150, -, 15000, 500, 1400

Wm. Drake, 250, 50, 30000, 200, 1250

S. Dickinson, 150, 10, 10000, 400, 1200

W. Deshane, 350, 250, 18000, 200, 1200

W. C. Crow, 250, 150, 15000, 400, 1000

U. Watts, 500, -, 20000, 500, 1400

J. A. Mears, 52, 10, 2500, 150, 400

J. Davis, 425, 175, 10000, 100, 200

L. Biddle, 300, 300, 10000, 100, 800

R. Carter, 200, 40, 7200, 400, 400

J. W. Bedwell, 100, 120, 8800, 200, 600

E. Rhodes, 260, 100, 15000, 450, 1400

E. J. Watson, 120, -, 13000, 350, 800

Joshua Biddle, 1375, 50, 20000, 200, 1570

Wm. Purdy, 170, 30, 10000, 300, 1000

B. W. Harris, 280, 60, 17000, 400, 1000

J. Griffith, 250, 50, 14000, 500, 1700

J. Mercer, 8, -, 500, 25, 200

J. W. Bouchell, 220, 40, 13000, 120, 350

Isaac Bouchell, 120, 30, 6500, 20, 100

J. Watson, 140, 30, 6800, 400, 800

A. Beaston, 140, 20, 5600, 250, 1000

John Bart, 100, -, 3000, 50, 300

J. Bowlski, 52, -, 25000, 100, 350

R. Cregg, 30, -, 1400, 100, 150

B. T. Pluiter, 500, 200, 35000, 500, 2000

S. W. DeCourcy, 110, -, -, -, -

J. Harris, 88, -, 5000, 150, 450

A. Cregg, 300, -, 12000, 100, 650

C. Cally, 190, -, 5700, 100, 300

D. A. Roberts, 110, -, 4400, 150, 850

B. Cregg, 130, -, 5200, 50, 400

C. Paradee, 200, -, 5000, 250, 500

P. Smack, 240, 60, 6000, 500, 1000

W. A. Rhoads, 100, -, 5000, 150, 600

W. A. Martin, 200, 50, 11500, 400, 1400

James Allen, 50, 150, 6000, 150, 800

J. Havelow, 60, 40, 4000, 50, 300

J. B. Jester, 300, 300, 18000, 300, 1000

J. S. Vandergrift, 150, 100, 6000, 150, 600

J. T. Hane (Kane), 100, 100, 2500, 100, 300

T. Buskirk, 300, -, 9000, 100, 1000

G. Kirk, 250, -, 12500, 400, 1400

Charles Rhodes, 160, -, 9600, 300, 1000

J. M. Pile, 180, -, 12000, 200, 700

C. Bryan, 200, -, 14000, 400, 1200

L. P. Ellison, 400, 50, 27000, 100, 2200

J. Brian, 70, -, 7000, 80, 350

Joseph Booth, 60, -, 6000, 100, 300

Jacob B. Ash owner, 30, 30, 3000, 200, 150

Thomas Watson owner, 12, -, 2500, 150, 450

Thomas Howard owner, 46, 16, 10000, 200, 500

Jacob Johnson owner, 56, 50, 3500, 50, 300

E. M. Hollingsworth owner, 200, 156, 20000, 150, 600

Charles Buckly owner, 150, -, 15000, 100, 600

Thomas Thachery owner, 80, 70, 7000, 400, 500

J. A. Criswell agent Grogan & Moffett, 250, 400, 24500, 450, 1100

George Earle owner, 30, -, 8000, 50, 120

Thomas Boulden owner, 300, 150, 20000, 500, 1500

Wm. Morgan owner, 80, 20, 7000, 100, 370

Thomas Stephens owner, 130, 20, 6000, 100, 250

Jessy Uptegrove, 45, -, 1800, 50, 200

Thomas Hempfield, 50, -, 1000, 50, 200

V. Clark, 80, -, 2000, 100, 100

James Scott, 84, 30, 7000, 150, 300

Susan Wilson, 25, -, 750, 50, 250

Lewis Owens, 50, 14, 1440, 100, 310

Mary A. C____, 10, -, 1000, 20, 60

Stephen Pluck, 100, 35, 6000, 500, 500

C. H. Owens, 30, -, 1000, 50, 150

Lewis Miller, 24, -, 700, 20, 100

Wilson Leister, 34, -, 750, 20, 100

Lemuel Whitely, 33, -, 1000, 50, 250

T. A. Vansant, 180, 60, 7000, 200, 820

Thos. Cotes, 31, -, 1500, 50, 150

Wm. H. McCullough, 100, 25, 7500, 200, 800

Hyland Marcus, 156, -, 7500, 300, 1000

Lewis Backloss, 140, 20, 8300, 200, 800

Andrew McIntire, 175, 35, 14700, 800, 1200

Mrs. B. A. Barr, 150, 100, 14000, 750, 1000

B. T. Biddle, 120, 30, 15000, 600, 1500

James Frazier, 300, 100, 2000, 500, 2000

David Ward, 160, -, 4800, 150, 1000

Thomas Dixon, 100, -, 2000, 50, 300

John T. Ward, 150, -, 6500, 150, 900

S. H. Mitchell, 160, 35, 8000, 700, 1500

Ann Reese, 250, 50, 10000, 300, 400

George Beecham, 125, 70, 12000, 300, 1100

Wm. M. Reed, 11, -, 1200, 200, 150

Ezekiel Wheatley, 150, 25, 5000, 150, 317

James E. Barrall, 165, 50, 18000, 300, 700

Nicholas Lothman, 28, 22, 1040, 100, 350

John Bryan, 55 25, 1400, 100, 500

A. Bathwell, 60, -, 1500, 100, 100

P. B. Pathwell, 25, -, 500, 50, 100

Jesse Stoops, 50, 12, 1400, 50, 400

Wm. Jeffries, 10, -, 500, 50, 100

W. F. Butler, 65, 100, 1600, 50, 150

Joseph George, 150, 50, 10000, 200, 600

Parks George, 100, -, 2000, 50, 200

N. F. Johnson, 80, -, 5400, 80, 500

Isaac Boulden, 100, -, 3000, 50, 400

Wm. Denny, 300, -, 10000, 200, 1700

Rev. J. McIntire, 150, 37, 10000, 500, 1600

Rev. J. McIntire, 100, 35, 9000, 200, 600

E. Pearson, 100, -, 2500, 100, 200

Conrad, 60, -, 3500, 150, 250

Mrs. M. Randolph, 220, 100, 18500, 300, 1500

M. M. Henderson, 100, 100, 14000, 700, 1100

Geo. McCullough, 175, 25, 14000, 1500, 400

Levi Ward, 32, 11, 1500, 100, 225

Stephen Pluck, 120, 30, 5000, 280, 385

John Gilpin, 325, -, 20000, 500, 1750

Osborn Reed, 70, 7, 3500, 200, 500

Jas. Yates, 70, 5, 3000, 200, 450

Geo. Simpson, 65, 5, 3000, 150, 300

John Underwood, 23, 3, 2000, 50, 2000

John N. Davis, 43, 2, 2000, 100, 300

Wm. Bowen, 105, 35, 7000, 350, 800

Thos. M. Neal, 60, 40, 5000, 200, 500

Jos. Jackson, 125, 125, 9000, 200, 920

John Davidson, 150, 44, 8000, 300, 500

Geo. Ricketts, 25, -, 2500, 100, 500

Jas. R. Brown, 37, -, 3000, 150, 420

Levi Tyson, 50, 15, 2000, 150, 350

Jessee Pierson, 25, 8, 1500, 75, 200

Wm. M. Tyson, 95, 35, 3000, 300, 500

Phillip Johnson, 60, 28, 3500, 200, 600

Jno. A. Johnson, 130, 20, 10000, 350, 1000

Mathew Naudain, 40, 59, 3000, 150, 300

John W. Moore, 70, 30, 2500, 150, 350

John T. Miller, 35, 11, 2000, 150, 250

Wm. Kershaw, 50, 16, 4000, 150, 225

Ebenezer Sutton, 25, 10, 800, 75, 225

David Greystone, 40, 29, 1500, 250, 300

John T. Smith, 80, 32, 1200, 300, 1065

David C. Parker, 23, 15, 2500, 200, 200

Jno. E. Davis, 128, 60, 6000, 40, 805

Jos. S. Ter___, 16, -, 1200, 50, 250

Cloud Carter, 16, -, 2000, 95, 200

Jos. Gibson, 18, 4, 1500, 100, 155

Peter Pierson, 85, 215, 12000, 300, 1200

Wm. V. Moore, 90, 15, 5000, 300, 525

Thos. Voxandine, 40, 6, 2500, 150, 325

Francis Green, 50, 43, 3000, 150, 685

Nalane McNamee, 30, 10, 2000, 100, 275

Hannah Harlan, 39, 81, 2000, 100, 280

Frisby Tull, 30, 20, 2300, 75, 200

John Gilpin, 50, 70, 5500, 400, 435

John Cantwell, 160, 20, 12000, 400, 810

J. D. Carter, 60, 40, 6000, 400, 500

John McCreary, 80, 85, 8000, 400, 650

Wm. Spence, 80, 20, 4000, 125, 600

Grael Valentine, 35, 50, 2000, 125, 400

Wm. Ottey, 70, 23, 4000, 350, 875

Thos. Flounders, 40, 13, 1600, 200, 150

John Campbell, 18, 2, 1600, 125, 400

Howard Davis, 60, 30, 2000, 150, 150

Wm. Campbell, 13, 12, 1600, 75, 150

Wm. Vanzant, 40, 16, 1400, 150, 250

Hugh F. Scarborough, 24, 6, 1000, 75, 235

Enos Scarborough, 33, 17, 1500, 100, 500

Watson Scarborough, 40, 20, 1000, 75, 250

Geo. Crow, 20, 17, 1000, 5, 165

Jos. Scarborough, 30, 8, 1000, 150, 165

John O'Daniel, 20, 4, 1600, 100, 150
Jerry Steel, 16, 24, 1500, 100, 320
Geo. Booth, 85, 25, 2000, 400, 380
Benj. Miller, 35, 50, 1500, 150, 310
A. Kirkpatrick, 40, 40, 1800, 150, 300
Samuel Hannah, 50, 52, 2000, 100, 420
Jethro McCauley, 20, 4, 1200, 100, 100
Elijah Mahoney, 20, 10, 1000, 100, 275
J. Harrigan, 20, 10, 1200, 50, 175
Wm. Lynch, 30, 8, 1200, 95, 225
Benj. Groves, 60, 50, 1200, 75, 100
Wm. Harding, 16, 14, 1000, 95, 185
Thos. McVey, 85, 55, 4500, 300, 830
Jno. S. Moffett, 60, 150, 4000, 300, 60
Eli Mendenhall, 60, 37, 3500, 125, 220
Mary Cantwell, 80, 80, 5000, 250, 1000
H. P. Bennett, 100, 42, 6000, 300, 650
Jacob Dawson, 200, 200, 8000, 350, 1250
Saml. Miller, 80, 20, 4000, 300, 465
Peter K. Wright, 133, 60, 6000, 15, 240
A. F. Smith, 120, 90, 8000, 300, 700
Thos. Brown, 60, 60, 8000, 250, 400
Kennard Blake, 100, 60, 5000, 200, 400
William Harvey, 60, 22, 3000, 200, 275
John T. Galleher, 12, 35, 1500, 50, 150
Wm. G. Carter, 100, 35, 6000, 100, 300
Benjamin Flounders, 60, 26, 1500, 50, 150
David Fulton, 70, 15, 2000, 15, 305
John F. Flounders, 60, 30, 1500, 100, 345
Wm. Ward, 18, 3, 1000, 100, 100

George C. Weaver, 29, 9, 1000, 150, 200
Robert McKaig, 22, 2, 1000, 100, 250
John W. Holt, 120, 55, 6000, 300, 450
Jame Burk, 80, 30, 2200, 250, 400
John Booth, 60, 15, 3000, 300, 420
Jesse H. Simpers, 60, 220, 4000, 250, 500
Thomas Miller, 40, 20, 2200, 200, 300
Levi T. Miller, 35, 40, 2000, 40, 125
John W. Miller, 65, 35, 3500, 300, 400
Wm. Simpers, 50, 30, 1200, 75, 225
Thomas Fitzsimons, 50, 9, 2500, 200, 225
Amos Scarborough, 35, 8, 1700, 100, 250
John Grant, 75, 25, 1500, 150, 300
Robert Cantwell, 80, 25, 1500, 150, 400
Samuel Knox, 60, 30, 1500, 75, 300
James H. Smith, 50, 49, 4000, 75, 300
James McCauly, 40, 195, 3500, 200, 420
Holland Hardgraves, 20, 22, 900, 30, 60
James Foster, 60, 100, 3000, 100, 400
Martin H. Biles, 50, 80, 2000, 100, 250
Henry Simpers, 50, 20, 1200, 100, 80
Jacob T. Collins, 130, 20, 6000, 250, 450
Daniel Lacy, 25, 60, 2000, 150, 120
Peter Johnson, 60, 77, 1600, 125, 175
Benjamin Woolf, 80, 60, 2000, 100, 350
Nicholas Manley, 140, 106, 7000, 100, 350

Thos. J. Reynolds, 100, 30, 8500, 325, 640

James P. Howell, 125, 10, 3500, 150, 300

George Fitzwater, 175, 44, 12000, 400, 900

Amos Alexander, 120, 25, 11250, 1000, 640

Joseph Alexander, 53, 15, 3500, 200, 350

George Alexander, 27, 3, 3000, 100, 175

Thomas Cunningham, 40, 16, 3000, 150, 300

George A. Garrett, 80, 10, 4000, 200, 670

Janis Matthias, 85, 15, 8000, 350, 800

Richard Simons, 25, 6, 1200, 100, 150

Wm. Benge, 80, 55, 4000, 100, 300

Jusbers Huggins, 95, 30, 9500, 50, 840

Jonathan Strahorn, 40, 20, 3000, 250, 425

Alexander Hill, 70, 30, 6000, 400, 630

Thomas Finley, 40, 35, 3000, 150, 650

Wm. Cannon, 20, 10, 1400, 80, 200

Robert Anderson, 50, 25, 3000, 100, 150

Andrew Jackson, 40, 16, 1800, 50, 220

Henry H. Kimble, 90, 21, 5000, 400, 652

Laurence Cintman, 36, 7, 1600, 100, 250

Michael Cintman, 14, 2, 1000, 60, 600

Amos Phillips, 35, 12, 1500, 75, 200

Thos. Ferguson, 50, -, 1400, 100, 250

John Samma, 35, 25, 1000, 50, 100

James Gregg, 50, 25, 3000, 35, 300

Jesse C. Buffington, 35, 15, 2000, 75, 200

Ninrod Minnor, 90, 35, 5500, 250, 900

Augustus Cann (Carr), 23, 5, 1500, 95, 600

Louis Wright, 50, 23, 2500, 125, 600

John L. Lynch, 27, 2, 1000, 75, 300

John L. Kingston, 90, 17, 2500, 125, 550

Abraham Dewett, 60, 40, 2500, 75, 350

George Hall, 44, 18, 2000, 100, 300

Theodore Warren, 90, 20, 3000, 150, 350

George Mullin, 20, 10, 1000, 75, 350

Jacob McConsel, 25, 21, 1500, 75, 175

John Mackel, 175, 65, 16000, 200, 1200

J. A. Mackie, 100, 50, 8000, 100, 800

Andrew White, 14, 1, 1400, 75, 300

Levi Hal, 200, 60, 13000, 600, 625

Wm. Mahan, 25, 24, 1500, 75, 250

John Gregg, 75, 30, 4000, 150, 600

William Holland, 70, 15, 4000, 100, 400

James McCartney, 70, 30, 2000, 100, 300

Henry Gatchell, 60, 20, 1500, 120, 300

Rachel Mackie, 120, 60, 10000, 350, 600

Walker Armstrong, 140, 27, 9000, 500, 1150

Jonathan Yerks, 100, 25, 6500, 400, 800

Nathan Tyson, 85, 14, 5000, 200, 500

Henry Billingford, 100, 50, 7500, 400, 300

Isaac Pass, 36, 12, 2500, 125, 200

Granville McCormic, 30, 7, 1500, 150, 400

Fred Lensgrov, 27, -, 1500, 100, 150

Samuel Rogers, 129, 16, 11500, 500, 900

Wm. Armstrong, 110, 20, 9900, 1000, 11500

Michael Tolling, 75, 11, 5000, 125, 340

John Ewing, 60, -, 4000, 200, 600

Edwin Mendenall, 100, 23, 6000, 300, 400

Israel Reynolds, 160, 30, 8500, 400, 600

Wm. B. Boyles, 220, 200, 16000, 500, 1200

Samuel Clements, 50, 75, 4000, 200, 500

Jesse Irwin, 25, 16, 2000, 100, 400

Wm. Roney, 70, 26, 7000, 100, 400

Reuben Gruffy, 60, 200, 2500, 125, 300

Gabriel Moore, 30, 10, 2000, 100, 250

Stephan Gilbert, 20, 3, 2000, 30, 400

John Conly, 12, 1, 1500, 75, 150

Elisha Ewing, 10, 1, 1300, 75, 150

Martha Hughes, 16, 8, 2000, -, 100

Wm. Gibson, 14, 5, 1200, 50, 150

Hiram Clement, 40, 10, 3500, 100, 268

Robert Alexander, 60, 15, 4000, 125, 350

Amos Gilbert, 75, 18, 3000, 100, 600

F. F. Ferguson, 80, 30, 2500, 100, 315

Catharine Boyles, 120, 33, 10000, 350, 600

Wm. H. Ewing, 70, 40, 4000, 200, 450

Ledsone Boyles, 150, 150, 6000, 300, 600

Alexander Steel, 25, 5, 800, 50, 178

Samuel McCauly, 50, 10, 1000, 50, 175

Caleb Kirk, 80, 63, 5000, 150, 800

W. T. Dawson, 90, 44, 4500, 125, 400

John Smith, 50, 60, 3000, 125, 400

George Cladon, 35, 15, 3500, 200, 350

John Thompson, 13, -, 1200, 75, 135

John Morison, 65, 35, 5000, 125, 500

James H. Mackey, 15, -, 1200, 50, 35

Mathew Bourland, 20, -, 1800, 50, 450

James Mackee, 37, -, 3000, 75, 300

David Collins, 65, -, 3500, 100, 400

Tobias Peterson, 70, 7, 2500, 50, 250

Richard Dean, 25, 20, 1500, 75, 300

James Cullen, 60, 30, 4000, 100, 500

John Campbell, 75, 47, 5000, 150, 400

Wm. M. Campbell, 50, 10, 3000, 100, 350

Alexander Curry, 80, 50, 4000, 200, 500

James McCane, 70, 10, 6000, 200, 500

James L. Foard, 85, 15, 7000, 500, 600

Aaron Minglin, 35, 23, 1500, 50, 150

John Devinna, 30, 22, 1500, 100, 250

Samuel Kelvington, 50, 14, 2000, 100, 300

James McGinn, 50, 14, 1500, 75, 250

John McCartney, 30, 72, 2000, 100, 500

James Duff, 45, 10, 2500, 100, 300

Richard Murphy, 28, 5, 1600, 50, 150

Ann Holland, 50, 10, 4000, 75, 200

John Mackee (Markee), 117, 14, 6000, 150, 600

Robert Taylor, 25, 4, 1000, 50,80

Isaac Gray, 100, 30, 5200, 100, 700

John Smith, 40, 35, 3000, 100, 400

Wm. Tuft, 50, 12, 2000, 100, 320

Rebecca Henderson, 70, 26, 3000, 75, 250

Samuel Maxwell, 30, -, 1600, 100, 200

Wm. Wayde, 32, 10, 2500, 150, 350
John Criswell, 40, 10, 1800, 100, 175
Thomas Ratliff, 100, 59, 6000, 175, 500
Thomas Owens, 25, 5, 1000, 75, 350
Benj. F. Zebley, 150, 7, 10000, 350, 1600
James Donnel, 80, 58, 10000, 250, 500
B. C. Cowan, 20, -, 3000, 100, 300
Hiram Walker, 90, 28, 6000, 300, 400
John Major, 60, 4, 2500, 150, 400
Rausler Beddle, 75, 25, 3000, 200, 400
Henry Hess, 40, 33, 3000, 120, 500
Levi Disart, 40, 65, 2500, 60, 200
George Sharpless, 70, 26, 6000, 80, 600
Jos. Warren, 60, 27, 3500, 100, 600
Jos. Oliver, 62, 14, 4000, 100, 500
Amos Blake, 56, 10, 2500, 100, 400
W. G. Drenon, 50, 10, 2000, 80, 300
Wm. Kirkwood, 40, 29, 3000, 150, 300
Jos. C. Hughes, 20, 8, 2000, 75, 200
Benj. Singles, 50, 17, 4000, 125, 400
J. T. Garrett, 70, 5, 4000, 100, 400
Joseph Weldman, 48, 14, 5000, 100, 500
George Johnson, 85, 20, 8000, 200, 600
Peter Peterson, 75, 12, 3500, 100, 500
Jesse Pyle, 30, 12, 2000, 100, 350
Robert Montgomery, 60, 30, 5000, 10, 380
Ann Garet, 80, 20, 5000, 125, 500
John M.Centman, 75, 34, 4000, 200, 540
Nathaniel Flounders, 25, 5, 1000, 75, 180
George Peterson, 40, 20, 2000,75, 50
Thos. Peterson, 48, 30, 2000, 75, 460
James Crow, 20, 5, 1800, 75, 350

James Willis, 30, -, 1500, 40, 80
Joseph Steel, 28, 3, 1000, 100, 200
Thomas Fulton, 20, -, 1000, 160, 160
David Scott, 180, 48, 13800, 600, 800
Hugh Roberts, 40, 10, 2500, 100, 200
Price Bullen, 100, 60, 5000, 150, 600
Lerman Drumon, 25, 8, 2500, 100, 350
John W. Moore, 45, 8, 2500, 200, 275
Silas Carter, 70, 90, 10000, 250, 600
Robert Queen, 13, 6, 1600, 75, 250
George W. Wright, 170, 41, 1000, 500, 1150
Thos. Tongs, 40, 10, 3000, 150, 350
Jos. Smith, 25, 25, 1500, 50, 160
James Moat, 50, 30, 2500,75, 200
Joseph Miller, 61, 66, 5500, 200, 700
Stewart Galleher, 10, -, 1800, 75, 150
Daniel Lord, 400, 550, 38000, 1000, 2875
J. L. Foard, 50, 11, 4000, 100, 350
F. A. Burnibe, 30, 5, 2000, 75, 125
Rebecca A. Getty, 15, 2, 2500, 75, 200
Wm. Hasson, 70, 135, 3000, 100, 250
Tobias Peterson, 20, 15, 1500, 75, 300
John W. Egner, 75, 25, 4000, 100, 400
John Smith, 75, 6, 4000, 125, 400
Park Smith (Park & Smith), 25, 17, 2500, 50, 425
Wm. G. Powell, 160, 40, 8000, 200, 750
Robert H. Gallaher, 40, 17, 1500, 100, 140
John E. Galleher, 40, 49, 1500, 75, 200
Ann Mahan, 24, 14 800, 50, 150
Jesse Taylor, 70, 26, 3000, 100, 300

Jonathan Centman, 40, 12, 1700, 75, 200

David Scott, 60, 35, 3000, 150, 400

Joshua Gatchell, 49, -, 3000, 100, 450

Joseph Pennock, 80, 10, 2000, 100, 500

Wm. Brocan, 110, 50, 7000, 200, 600

John McCleary, 50, 5, 2500, 100, 350

Wm. Lane, 80, 30, 3500, 160, 400

Ephraim Jourdan, 30, -, 2000, 100, 100

Joseph McCullough, 45, 15, 3500, 100, 250

Wm. Meridith, 13, -, 1000, 51, 150

Amos Lois (Lewis), 50, 25, 3000, 100, 300

Mathew Galleher, 40, 20, 4000, 100, 300

Daniel Sutton, 25, 5, 1800, 100, 300

Josiah Green, 16, 1, 1200, 50, 200

T. W. Downing, 500, 600, 22000, 500, 2400

James McCown agent Deaker, 7, 7, 3500, 50, 200

John Yeoman, 12, 12, 1200, 100, 100

J. B. Graham, 10, 10, 2000, 200, 250

John Ford, 160, 130, 8500, 100, 600

H. A. Land, 50, -, 3000, 100, 400

E. W. Thomas, 100, 50, 5000, 200, 600

S. S. Thomas, 100, 600, 8000, 200, 900

Miss Ann Russler, 250, 150, 12000, 200, 300

William Seatten, 32, -, 3500, 100, 300

J. T. Scott, 50, -, 5000, 150, 300

John S. Siner (Sines), 80, -, 6000, 200, 400

George Chanden, 6, 7, 3000, 50, 300

Johnson Simpers, 50, -, 3500, 150, 400

Henry Bernett, 40, -, 1170, 120, 320

John N. Block (Black), 30, -, 2000, 150, 500

Jeremiah Watson, 50, -, 3000, 400, 400

William Grant, 25, -, 1500, 100, 400

C. C. Graham, 16, -, 2000, 150, 440

S. A. West, 40, 30, 1600, 100, 200

Edward Jackson, 160, 30, 8000, 200, 600

John Thompson, 80, 120, 4000, 100, 400

Joseph Harris, 90, 200, 9000, 200, 500

William Gibson, 120, 80, 9000, 50, 500

Victor Craig, 170, 90, 12000, 300, 800

T. W. Baker, 80, 60, 4000, 200, 500

John Jackson, 100, 80, 6000, 200, 800

Y. S. Kerby, 175, 25, 12000, 250, 1300

G. W. Barns, 270, 140, 30000, 400, 1300

George Winchester, 80, 50, 4000, 100, 400

John Pennington, 20, 60, 840, 50, 150

W. T. Owens, 50, 30, 1000, 50, 250

John Shaw, 100, 100, 6000, 250, 900

J. B. Tyson, 80, 100, 4000, 150, 500

W. M. Brown, 80, 25, 5300, 200, 860

James Lackland, 20, 6, 3000, 50, 150

Thos. S. Abrams, 120, 30, 6000, 50, 200

James Gilford, 70, -, 4000, 150, 300

John M. Gilford, 60, -, 1300, 100, 520

Benjamin Gilford, 88, 20, 2100, 150, 1000

David Cameron, 60, 8, 3500, 200, 500

Dr. J. S. Stiles, 100, 60, 8000, 200, 1000

William Crosgrove, 53, -, 750, 50, 200

T. B. Palmer, 75, -, 3700, 100, 300

William Grant, 11, -, 1000, 20, 70

Thomas Williams, 12, 3, 1000,70, 300

Hasson Lynch, 33, -, 1300, 50, 300

John Williams, 21, -, 1000, 25, 200

J. Reeder, 60, 40, 4000, 200, 500

E. Falls, 20, -, 1000, 150, 500

A. Brown, 120, -, 4000, 100, 700

R. Boucher, 66, -, 3000, 150, 800

Samuel Thompson, 30, 60, 2000, 200, 300

R. E. Oldham, 100, -, 5000, 200, 700

B. R. Lair, 85, 34, 5000, 350, 400

William Brickley, 50, 70, 3000, 250, 500

William Sedon, 52, -, 1500, 150, 350

Jacob Walton, 100, 60, 4800, 150, 400

Ely Janney, 96, -, 2800, 100, 300

John Janney Jr., 25, -, 1000, 100, 100

Thomas Riddle, 40, -, 1000, 150, 320

John Tyson, 25, -, 500, 50, 120

Thomas Tyson, 62, -, 1500, 200, 350

E. S. Lewis, 100, 40, 2800, 100, 600

Jesse Janney, 110, 50, 8000, 150, 700

William Lynch, 26, -, 1000, 50,3 50

James Trimble, 52, -, 1600, 40, 180

Stephen Armore (Armour), 23, -, 1200, 40, 200

John S. Hoster (Yoster), 40, 74, 2500, 50, 120

John Abrams, 70, -, 1500, 50, 200

William Abrams, 40, 20, 750, 10, 125

Margarett Terry, 200, 90, 4000, -, 700

E. T. Mask, 65, 30, 4000, 150, 420

James Shote, 100, -, 3000, 100, 450

William Thompson, 20, 5, 800, 10, 250

James Thompson, 65, -, 1000, 100, 200

D. H. Paul, 75, -, 750, 100, 80

E. Sentman, 17, -, 500, 50, 150

David Tucker, 56, -, 2000, 100, 250

Thomas Cooper, 50, -, 2000, 100, 250

L. H. Thompson, 50, -, 1500, 50, 100

Alfronso A. DeFigore (DeFagone), 80, 40, 8500, 150, 400

Charles Grier, 150, 50, 5000, 50, 400

David Bial, 30, 65, 2000, 20, 200

William Denison, 35, -, 500, 20, 100

Stephen Atkinson, 80, -, 800, 40, 50

William Reade, 100, 200, 4000, 150, 450

B. Frelds (Fields), 100, 60, 2000, -, 500

Wm. Benjamin, 150, 50, 4000, 100, 400

Jacob Hague, 80, 30, 5000, 150, 300

Armour Cameron, 15, 40, 1500, 50, 200

Samuel Moffit, 50, 80, 2600, 200, 400

James Russell, 80, 60, 3500, 150, 500

John Riddle, 43, -, 960, 100, 350

Henry Burns, 35, 120, 3000, 100, 300

W. Reed Jr., 26, -, 500, 40, 200

John Welsh, 30, -, 500, 40, 200

Ben Reed, 80, -, 800, 40, 300

James Furguson, 40, -, 800, 20, 300

Westly Brown, 50, 60, 3000, 50, 300

M. Milbourn, 24, 6, 1000, 50, 150

John Furguson, 44, -, 1000, 150, 500

Samuel Burns, 100, 75, 3000, 100, 400

Joseph Benjamin, 25, -, 600, 50, 200

Robert Logan, 40, 10, 400, 200, 500

James Crawford, 40, 50, 6000, 250, 700

T. T. Benjamin, 25, -, 800, 50, 170

P. A. Ricards, 140, 97, 9000, 250, 1200

J. T. Reed, 20, -, 800, 20, 150

John Johnson, 65, -, 1500, 100, 200

John Tyson, 60, 20, 4000, 200, 300
P. M. Coulter, 10, -, 1800, 50, 100
John McClan, 60, 30, 3000, 100, 300
Sarah Paul, 30, 40, 1000, 100, 200
G. S. Jeffries, 35, 125, 10000, 70, 300
D. Y. Cooly, 75, 125, 4000, 100, 300
T. V. Rose, 80, 40, 6000, 150, 900
T. Terry, 60, 60, 3600, 100, 400
W. Vandergrift, 100, 15, 3400, 100, 450
B. Walsey, 50, -, 1500, 100, 100
J. Crumy, 30, -, 800, 50, 300
Russell Thomas, 100, 150, 6000, 250, 700
B. Reynolds, 100, 100, 600, 100, 200
W. P. Howard, 200, 470, 10000, 150, 800
Edgar Moon, 75, 25, 5000, 300, 400
S. Maheny, 50, -, 2000, 100, 400
John Currey, 100, -, 3000, 50, 150
Joseph Lost, 80, 40, 2000, 200, 300
E. Atkinson, 70, -, 1400, 50, 400
A. Lambert, 50, -, 1000, 30, 300
Thomas Russell, 80, 20, 7000, 100, 600
T. R. Beddle, 70, 70, 6000, 100, 900
William B. Morris, 70, 80, 10000, 200, 800
J. W. Huss, 60, 60, 5000, 200, 400
Hugh Armour, 50, 250, 9000, 150, 150
G. W. Forde, 100, 30, 400, 200, 200
Marjery Butler, 40, 40, 4000, 100, 500
G. W. Chapell, 85, 20, 6000, 200, 800
Edward Pearch, 75, 60, 5000, 150, 400
Edward Pearch, 70, 80, 5000, 100, 600
E. P. Grace, 100, 20, 3600, 150, 640
W. H. Arance, 180, -, 9500, 400, 1100
J. Phelps, 50, 30, 4000, 150, 350
A. George, 80, 15, 3800, 150, 500

E. H. Rilay, 70, 100, 2500, 150, 300
J. Y. Collins, 250, 740, 20000, 700, 1100
J. A. Rutter, 150, 100, 700, 100, 500
John T. Veasey, 100, 100, 4000, 100, 400
Joseph McKinney, 50, 50, 1000, 50, 600
Robert Thackey, 80, 100, 5000, 100, 600
A. Pennington, 20, 60, 2000, 50, 80
Robert Hart, 65, 12, 1800, 200, 120
H. Richardson, 60, 160, 7000, 300, 300
J. F. Wilson, 100, 100, 6000, 150, 500
W. J. Wilson, 170, 100, 7000, 200, 750
Samuel White, 120, 40, 5000, 200, 700
A. J. Rancant, 100, -, 2000, 50, 200
William Thomas, 130, 80, 2000, 150, 670
J. F. Ford, 60, 30, 4000, 50, 100
T. B. McKinney, 70, 50, 4000, 100, 300
G. G. Leiper, 75, 350, 8000, 100, 300
W. R. Hyland, 8, 10, 500, 20, 120
John McCrack, 70, 130, 6000, 150, 500
J. S. Wingate, 120, 180, 5000, 200, 750
Jacob Ricard, 60, 130, 7000, 50, 600
M. Mousall, 70, 60, 2000, 50, 130
John A. Wilson, 40, 60, 2500, 200, 540
A. Hyland, 12, 2, 700, 200, 200
M. Larydear, 100, 65, 3500, 500, 800
C. B. Snuyder, 70, 50, 5000, 360, 240
B. Brock, 100, 40, 2800, 100, 300
Hall Boodough, 175, 70, 7000, 200, 200
G. F. Glading, 160, 40, 5000, 150, 650

G. F. Glading, 160, 40, 5000, 150, 650

W. Foard, 58, -, 2000, 50, 200

Wm. McMeme, 170, 90, 7800, 200, 9780

J. F. Rattemburg, 16, -, 800, 50, 100

H. Shellcross, 30, 70, 700, 20, 100

John Arison, 35, 10, 2000, 200, 1000

W. Barr, 40, 60, 1400, 100, 100

Charles James, 30, 20, 1800, 50, 350

James Goul, 30, 5, 500, 50, 30

Capt. W. Pryor, 50, 50, 2000, 100, 350

Thomas Simpers, 15, 35, 800, 50, 200

Rev. J. Merry, 170, 110, 2000, 200, 350

G. W. Merry, 10, 70, 1500, 100, 100

Thomas Lake, 30, 100, 2000, 50, 400

W. Holt, 30, 70, 100, 50, 200

J. S. Bouldon, 120, 90, 4000, 150, 800

Joseph Glading, 130, 150, 6000, 350, 700

Capt. W. Robinson, 20, 20, 800, 50, 100

E. H. Hyland 70, 10, 3000, 150, 200

John Aldridge, 100, 100, 5000, 400, 800

George R. Howard, 125, 80, 6000, 300, 780

Jefferson Purnee, -, -, -, -, 500

George Alexander, 8, -, 400, 50, 400

John R. George, 80, 150, 11500, 1000, 3000

Samuel Aldridge, 45, -, 11500, 400, 1000

William D. Tully, 50, -, 2000, 50, 100

H. Pearrey, 200, 100, 15000, 150, 12000

Isaac Lort (Lost), -, -, -, 100, 500

F. W. Lort, 110, -, 3000, 100, 600

Mrs. J. B. Aldridge, 100, 65, 5000, 500, 700

Mrs. E. Fister, 100, 70, 3000, 80, 450

R. Currey, 100, 50, 3000, 100, 400

Andrew Barrett, 35, 35, 1400, 50, 150

J. T. Alexander, 35, 65, 1500, 80, 20

Thos. Culterton, 40, 35, 800, 25, 300

J. F. Beaston, 140, 200, 12000, 200, 550

H. Geters, 35, 70, 3000, 150, 500

Richard Roach, 60, 40, 3000, 100, 300

Samuel Hitchcox, 25, 30, 4000, 100, 200

D. L. H. Adams, 60, 150, 6000, 50, 200

E. Mahoney, 45, 20, 2500, 400, 300

And. Alexander, 18, -, 1800, 20, 250

Jacob Johnson, 60, 50, 2500, 50, 200

Asariah Rittenhaus, 35, 9, 3000, 150, 300

Job Haines, 40, 2, 4000, 100, 200

James M. Evans, 90, 90, 6000, 300, 750

Edwin Haines, 100, 64, 7000, 300, 600

Jas. H. Haines, 70, 23, 4000, 250, 400

Barclay Reynolds, 60, 25, 6000, 200, 600

Jacob Reynolds, 100, 20, 8000, 200, 600

Jacob Kirk, 18, 7, 1500, 100, 200

G.(S) B. Stubbs, 75, -, 8000, 200, 600

M. J. Hunt, 130, 21, 6000, 200, 700

Jonathan Reynolds, 80, 45, 4000, 150, 675

John Lincoln, 200, 105, 10000, 300, 1100

Benj. H. Buckley, 20, 10, 1500, 80, 200

Benj. Hannah, 70, 5, 4000, 100, 600

Wm. Shelby, 25, 10, 1500, 75, 120

Henry Reynolds, 70, 14, 5000, 200, 500

Hains Reynolds, 35, 10, 1400, 100, 320

Edwin Reynolds, 50, 34, 3000, 80, 425

Wm. Wiley, 50, 50, 2000, 80, 350

Gibbons Moore, 50, 25, 2500, 125, 350

Rebecca Jackson, 30, 35, 3000, 130, 300

David Phillips, 25, 15, 2500, 100, 650

John T. Reynolds, 20, 5, 1600, 75, 300

Jacob Reynolds, 40, 15, 3500, 100, 375

Isaac Reynolds, 50, 50, 2000, 100, 800

James Wiley, 20, 10, 1000, 40, 300

Joseph Reynolds, 30, 18, 2000, 80, 175

Elis Passmore, 60, 44, 3500, 200, 800

Thomas Gardner, 100, 100, 6000, 250, 750

Jacob T. Reynolds, 45, 15, 3500, 100, 300

Robert Evans, 130, 60, 7000, 300, 700

Wm. J. Evans, 114, 100, 10000, 200, 1300

Andrew Pisver (Pierce), 50, 15, 3500, 50, 260

Samuel Torsh, 53, 3, 5000, 150, 450

James A. Coulson, 30, 10, 1500, 75, 200

John P. Coulson, 50, 9, 4000, 100, 300

Stephen Reynolds, 30, 4, 3000, 75, 300

Stephen Richards, 75, 25, 5000, 250, 500

Stephen Reynolds, 40, -, 3000, 100, 225

Reuben Reynolds, 50, 25, 5000, 175, 500

John P. Evans, 120, 80, 10000, 200, 1000

Thos. Krauss, 28, 3, 1000, 50, 175

Jacob Krauss, 33, -, 1200, 50, 200

Isaac R. Taylor, 7, -, 1400, 50, 150

Joseph Lincoln, 80, 20, 6000, 150, 450

David Mullen, 8, 8, 2500, 50, 300

Agnes Touchstone, 21, -, 1000, 50, 75

Henry Haines, 85, 35, 8000, 200, 600

Alfonces Kirk, 35, 15, 1500, 50, 350

Joseph L. Stephens, 90, 60, 6000, 175, 500

Job Gatchell, 18, 6, 1000, 75, 160

Wm. Cherry, 20, -, 1000, 75, 175

Sarah Coulson, 40, 30, 2500, 100, 200

Wilson Brown, 15, 15, 1200, 100, 200

Samuel Brown, 60, 41, 3000, 150, 350

David Gallaher, 63, 9, 1500, 100, 250

Wilson Sedwell (Ledwell), 80, 60, 4000, 150, 575

Wm. McRath, 45, 12, 2500, 125, 350

Joseph Laird, 60, 20, 2000, 100, 400

Jonathan Headley, 55, 15, 6000, 150, 350

John L. Nice, 85, 40, 7000, 500, 900

Hampton Langdon, 50, 30, 4000, 150, 300

Luke Brown, 50, 14, 4000, 200, 350

Baziel Haines, 100, 55, 8000, 300, 750

Henry J. Briscoe, 120, 20, 7000, 200, 750

A. H. Briscoe, 80, 25, 4000, 150, 600

John Armour, 20, -, 1000, 75, 150

John A. Mount, 140, 44, 11000, 500, 900

B. F. Thomas, 60, 30, 4500, 125, 300

James Quigley, 60, 40, 4000, 75, 300

Thos. Maxwell, 100, 44, 6000, 175, 500

Elizabeth Maxwell, 35, 15, 2500, 125, 250

Joakim Brickley, 85, 15, 5000, 175, 500

John D. Mitchner, 60, 20, 4000, 200, 500

Samuel Campbell, 20, 3, 1000, 75, 200

James Healdley, 60, 10, 4000, 125, 400

Andrew Brickley, 75, 15, 3500, 175, 650

Elisha Kirk, 97, 20, 3000, 75, 300

John Alison, 40, 18, 2500, 75, 250

Leonard Reynolds, 8, -, 2000, 75, 150

Benj. Tyson, 90, 40, 3000, 100, 400

John McCullough, 75, 82, 90000, 200, 600

John W. Coldwell, 34, 6, 3000, 75, 300

James Egan, 20, 6, 1600, 75, 300

Andrew Brickley, 50, 20, 2000, 100, 375

Andrew Ramsey, 220, 50, 4000, 100, 710

William McCullough, 100, 60, 6000, 100, 500

Elijah Kister, 335, 15, 2000, 125, 210

John Barnes, 55, 25, 3500, 125, 600

John Hammersmith, 100, 125, 6000, 150,700

Joseph B. Taylor, 154, 60, 9000, 350, 1280

John Wason, 40, 14, 1500, 150, 250

John Krauss, 24, 4, 1500, 100, 225

Thos. Richards, 100, 100, 6000, 150, 1355

Jacob Richards, 60, 42, 7000, 300, 675

Amos Moore, 25, 12, 2200, 30, 505

Samuel Hindman, 60, 56, 6000, 150, 625

Robert Hindman, 150, 110, 8000, 200, 750

Wm. Gillispie, 60, 50, 2000, 150, 250

Wm. Coale (Cook), 72, 10, 4000, 200, 80

Jeremiah Swisher, 60, 20, 3500, 150, 510

James Swisher, 60, 10, 2500, 100, 300

James Gerry, 100, 40, 5000, 200, 670

John Gery, 60, 20, 2000, 30, 150

John Graham, 20, 13, 1000, 60, 150

Reese P. McDowell, 18, 10, 1200, 75, 150

Stephen Woodrow, 90, 31, 3500, 100, 350

Mathew Morrison, 75, 75, 7000, 150, 400

Robert H. Nesbit, 60, 13, 2800, 150, 300

Robert Nesbit, 51, 10, 3000, 100, 350

Thomas Kennard, 50, 7, 4000, 300, 350

Alex Ewing, 16, -, 1000, 10, 125

Wm. A. Torsh, 18, 7, 1200, 35, 150

Moses Nesbit, 68, 30, 6000, 250, 625

Thos. Hopkins, 40, 11, 2000, 72, 200

John Nickel, 55, 38, 4000, 125, 400

Wm. Woodrow, 65, 31, 4000, 175, 600

Ira White, 60, 14, 4000, 200, 450

George W. Poyst, 30, 12, 2000, 200, 300

Amos Clendenan, 50, 15, 3000, 100, 350

Ambrose Ewing, 75, 23, 6000, 150, 600

W. H. Balderston, 25, 5, 3000, 140, 2000

John P. Balderston, 30, 10, 4000, 25, 250

Ellis P. Brinton, 50, 21, 4500, 100, 510

Lebetic Jackson, 80, 10, 6000, 150, 500

John Moore, 100, 20, 6000, 150, 600

Peter Brady, 35, 11, 3000, 100, 160

Wm. Waring, 65, 30, 4300, 50, 400
Thos. Hay, 140, 22, 8000, 250, 500
G. W. Moore, 100, 30, 3000, 150, 600
Wm. W. Moore, 200, 34, 11000, 500, 1000
John Brown, 20, 2, 2000, 75, 125
James Torsh, 80, 35, 4000, 200, 600
Sarah Howell, 40, 47, 2000, 50, 150
Loyd Balderston, 80, 30, 7000, 600, 1000
John More, 15, -, 760, 30, 100
James S. Nickell, 100, 25, 8000, 100, 300
William Nickell, 50, 16, 3000, 100, 400
Wm. P. Coulson, 70, 60, 3000, 150, 500
J. W. Buckley, 50, 40, 3000, 150, 400
Leonard Krauss, 50, 20, 2000, 100, 200
Jacob Slicer, 45, 18, 3000, 150, 200
James McCullough, 36, -, 2500, 150, 250
Anthony Thompson, 22, -, 1500, 50, 250
John Taylor, 44, 4, 2500, 100, 450
Thos. J. Gillespie, 120, 55, 7000, 175, 400
John Cameron, 80, 80, 4000, 150, 400
Moses K. Purson, 60, 8, 2000, 75, 450
William Wilds, 15, 5, 2500, 75, 175
Robert Cother, 100, 20, 4000, 150, 600
Alen Kirk, 45, 31, 5000, 100, 350
James Cameron, 50, 15, 2000, 160, 300
George Gillispie, 70, 30, 2000, 160, 300
George Gillispie, 45, 10, 1500, 75, 150
John Kultey, 15, -, 1000, 50, 125

Benjamin Wilson, 100, 90, 5000, 200, 600
Samuel Cooper, 38, -, 3000, 100, 250
Saml. Hay, 60, 15, 3000, 150, 375
Agnes Brickley, 80, 25, 8000, 300, 450
Roda Barnes, 110, 30, 6000, 250, 800
John J. Fasher, 22, 5, 1200, 100, 150
Saml. Fasher, 20, 6, 1000, 75, 150
Henry Alen, 40, 6, 2500, 75, 200
William McCullough, 50, 26, 4500, 100, 400
James Boyd, 35, 25, 2000, 75, 200
James Beard, 20, 2, 1600, 75, 150
Arvin Orr, 80, 20, 4000, 100, 600
Dugirth Hall, 50, 10, 4000, 100, 400
Daniel Clendenon, 70, 6, 4500, 150, 400
James Ewing, 25, -, 2500, 75, 150
Wm. P. Ewing, 25, -, 2500, 100, 300
Elijah Haines, 28, 7, 1000, 35, 165
Elijah Walker, 95, 13, 3500, 75, 300
William Kirk, 100, 30, 3500, 300, 70
Robert Williams, 25, 5, 1000, 100, 250
John Keelhultz, 40, -, 3000, 100, 400
James Nickel, 75, 25, 4000, 150, 450
William Little, 60, 40, 6000, 300, 600
Catharine Nickel, 70, 30, 5000, 250, 400
Samuel Mathews, 127, 25, 8000, 2000, 975
Aaron Mitchner, 42, 8, 4000, 300, 350
David Nesbit, 66, 31, 2800, 40, 250
Joseph Golebart, 135, 30, 10000, 400, 1150
James McCoy, 157, 50, 12000, 300, 5500
Theodore Marshall, 50, 2, 2000, 100, 600
Townsend Brown, 60, 7, 3000, 150, 500

Jonathan Nesbit, 26, 10, 1500, 100, 350

John L. Price, 70, 38, 6000, 150, 350

James O. McCormer, 100, 60, 12000, 400, -

Nevin W. McCormer, 70, 10, 7000, 350, 675

Robert Thompson, 80, 24, 3000, 150, 400

Samuel Harris, 40, 27, 2000, 75, 250

John Burlin, 120, 120, 5000, 100, 500

Jas. Ramsey, 80, 20, 3000, 100, 800

Saml. Gayty, 21, 9, 2000, 100, 400

Amos Ewing, 240, 60, 180000, 600, 2010

Sylvester Coulson, 35, 35, 3000, 125, 350

Christopher Ward, 200, 100, 15000, 250, 500

J. J. Heckart, 40, 8, 8500, 200, 500

Henry Smithson, 130, 47, 8000, 125, 450

John Nibblock, 120, 40, 2000, 200, 800

Andrew Lyon, 90, 10, 8000, 250, 750

Sheldon Rutter, 40, 10, 2000, 120, 150

Robert Trump, 18, 15, 3000, 75, 400

James Todd, 40, 55, 2000, 100, 250

James Gillispie, 335, 15, 2000, 125, 300

Jonathan Willet, 30, -, 3000, 150, 500

Hugh Wright, 20, 20, 900, 100, 200

Edward Glacken, 50, 12, 3000, 200, 450

William Rowland, 70, 70, 8000, 200, 620

David Craig, 70, 30, 4000, 120, 400

Edward Russell, 70, 37, 4000, 150, 500

James McCauley, 200, 30, 8000, 150, 1000

Jacob B. Thomas, 55, 25, 2500, 200, 325

Summerfield Kidd, 80, 20, 3500, 180, 400

James McCoy, 55, 7, 2000, 75, 250

John B. McCoy, 70, 20, 2500, 125, 300

Madison Rowland, 50, 14, 300, 25, 130

Wm. J. Stebbins, 12, -, 1000, 40, 225

Granville Brown, 64, -, 2000, 100, 500

Joshua McKeaver, 100, 20, 3500, 250, 400

Elisha Harris, 66, -, 1000, 100, 200

Joseph Stebbings, 12, -, 1700, 75, 150

James Frist, 100, 70, 9000, 250, 800

Robert Hanes, 15, 1, 1500, 75, 75

James Way, 130, 20, 8000 250, 125

Richard Griffy, 25, 10, 1000, 75, 150

Wm. Tamerry, 25, 13, 1200, 75, 250

David Rea, 120, 30, 12000, 500, 1000

James Barns, 40, 12, 2000, 150, 300

Edward Kelly, 50, 20, 2000, 75, 275

James Toulson, 30, 8, 2500, 125, 200

David Archibald, 25, 12, 1500, 100, 200

John Linton, 58, -, 3000, 150, 300

R. E. Brownwell, 50, 17, 3000, 100, 450

F. M. Rawlins, 100, 17, 6000, 200, 450

Robert Montgomery, 70, 20, 8000, 200, 500

William Griffy, 50, 40, 9500, 100, 600

William Hanes, 24, 16, 2000, 75, 300

Samuel Hall, 18, 20, 1000, 75, 200

Corbin Cooley, 100, 25, 7000, 300, 1000

Cathrine Broughter, 150, 50, 8000, 500, 600

Frances Clark, 50, 87, 6000, 100, 250

Alexander Boyd, 56, 40, 3500, 125, 500

William Patton, 20, 68, 14000, 500, 600

James Evans, 135, 55, 14000, 500, 1100

Hugh Steele, 75, 37, 7000, 500, 1000

George Hines, 40, 3, 2000, 100, 200

Peter Tome, 100, 20, 5000, 150, 500

John Starrett, 70, 54, 7000, 175, 700

John Carson, 80, 4, 5000, 250, 700

Edmond T. Brown, 100, 70, 7000, 200, 600

John B. Campbell, 30, 40, 1000, 100, 300

John B. Cambell, 10, 4, 1000, 100, 200

William McMullin, 90, 44, 3000, 200, 500

David Jermess, 50, 8, 2000, 100, 500

Samuel Jermess (Jenness), 50, 8, 2000, 150, 400

John Campbell, 20, 25, 1500, 100, 300

Rebecca Williams, 65, 35, 4000, 200, 400

John Weir, 50, 90, 4000, 75, 300

Samuel Hines, 70, 20, 2000, 100, 400

Elisha Brown, 50, 15, 3000, 200, 600

John Steele, 100, 60, 14000, 400, 700

E. D. McClenehan, 135, 8, 13000, 500, 900

John Moon, 1, -, 1500, 150, 200

John Patton, 150, 50, 15000, 500, 600

James M. Gay, 60, 46, 4000, 250, 450

John Alexander, 10, 20, 2000, 100, 200

John S. Everist, 47, -, 4500, 150, 450

Jefferson Ramsey, 160, 20, 8000, 300, 900

Samuel Gay, 40, 20, 2000, 100, 200

Edmond Physic, 12, -, 2000, 100, 250

John McMullin, 26, 41, 1500, 125, 200

Thomas McMullin, 22, 8, 1000, 100, 200

Samuel Whitaker, 4, 19, 600, 75, 175

John Craig, 8, 12, 600, 75, 200

Jacob Dean, 100, 62, 8000, 200, 600

John Sheridan, 35, 40, 800, 50, 160

John T. Rutter, 80, 40, 400, 100, 300

Francis Boyd, 80, 30, 2500, 75, 200

Calender Patterson, 12, 1, 2000, 75, 225

Joel Reynolds, 16, -, 1500, 75, 200

George Thompson, 60, 50, 4500, 100, 350

Eli Cosgrove, 75, 52, 4500, 150, 790

Elizabeth Ryon, 50, 20, 4000, 75, 200

Joel Ryon, 30, 10, 800, 50, 200

John Gorrell, 20, 10, 800, 176, 300

Jonathan Corier, 30, 21, 1000, 100, 225

James McGonegal, 25, 10, 1200, 100, 200

William Rutter, 30, 20, 1000, 100, 175

Alexander Jackson, 18, -, 1000, 75, 200

Absolum Jackson, 20, 2, 1000, 75, 250

James Tucker, 12, 1, 1000, 75, 250

John Russell, 13, 1200, 150, 100

Edward Jackson, 16, -, 1400, 75, 250

Richard Rutter, 22, -, 1500, 57, 150

William Ward, 180, 60, 6000, 200, 500

Thomas Lamar, 60, 10, 4000, 125, 450

William Wilson, 150, 50, 8000, 500, 1000

Samuel Wilson, 150, 20, 8000, 500, 800

John Maulden, 150, 50, 9000, 500, 900

Joseph Couden, 100, 50, 7000, 500,700

Henry S. Couden, 225, 87, 12000, 500, 1800

Thomas Watson, 20, 4, 800, 75, 150

Samuel Aiken, 16, -, 700, 25, 75

Henry Jackson, 300, 100, 15000, 500, 1500

John Branfield, 160, 40, 8000, 400, 1000

John Stump, 200, 100, 20000, 500, 1500

W. W. Black, 100, 125, 4000, 300, 600

George P. Whitaker, 300, 754, 10000, 200, 2175

E. M. Seeley, 150, 50, 16000, 500, 1000

James McMullen, 40, 23, 2000, 200, 900

John Aiken, 100, 20, 3500, 200, 400

William Patterson, 40, 50, 1000, 50, 300

William Dennison, 25, 27, 2000, 100, 350

Samuel Chamberlin, 70, 56, 4000, 100, 300

Thomas Smith, 12, -, 500, 50, 200

George Davidson, 200, 160, 9000, 180, 650

Robert Kerr, 100, 37, 3500, 160, 600

William Craig, 100, 40, 4000, 100, 700

Henry L. Physic, 60, 24, 5000, 100, 300

John T. Smith, 100, 200, 5000, 200, 450

James Quigley, 100, 200, 5000, 200, 400

Wiliam Paca, 100, 40, 2000, 100, 500

Henry Chamberlin, 180, 20, 8000, 150, 1000

John H. Harlan, 100, 100, 8000, 1000, 500

David Brown, 65, 32, 400, 175, 665

Joseph Preston, 140, 75, 12000, 800, 900

John Derixon, 100, 100, 6000, 400, 350

Benj. Brown, 40, 27, 3000, 100, 200

Jacob Reynolds, 60, 40, 3000, 200, 450

Wm. Keittry, 35, 15, 1500, 100, 350

Wm. Keittey, 40, 40, 1000, 80, 275

Jacob McVey, 25, 25, 1000, 80, 150

Wm. G. Moore, 20, 75, 1800, 100, 300

Wm. Richards, 100, 43, 7000, 250, 500

John Peters, 110, 30, 7000, 500, 600

John McKinney, 60, 38, 4000, 200, 575

Frederick McNamee, 55, 15, 2500, 160, 350

David Harlan, 100, 100, 2000, 50, 300

John A. Thompson, 100, 20, 8000, 200, 615

James A. Hannah, 60, 30, 1000, 125, 250

Washington Hill, 26, 4, 500, 50, 300

Stephen Hannah, 70, 10, 1000, 50, 500

Alphoneus Crothers, 100, 20, 4000, 50, 425

Wm. Hathaway, 16, -, 1000, 75, 175

John Fulton, 42, 15, 1500, 100, 390

George P. Stewart, 60, 20, 3000, 150, 450

Isaac Hill, 15, 35, 1500, 25, 260

Chalkly B. Cutter, 100, 80, 5000, 500, 800

John L. McGuigan, 47, 7, 2000, 100, 275

Thos. Grubb, 280, 130, 12000, 600, 1200

Absolom Romer, 115, 59, 6000, 500, 1385

Thos. Fulton, 25, 17, 1000, 75, 175

Reuben Alexander, 17, -, 2000, 50, 200

Alexander Fulton, 50, 27, 200, 50, 250

Thos. Fulton, 25, 15, 1000, 75, 300

John H. Wilson, 150, 75, 4000, 150, 750

George Brown, 35, 25, 1000, 100, 250

Jonas Preston, 100, 75, 7000, 250, 670

Francis P. Smith, 75, 65, 5000, 200, 915

Wm. Gillespie, 100, 115, 4000, 200, 680

Amos Preston, 95, 38, 4000, 200, 560

Robert Rawlins, 95, 55, 4000, 175, 650

Geo. Gillespie, 100, 40, 6000, 250, 885

Thos. A. Hanshaw, 60, 30, 4000, 200, 500

Patrick Ewing, 170, 69, 16800, 500, 1525

Joseph Peoples, 80, 34, 4000, 250, 600

Benj. McVey, 25, 5, 1200, 175, 400

Jacob B. Kennard, 10, 10, 2500, 175, 237

Jno. C. Way, 90, 27, 6500, 500, 925

R. D. Hall, 120, 30, 7000, 280, 550

Charles Hall, 125, 55, 10000, 250, 600

John Rowland, 60, 55, 6800, 150, 500

Theodore Barrett, 75, 35, 8000, 250, 400

Charles Boddy, 20, 40, 800, 75, 150

John Sheur, 100, 70, 7000, 200, 250

Wm. J. West, 100, 12, 6000, 400, 850

Jas. Blackburn, 80, 30, 4000, 180, 300

Lewis C. Martindale, 120, 81, 16000, 800, 835

Andrew Nickel, 50, 100, 4000, 200, 700

Samuel Gillespie, 75, 76, 4000, 250, 650

Patrick Boyle, 40, 82, 2000, 20, 180

Elihu Gillespie, 8, 4, 1000, 75, 155

James Smith, 65, 46, 5000, 200, 1000

Everson McGraw, 20, 40, 1500, 60, 125

David M. Shearer, 16, 20, 800, 60, 400

Wm. Warberton, 100, 340, 8800, 300, 1000

J. T. Coothers, 40, 53, 2000, 125, 200

E. W. Mahoney, 30, -, 90, 100, 300

Hugh Cameron, 80, 40, 6000, 200, 800

Benonia Nowland, 40, 28, 2000, 100, 150

Wm. H. Reader, 75, 112, 3750, 150, 300

James Hall, 17, -, 750, 50, 150

George W. Janney, 60, 20, 2400, 50, 250

John Nowland, 50, 40, 2000, 50, 200

Wm. P. C. Stroud, 125, 87, 6000, 225, 440

James Steel, 40, 23, 2000, 25, 775

David Moon, 30, 70, 1000, 40, 350

Saml. Hadock, 60, 27, 3000, 150, 400

Saml. Brown, 40, 20, 2000, 100, 300

Isad Anderson, 40, 8, 2200, 125, 370

Saml. Mearns, 100, 60, 6000, 500, 780

Andrew F. Mearns, 50, 42, 2700, 200, 550

Absolum McVey, 50, 31, 5000, 250, 375

Geo. W. McVey, 20, 10, 1200, 50, 130

George W. Smith, 44, 14, 3500, 200, 440

Andrew M. Cameron, 35, 25, 3500, 200, 415

Mary Mearns, 90, 40, 3500, 75, 220

Jonathan Harigan, 13, 2, 600, 10, 600

Jacob Woodrow, 80, 45, 3000, 75, 225

James Brittian, 70, 12, 3000, 125, 340

James K. McVey, 40, 10, 1600, 50, 250

James Mearns, 100, 52, 5000, 400, 550

John B. Cathers, 35, -, 2000, 100, 230

David Conley, 60, 34, 3000, 300, 700

Montteleon Brown, 138, 10, 5000, 200, 90

Hannah Kirk, 40, 20, 2000, 10, 125

Elisha E. Kirk, 40, 15, 1600, 500, 250

John Rogers, 60, 45, 2800, 200, 400

Israil Rogers, 70, 36, 2000, 140, 600

Ambrose Holcomb, 75, 15, 3000, 50, 400

Jacob Bonham, 22, 8, 800, 50, 100

Henry Long, 65, 10, 3800, 50, 350

Jacob Jobe, 40, 70, 2500, 150, 400

John P. Smith, 80, 14, 4000, 300, 485

Ambrose Owens, 60, 40, 4000, 300, 325

Cornelius Barret, 70, 25, 3500, 150, 350

Ruth Shearer, 13, 9, 600, 20, 40

Edwin J. Brown, 43, 18, 2500, 150, 350

Wm. Gruserd (Grussan), 60, 40, 4000, 200, 600

John Algard, 40, 20, 1500, 80, 250

Jos. Moon, 40, 25, 3000, 150, 300

Milton Wilkinson, 14, -, 700, 40, 150

Wm. Williams, 14, -, 1200, 40, 200

Edward Yerkes, 16, -, 1000, 50, 150

Andrew Yerkes, 34, 5, 2500, 200, 250

Robert McMullin, 46, 10, 2000, 200, 400

James McVey, 25, 15, 1200, 20, 100

Nathan Griffith, 24, 4, 1400, 50, 150

Jos. Pierson, 40, 35, 2000, 150, 350

Jos. S. Rile, 13, 4, 4000, 50, 150

Daniel Brown, 50, 5, 3500, 50, 200

Jacob Tyson, 9, 32, 800, 20, 150

Abner Rogers, 25, 55, 1500, 80, 250

Wm. T. Brown, 60, 20, 2500, 150, 500

Jesse Tyson, 20, 12, 850, 75, 150

John R. Abrams, 55, 10, 3500, 250, 600

George Chambers, 24, 4, 60, 25, 80

Jos. Tyson, 30, 10, 1800, 75, 350

Thos. Rogers, 25, 10, 1000, 75, 250

Evan Reisler (Rusler), 100, 60, 3300, 150, 600

John Dawson, 16, 8, 1000, 60, 250

Clemson Brown, 100, 36, 3300, 100, 400

David Wherry, 25, -, 1600, 25, 125

Elizabeth Hoops, 76, 30, 5000, 100, 175

Saml. Smith, 50, -, 4500, 350, 400

Samuel Lytte, 110, 30, 5000, 500, 600

Wm. Glenn, 145, 30, 7000, 500, 1500

Henry Rile (Rill), 47, 24, 3500, 250, 450

Otho Nowland, 40, 30, 2000, 50, 300

Wm. Cameron, 75, 125, 5000, 250, 600

Christian Reader, 80, 40, 3800, 150, 400

Isaac F. Vanarsdale, 45, -, 2600, 200, 300

John Carhart, 200, 85, 10000, 400, 600

Ducit Charles Carhart, 80, 15, 3000, 150, 565

Jos. Cameron, 35, 10, 2000, 150, 275

Sutton Scarborough, 80, 10, 5000, 800, 700

Howard Scarborough, 40, 20, 2800, 150, 400
A. R. Smith, 100, 22, 4500, 300, 700
Carey Isaacs, 22, 3, 1200, 75, 110
G. W. Oldham, 68, 7, 5000, 200, 600
John W. Oldham, 35, 40, 2500, 100, 125
Elen Martindale, 90, 50, 6500, 400, 600
Mary Gale, 48, 37, 1600, 25, 20
James R. Rainer, 70, 100, 300, 80, 250
Thos. Janney, 70, 38, 3000, 180, 500
John C. Murry, 65, 45, 3500, 250, 800
John T. Janney, 11, -, 600, 15, 100
Saml. Harris, 40, 40, 2000, 35, 325
Jonathan Burns, 50, 30, 2000, 200, 150
John Gambol, 20, 36, 2000, 25, 250
Robert Tromble, 35, 30, 1800, 150, 325
Levi Johnson, 225, 95, 14000, 300, 1600
Wm. F. Buckland, 20, 40, 1800, 80, 375
James Crothers, 130, 30, 7000, 150, 700
Charles Matthews, 60, 39, 8000, 150, 750
Isaac England, 70, 30, 5000, 200, 250
Jos. England, 22, 11, 1500, 100, 225
Enoch Johnson, 60, 20, 3000, 300, 450
John Kirk, 100, 30, 4000, 250, 450
Saml. Slicer, 46, 39, 300, 150, 425
Hugh Russell, 25, 9, 1800, 150, 125
John Slicer, 60, 14, 2000, 200, 350
Wm. Haines, 150, 50, 15000, 1000, 2000
Horris Deykins, 100, 40, 6000, 250, 625
Mark Brown, 95, 8, 5500, 350, 750
Wm. Kirk, 20, 20, 800, 100, 180
John England, 40, 20, 2000, 100, 500

Charles Gamble, 140, 40, 4000, 500, 800
Milton White, 50, 30, 4000, 75, 200
James Turner, 30, 14, 4000, 75, 300
Elisha J. Brown, 50, 22, 2000, 80, 275
Ann Kirk, 25, 28, 3000, -, 75
Owen Reisler (Rusler), 100, 105, 6000, 400, 1275
Jos. Hopkins, 35, 5, 3000, 200, 300
Reuben Kirk, 100, 75, 5000, 350, 700
Jos. Hamilton, 160, 100, 6500, 300, 1235
Abner Kirk, 30, 23, 3000, 125, 275
Amassa Churchman, 50, 25, 2500, 200, 550
Saml. England, 97, 70, 5000, 350, 600
Edwin J. Sheppard, 58, 10, 2000, 100, 240
Charles Barker, 113, 10, 5000, 350, 550
John Marshal, 70, 41, 5000, 425, 200
A. B. Newton, 40, 60, 3500, 200, 170
Jos. B. Carhart, 130, 50, 5000, 300, 550
Lorie Mearns, 75, 15, 2000, 250, 550
Elisha H. England, 25, 25, 2000, 150, 250
Thomas P. Marshon, 75, 30, 4000, 400, 700
Onias C. Marshon, 60, 20, 5000, 200, 500
John Paschall, 125, 12, 8200, 250, 620
Emsor Touchstone, 12, -, 1000, 50, 100
Wm. Cameron, 100, 125, 7500, 300, 650
Jos. England, 50, 7, 2000, 200, 250
John Dunahoo, 15, 25, 1500, 75, 100
Zimri Taylor, 30, 10, 2500, 100, 165
Anthony Krauss, 22, -, 800, 100, 160
Wilson Marshall, 50, 20, 3000, 200, 350

Jos. Sidwell, 65, 80, 4000, 100, 500
Saml. D. Wilson, 65, 9, 2500, 150, 400
Isaac Bitte, 40, 40, 1200, 75, 200
Joseph Lechler, 130, 25, 5000, 100, 650

Job Kirk, 50, 82, 3000, 150, 580
Eli Hulford, 150, 20, 10000, 200, 500
Jos. Martindale, 100, 60, 8000, 500, 1000

Charles County Maryland
1860 Agricultural Census

The University of North Carolina Library under a grant from the National Science Foundation microfilmed agricultural Census records. Records were filmed at the University of North Carolina from original records at the Maryland State Library.

Columns 1, 2, 3, 4, 5, and 13 represent the following information on the census:
1. Name of Owner, Agent or Manager of Farm
2. Acres of Improved Land
3. Acres of Unimproved Land
4. Cash Value of the Farm
5. Value of Farming Implements and Machinery
13. Value of Livestock

C. Lancaster, 175, 25, 6000, 80, 400
Ignatius Lancster, 220, 100, 6000, 150, 800
John Henersly, 200, 300, 13000, 150, 600
William Matthews, 200, 300, 18000, 300, 1500
J. M. Williams, 180, 18, 5000, 100, 500
A. B. Simms, 740, 260, 25000, 500, 2000
C. F. Lancaster, 120, 85, 2900, 100, 600
M. Lancaster, 25, 75, 5000, -, 150
Rebecca Barber, 750, 20, 2000, -, 4
William A. Wills, 100, 48, 4000, 200, 400
J. F. Wills, 120, 80, 3000, 100, 1000
F. Matthews, 200, 184, 10000, 300, 800
A. E. Ramsey, 160, 30, 7000, 300, 800
Z. Bailey, 200, 100, 9000, 75, 225
R. Thomas, 15, 20, 1500, 150, 300
James Hayden, 100, 50, 4000, 100, 500
_. R. Thompkins, 100, 75, 8000, 75, 300
J. F. Hayden, 100, 100, 2000, 100, 500
J. E. Hayden, 120, 20, 6000, 50, 500

_. F. Simms, 250, 78, 8000, 100, 500
R. Bateman, 60, 30, 1500, 20, 600
A. Bailey, 90, 92, 1500, -, 100
H. _. Perry, 25, 15, 600, 50, 150
Samuel Smoot, 25, 12, 800, 50, 250
T. Posey, 222, 62, 5000, 125, 1000
William A. Maddox, 137, 36, 5000, 500, 475
_. A. Bourroughs, 300, 216, 9000, 250, 800
A. Dyson, 4, 6, 500, 100, 600
John T. Stadtart, 1050, 720, 45000, 300, 220
J. A. Ching, 60, 40, 1500, 75, 400
V. Turner, 250, 150, 8000, 150, 500
A. J. Smoot, 300, 70, 8000, 100, 1000
J. H. Dutton, 15, 5, 1000, 50, 166
_. H. Morgan, 200, 100, 10000, 600, 1975
H. Barber, 250, 90, 10000, 50, 650
J. P. Feson, 160, 45, 4000, 80, 300
M. Robertson, 200, 100, 7000, 250, 375
John H. Wingat, 200, 70, 7000, 100, 800
E. Shoeburn, 150, 50, 6000, 100, 700
Nally Dutton, 430, 230, 15000, 700, 1200
T. A. Jones, 300, 240, 8000, 200, 100

W. R. Rems, 100, 48, 500, 200, 650
R. Johnson, 120, 75, 1600, 50, 400
W. H. Higgs, 12, 100, 2000, -, 300
J. Calton, 150, 102, 8000, 100, 450
Benjamin Jamison, 500, 400, 20000, 300, 1725
J. D. Gardiner, 300, 200, 8000, 100, 1320
L. Lancaster, 150, 450, 3400, 60, 500
C. Lancaster, 150, 600, 5000, 100, 500
B. S. Higdon, 700, 500, 19000, 300, 3500
T. Hurbert, 400, 150, 15000, 300, 2000
T. L. Budd, 400, 200, 11000, 250, 1500
J. A. Bosey, 300, 200, 12000, 200, 600
W. A. Lyons, 300, 250, 10000, 85, 700
J. W. Burch, 68, 54, 7500, 100, 645
J. Crismond, 900, 275, 25000, 350, 1000
A. Simpson, 215, 150, 8000, 100, 500
L. Lancaster, 250, 100, 10000, 150, 150
L. E. Edelen, 200, 100, 8000, 75, 700
R. M. Smoot, 300, 150, 7000, 200, 375
R. H. Boarman, 800, 400, 30000, 100, 200
R. A. Miles, 175, 55, 4000, 175, 375
J. W. Boarman, 250, 250, 6000, 100, 500
S. E. Simpson, 200, 100, 5000, 75, 375
E. Edelen, 175, 75, 3000, 75, 300
B. Edeln, 200, 80, 4000, 320, 600
T. T. Latimer, 100, 100, 5000, 100, 750
J. M. Latimer, 500, 270, 14500, 300, 1000
J. S. Boarman, 146, 40, 4000, 100, 1000

E. S. Gardiner, 550, 350, 15000, 500, 1900
G. A. Hunt, 150, 60, 5000, 50, 800
F. L. Boarman, 250, 150, 5000, 50, 400
J. Ware, 300, 340, 10000, 250, 1900
S. Ferrell, 200, 140, 7000, 150, 1500
E. Robertson, 300, 350, 9000, 75, 300
J. W. Smoot, 400, 200, 10000, 300, 1500
E. A. Middleton, 254, 130, 12000, 200, 1000
D. Middleton, 100, 100, 1000, 10, 500
J. J. Hughes, 300, 200, 12000, 150, 1500
Mary Crain, 120, 20, 2800, 30, 300
L. Posey, 300, 60, 8000, 250, 1150
W. A. Posey, 320, 90, 17000, 350, 1200
F. L. Mills, 58, 58, 500, 50, 250
C. Shaw, 350, 150, 15000, 500, 600
H. Oliver, 100, 50, 1500, 50, 100
H. W. Wathen, 200, 150, 3000, 50, 800
C. Bailey, 60, 40, 5000, 30, 250
J. H. Wathen, 60, 40, 2000, 50, 250
J. H. Kinnehan, 30, 20, 1500, 50, 500
S. Godwin, 600, 400, 18000, 300, 800
T. B. Turner, 220, 200, 4500, 250, 620
J. B. Lyons, 300, 241, 5500, 250, 870
H. B. Swan, 75, 55, 1200, 30, 400
T. H. Burch, 100, 150, 3000, 50, 650
H. Simpson, 60, 190, 1500, -, 150
K. Cartwright, 130, 60, 100, 30, 150
J. O. Maddox, 93, 50, 3000, 150, 1000
J. B. Maddox, 100, 100, 4000, 50,700
G. Dugger (Digges), 300, 150, 10000, 50, 950
S. Cox, 700, 600, 20000, 500, 5250

Wm. Nevett, 320, 50, 1400, 200, 2000

T. R. Robertson, 200, 200, 4500, 150, 600

T. C. Howard, 400, 250, 10000, 20, 500

F. H. Edelin, 230, 100, 2000, 75,600

Wm. Gough, 350, 200, 150000, 200, 1720

J. A. Keach, 250, 150, 8000, 150, 1360

C. A. Bower, 150, 150, 4000,75, 500

T. F. Darnell, 200, 100, 6000, 65, 700

M. A. Dyson, 300, 100, 6000, 50, 450

S. T. Swane, 300, 100, 6000, 100, 1200

W. H. Freeman, 80, 80, 1800, 40, 300

T. O. Bean, 500, 500, 15000, 50, 250

F. Wills, 600, 460, 25000, 500, 3000

Mathias Cooksy, 75, 75, 1500,100, 350

T. J. Ward, 125, 50, 2000, 50, 300

G. C. Swan, 50, 86, 1360, 5, 280

Robert Monod, 25, 300, 2000, 10, 75

Dr. S. Derett (Dentt), 100, 50, 2500, 150, 500

B. T. Tenison, 130, 13, 3000, 25, 500

Wm. Thompson, 150, 150, 1500, 25, 60

J. F. Swan, 60, 40, 500, 30, 250

Peter Roby, 100, 100, 3500, 15, 200

J. P. Marshall, 125, 80, 3000, 800, 900

C. Martin, 200, 100, 3000, 50, 375

Thomas Simpson, 100, 100, 3000, 50, 36

F. M. Lancaster, 150, 100, 8000, 50, 650

Rebecca Boteman, 140, 50, 4000, 20, 275

C. Hamilton, 300, 200, 15000, 200, 1600

G. Dent, 600, 400, 15000, 500, 1160

J. Harrison, 250, 250, 5000, 300, 1075

J. W. Freeman, 200, 700, 8000, 50, 1000

S. Swann, 150, 300, 1500, 100, 1200

C. Tarner, 120, 120, 1000, 75, 500

H. Good, 70, 100, 800, 75, 400

T. Rencher, 45, 50, 600, 50, 75

R. St. Clair, 100, 200, 3000, 100, 800

C. Stincock, 75, 100, 2000, 100, 800

C. Johnson, 60, 20, 3000, 100, 600

J. H. Simpson, 60, 20, 2000, 20, 500

G. Simpson, 900, 150, 6000, 300, 1300

R. Lloyd, 350, 150, 12000, 5000, 2500

F. H. Digges, 400, 250, 18000, 400, 2000

T. Stone, 200, 40, 8000, 500, 2000

E. Edelin, 300, 350, 20000, 2000, 3300

Benj. Jamison, 350, 50, 8000, 150, 1500

R. Watson, 250, 150, 14000, 300, 1500

E. Marshal, 100, 100, 8000, 150, 600

H. R. Slanis, 700, 400, 35000, 2000, 200

J. R. Reeder, 8, 20, 1000, 100, 400

F. Weems, 75, 20, 2000, 300, 800

M. Hawkins, 350, 120, 12000, 500, 1000

H. Furguson, 300, 160, 20000, 500, 1500

R. Furguson, 400, 400, 20000, 300, 2000

R. Digges, 400, 50, 10000, 300, 2000

A. Willi, 400, 400, 18000, 300, 950

Keech, 100, 50, 1500, 300, 120

Wm. Bateman, 150, 125, 8000, 200, 1200

J. G. Harris, 150, 150, 8000, 400, 120

E. J. Harris, 250, 200, 15000, 500, 2000

G. W. Hunbertford, 400, 100, 12000, 400, 1800

J. H. Borroughes, 375, 170, 16000, 400, 1600

J. Canter, 200, 400, 15000, 400, 1900

J. Todd, 40, 10, 3000, 200, 700

E. Floyd, 200, 130, 10000, 300, 800

L. Floyd, 175, 100, 6000, 200, 800

E. Baily, 75, 75, 3000, 200, 500

C. Pery, 100, 50, 4000, 150, 800

R. McClain, 175, 50, 5000, 200, 800

Dr. R. Crain, 25, 20, 3000, 50, 400

Thomas Wood, 100, 250, 6000, 150, 800

J. H. Neal, 350, 150, 16000, 1500, 2200

H. Neal, 200, 350, 12500, 200, 1500

A. Boarman, 150, 70, 12500, 200, 1200

L. Ferrell, 300, 70, 8000, 300, 1500

Osmal Swan, 10, 40, 400, 20, 150

Henry Lucas, 200, 243, 6000, 30, 700

Edward Boone, 120, 180, 3500, 90, 816

James Turner, 120, 180, 2000, 120, 650

Smith Butler, -, -, -, 25, 250

Richard Harlin, 40, 46, 500, 12, 90

Augustus Langley, 100, 40, 1200, 25, 530

Charles Moran, -, -, -, 5, 250

Leonard Roby, 100, 20, 1800, 25, 324

Benedict Montgomery, -, -, -, 20, 260

William Bowman, 212, 232, 4000, 40, 450

Joseph Thompson, 320, 167, 5000, 30, 450

Leonard Edelin, -, -, -, 50, 190

Munce McDaniel, 125, 100, 1200, 30, 200

Stanislaus Tucker, -, -, -, 40, 400

John Proctor, 100, 25, 600, 30, 350

Saulsbury Covington, 15, 35, 300, 30, 120

Mary Thomas, 200, 300, 3000, 40, 555

Martha Cooksey, -, -, -, 25, 225

Elizabeth Wright, 3, 15, 75, 6, 120

Caleb Rawlings, 52, 10, 350, 20, 230

Joshua Naylor, 20, 10, 300, 20, 215

Thomas Smith, 60, 24, 450, 5, 195

Joseph Watson, 44, 60, 600, 25, 160

Joshua Naylor, 30, 50, 600, 25, 500

Peter Wood, 300, 200, 20000, 150, 487

John McKey, 15, -, 150, 20, 160

Edward Waters, 18, 5, 150, 30, 125

Ann Bowling, 260, 200, 5000, 20, 360

John Gardiner, 400, 200, 12000, 200, 1820

John Acton, -, -, -, 15, 190

Peter Hawkins, 200, 178, 400, 25, 850

John Hawkins, 300, 100, 8000, 1000, 1500

Thomas Davis, -, -, -, 15, 240

Nancy Murray, -, -, -, 15, 180

Noble Richards, -, -, -, 25, 195

George Berry, 400, 300, 10000, 100, 1400

Malcom Horton, 20, 185, 2500, 260, 1000

John Cluff, -, -, -, 20, 575

George Acton, -, -, -, 21, 135

William H. Berry, 400, 200, 9500, 75, 1200

Caroline Hawkins, 600, 200, 10000, 250, 2000

John Watson, -, -, -, 20, 200

John Robinson, 400, 200, 15000, 500, 1200

William Hall, 120, 20, 8000, 65, 400

William Robey, -, -, -, 5, 60

Josias H. Hawkins, 500, 100,6000, 200, 2200

John Hamilton, 1400, 900, 75000, 1200, 6300

George Richards, -, -, -, 5, 125
Luke Hawkins, 500, 100, 12000,
200, 2000
Hatty Richards, -, -, -, 30, 250
Cecelia Edelin, -, -, -, 20, 450
Valentine Miles, 300, 300, 12000,
250, 2000
Edward Edelin, 500, 500, 20000,
300, 3000
Richard Wheatley, -, -, -, 25, 275
Alexander Smoot, 200, 150, 3900,
100, 800
Smith Turner, -, -, -, 20, 300
Robert Brayfield, -, -, -, 20, 250
Thomas Jamison, 200, 75, 5000, 125,
900
Eliza Gardiner, 350, 450, 10000,
200, 1200
Walter Jamison, 200, 300, 5000, 75,
1100
John Jackson, -, -, -, 35, 275
Wilfred Moore (Moon), 90, 110,
1500, 35, 500
Ann Beam, 100, 160, 2000, 35, 600
Ignatius Gardiner, 1000, 1000, 7000,
50, 1500
John Harbin, 100, 37, 1000, 20, 235
John Brannan, -, -, -, 20, 175
Marcellus Jamison, 300, 164, 5000,
25, 600
Dr. W. Queen, 400, 300, 10000, 200,
1400
Alexander Queen, -, -, -, 40, 515
William Wheatley, 150, 150, 2000,
20, 540
Cincinnatius Murphy, -, -, -, 20, 150
Alpheusus Murphy, -, -, -, -, 100
Oliver Burch, 100, 125, 2000, 30,
225
Charles Burch, 100, 50, 1500, 25,
135
John Parker, 25, 40, 500, 20, 200
Franklin Burch, 110, 150, 3000, 25,
420
John B. Langley, 200, 120, 2600, 40,
430

Nancy Langley, -, -, -, 100, 400
James Goldsmith, -, -, -, 14, 130
Wm. J. Middleton, 100, 100, 1600,
2, 400
William Alvey, 300, 150, 20000,
175, 1500
Edward Gardner, 150, 35, 2000, 50,
700
Samuel Harrison, -, -, -, 40, 600
Elizabeth Middleton, 200, 100, 6000,
175, 450
Louisa Smith, 250, 350, 6000, 40,
550
Marcellus Edelin, 300, 200, 12000,
300, 1950
Jno. Gibbons, 100, 80, 1600, 30, 350
Jesse Burch, 100, 75, 100, 20, 450
Theoplilus Dent, 300, 596, 8000, 40,
650
Henry Truman, 450, 350, 25000,
300, 1140
William Keech, 250, 110, 10000,
175, 1200
Clinton Dent, -, -, -, 85, 1050
Dr. Albert Carrico, 152, 151, 5000,
175, 1350
Mary Matthews, 400 500, 6000, 100,
1200
Thomas Lathron, 71, 120, 1500, 20,
210
Chas. Harrison, 100, 105, 1600, 25,
165
Theophilus Dent, 200, 300, 7000, 50,
560
John Wiley, -, -, -, 3, 200
Thomas Carrore (Carnore), 450, 55,
12000, 200, 1210
Lloyd Johnson, 50, 100 1000, 15,
275
George Carrico, 260, 110, 8000, 75,
1400
Jno. B. Parker, 50, 50, 1000, 20, 140
Thos. Lyon, 60, 60, 500, 20, 210
Josias Hancock, 60, 90, 1000, 20,
100
Henry Canter (Carter), -, -, -, 10, 250

Chas. Reidd (Ridd), -, -, -, -, -

Zachariah Webster, 125, 75, 5000, 57, 550

Zachariah Swan, 215, 215, 8000, 75, 428

Ellen Woodburn, 200, 60, 3500, 30, 370

Benj. Goldsmith, 50, 50, 1500, 30, 320

Joseph Canter, 140, 80, 3000, 25, 380

William Coomes (Coomer), 115, 115, 800, 35, 380

Levi Nutwell, 150, 98, 3000, 20, 300

Benj. Swan, 25, 50, 500, 8, 200

Ignatius Hunter, 40, 35, 500, 60, 75

Anna Deakins, 125, 50, 2000, 20, 435

Hoviah Burch, 150, 150, 2000, 150, 700

Richard Farrall, -, -, -, 30, 350

Thomas Cage, -, -, -, 8, 95

William Stonestreet, -, -, -, 35, 275

Edgar Brawner, 300, 150, 700, 175, 1600

Alfred Gardner, 450, 550, 20000, 1150, 2450

Winifred Martin, 150, 150, 1500, 100, 450

Joseph Heran (Horan), 80, 20, 1000, 20, 315

Harriet Horon, 100, 100, 1200, 25, 175

Elizabeth Heran, 60, 40, 1000, 40, 280

William Wood, -, -, -, 30, 115

Rosina Turner, 75, 25, 1000, 30, 210

Thomas Proctor, -, -, 30, 400

William Richards, 55, 55, 1000, 60, 350

William Proctor, -, -, -, 8, 180

Catherine Montgomery, -, -, -, 12, 110

William Worling, -, -, -, 25, 175

William Boswell, 330, 150, 4000, 30, 275

Lemuel Wilmer, 100, 20, 4000, 30, 400

Alfred Roby, -, -, -, 20, 360

Betsie Bean, -, -, -, 30, 230

Hendly Bean, 75, 98, 1800, 15, 225

Aquilla Turner, 500, 200, 9000, 45, 1100

Jerry Dyer, 275, 125, 6000, 275, 1754

Pamela Dent, 200, 130, 3000, 40, 690

John Dixon, 80, 120, 2000, 20, 200

Jas. Montgomery, -, -, -, 100, 860

Hiscah Padgett, -, -, -, 15, 275

Jas. Farrall, -, -, -, 20, 300

Wm. Bartles, 250, 150, 3000, 40, 700

William Robey, 106, 107, 2000, 25, 340

Hannibal Downing, 250, 100, 3000, 100, 500

Joseph Bowerman, 18, 7, 500, 10, 125

William Boswell, -, -, -, 20, 150

Wm. Hancock, -, -, -, 5, 50

Danl. Thomas, 100, 100, 1500, 10, 150

Henry Covington, -, -, -, 10, 30

Henry Proctor, -, -, -, 20, 220

Josin Butler, -, -, -, 3, 75

Benj. Smoot, 75, 50, 1000, 20, 175

Jas. Goldsmith, -, -, -, 20, 115

Robert Dyson, -, -, -, 18, 330

Catharine Smith, -, -, -, 30, 225

George Willett, 150, 150, 2500, 25, 900

Joseph Gardner, 340, 321, 9500, 100, 1200

Jere Mudd, 275, 150, 5000, 300, 1000

Warren Adams, 80, 100, 2000, 20, 350

Henry Moore, 350, 450, 10000, 450, 1200

Maria Berry, 30, 60, 600, 20, 100

Francis Montgomery, 250, 550, 5000, 100, 200
George Willett, -, -, -, -, 60
Dennis Spalding, 800, 1200, 20000, 500, 2000
Sarah Robey, 100, 100, 1000, 50, 300
George Robey, 175, 50, 3000, 20, 200
Charles Hardy, -, -, -, 20, 150
William Shaw, 50, 106, 1500, 30, 250
Henrietta Wherts, 100, 100, 150, 50, 378
Samuel Robey, 60, 60, 350, 50, 200
Alfred Murray, 150, 50, 1200, 50, 350
Hannibal Acton, 150, -, 200, 25, 170
Robert Tucker, 75, 125, 2000, 25, 200
Thomas Tucker, 12, -, 100, 11, 50
Morris Hamilton, 100, 50, 1000, 20, 250
Thomas Padgett, 80, 67, 1000, 20, 250
Charles Harper (Harkes), 75, 30, 1000, 100, 120
John Garner, 50, 50, 400, 20, 150
Theodore Monroe, -, -, -, 20, 300
Samuel Maddox, -, -, -, 20, 200
James Osborne, -, -, -, 20, 100
Jane Clements, -, -, -, 20, 100
Charles Garner, -, -, -, 40, 250
Francis Pye, -, -, -, 300, 400
Amanda Mathias, 300, 200, 12000, 200, 600
John Chapman, 600, 300, 14000, 300, 2500
John Albrittian, 150, 78, 3500, 25, 600
Ignatius Morais, 4, -, 100, 1, 60
John Owen, 100, 100, 1800, 25, 200
Mary Owen, 150, 50, 2000, 25, 200
James Garner, -, -, -, -, 6
James Freeman, -, -, -, 10, 50
Richard Murdock, -, -, -, 25, 300

Asa Jenkins, -, -, -, 5, 150
Maria Hancock, 10, -, 150, -, -
William Cox, 435, 436, 33000, 800, 2500
Benj. Stonestreet, 250, 50, 10000, 300, 400
Joseph Stonestreet, 400, 160, 14000, 300, 2000
Frederick Stone, 130, 40, 10000, 150, 650
Frederick Stone agt., 400, 900, 13000, -, -
Frederick Stone agt., 350, 480, 13000, -, -
Mary Davis, 550, 550, 15000, 350, 1500
George Jenkins, 350, 380, 25000, 500, 3000
William Hamilton, 20, 150, 12000, 200, 1000
Charles Williams, 120, 320, 7000, -, -
Elizabeth Davis, 200, 130, 12000, 300, 1500
Edgar Furguson, 15, -, 5000, 50, 500
Mary Sanders, 500, 250, 1000, 300, 2000
William Roby, 295, 250, 12000, 150, 880
John Hancock, 45, 150, 2000, 25, 150
John Delozier, 83, 40, 300, 25, 150
William Garner, 50, 60, 2000, 25, 300
Francis Green, 200, 280, 9000, 100, 800
Thomas Edelin, 125, 155, 360, 30, 400
Richard Wade, 40, 160, 1000, 25, 500
John Snow, 60, 60, 1200, 10, 400
Samuel Adams, 100, 100, 3000, 25, 358
Richard Tubman, 200, 200, 6000, 300, 1500
William Swan, -, -, -, 20, 80

Robert Hunt, -, -, -, 5, 35
Richard Wade, 90, 112, 1000, 50, 650
William Smalwood, 162, 65, 1000, 50, 450
Judson Meding, 100, 152, 1200, 50, 500
William Padgett, -, -, -, 50, 250
Thomas Willett, 100, 40, 1500, 30, 8
Bennett Wilkinson, 73, -, 800, 30, 125
Asarias Padget, 160, 40, 5800, 30, 500
James Burch, 150, 150, 4000, 50, 620
Alfred Batter (Butler), 52, 90, 2000, 30, 150
Catherine Stewart, 200, 200, 3000, 100, 1000
Jacob Stewart, 180, 82, 1300, 50, 500
Charles Stewart, 250, 250, 3500, 50,700
Warren Willett, 150, 90, 3000, 60, 400
Richard Willett, 300, 113, 2500, 65, 450
William Doment, 150, 140, 2000, 60, 400
Claggett Page, 250, 200, 6000, 250, 200
John Higdon, -, -, -, 100, 845
Cornelius Willett, -, -, -, 40, 350
Francis Robey, 300, 260, 5500, 125, 2000
Saml. Robey, -, -, -, 50, 700
Samuel Cox, 100, 100, 3000, -, -
Hugh Roby, -, -, -, 25, 130
William Willett, 65, 60, 800, 40, 400
Mary Virtman, 20, 240, 3000, 15, 100
George Tubman, 200,70, 4000, 125, 200
Nathaniel Mally, 110, 37, 2000, 100, 500

William Claggett, 250, 260, 5000, 150, 1200
Henry Cawood, 200, 200, 2000, 100, 1200
Edward Jones, 100, 190, 6500, 100, 400
Domenick Oliver, -, -, -, 40, 250
Richard Bryan, 310, 190, 10000, 150, 1200
Seaton Morris, 300, 75, 20000, 100,750
Oliver Bryan, 150, 100, 4000, 200, 1000
Elijah Brown, 50, 100, 750, -, -
Joseph Downs, -, -, -, 25, 200
William Harr (Haw), -, -, -, 25, 450
Robert Wade, -, -, -, 20, 375
Ann Brown, -, -, -, 75, 1120
Henry Berry, -, -, -, 50, 358
Henry Butler, 100, 33, 1600, 30, 260
Noel Hannon, 250, 150, 4000, 100, 1050
Alexander Hamilton, 115, 75, 7000, 150, 1000
Benj. McPherson, 250, 100, 2000, 75, 1180
John Cox, 332, 246, 8000, 300, 1800
Miley Savey, 65, 65, 800, 25, 180
Washington Roster, 100, 40,800, 23, 180
Henry Butler, -, -, -, 20, 220
Benj. Gardiner, 250, 125, 2000, 40, 537
Michael McSherry, 20, 5, 200, 15, 190
Cornelius Maley, 10, 15, 200, 20, 60
Robert Mudd, 100, 100, 800, 25, 248
Catherine Tippett, 100, 218,3000, 125, 370
Joseph Hunt, 150, 260, 3500, 125, 750
Elizabeth Bowling, -, -, -, 4, 120
Underwood Sassan, 40, 10, 300, 40, 200
Wheeler Hamilton, 10, 40, 250, 20, 160

George Hamilton, 30, 20, 3000, 15, 180

James Wheatley, 175, 139, 3000, 50, 300

Joseph Wheatly, 80,70, 150, 30, 380

Catherine Mudd, 150, 150, 4000, 50, 825

Alfred Mudd, 8, 37, 300, 2, 195

Washington Marlow, -, 86, 350, 60, 190

James Hicks, 150, 120, 1200, 30, 400

Lemuel McDaniel, -, -, -, 30, 180

Lenia Moore, 75, 75, 1002, 15, 230

Lenard Adams, 170, 250, 2500, 20, 350

Hezekiah Monroe, 250, 350, 8000, 100, 1000

Richard Williams, -, -, -, 3, 140

Gran Clemments, -, -, -, 15, 200

Allison Hicks, 100, 200, 2000, 25, 600

Benj. Blandford, 82, 100, 2300, 50, 580

William Wolfe, - -, -, 2, 550

Caleb Pickerd, -, -, -, 12, 260

William Mathews agt., 1600, 1000, 60000, 3000, 8500

William Mathews, 800, 600, 20000, 1000, 3000

Yates Roby, 350, 308, 12000, 150, 1286

Walter Barkley, 137, 100, 1500, 20,180

John Huntt, 75, 36, 1000, 20, 400

Andrew Norris, 100, 30, 2000, 50, 560

Frederick Maddox, 15, -, 300, 20, 200

Elizabeth Hodges, 100, 63, 300, 50, 600

Charles Pye, 150, 125, 3000, 200, 700

Anna Beale, 200, 100, 3000, 25, 365

Thomas Clements Jr., 90, 20, 1500, 20, 300

Thomas Clements, 100, 50, 1500, 20, 350

William Lyon, 175, 150, 2000, 30, 478

John Wise, 130, 95, 1500, 30, 315

Sarah Norris, 43, 75, 1500, 25, 165

Marcellisu Thompson, 350, 150, 10000, 323, 1300

Nicholas Stonestreet, 285, 285, 8000, 545, 1124

Sophia Gardiner, 400, 470, 5000, 50, 792

Thomas Wilkinson, 40, 100, 700, 25, 180

James Latimer, 100, 115, 1500, 30, 300

Wm. Hamilton, 25, 70, 300, 10, 150

Dennis Wilkinson, -, -, -, 21, 135

Samuel Berry, 188, 40, 3500, 40, 670

Thomas Berry, 300, 400, 12000, 185, 1265

Richard Smallwood, 80, 15, 800, 15, 230

Francis Gardner, 380, 536, 20000, 200, 1287

James Moore, 110, 70, 1000, 20, 260

Hezekiah Willett, 150, 430, 3000, 30, 400

Mary Adams, -, -, -, 15, 100

Fielden Willett, -, -, -, 20, 175

Elizabeth Duffy, 100, 200, 3000, 30, 425

Patrick Duffy, 50, 50, 500, 60, 350

John Ford, 500, 343, 5000, 300, 1865

Ann Gay, 175, 125, 5000, 40, 560

Benj. Tubman, 300, 500, 7000, 300, 1600

John Carrington, 250, 150, 6000, 60, 200

Charles Hannon, -, -, -, 20, 160

Thomas Harris, 150, 50, 3000, 50, 380

John Harris, 260, 260, 8000, 120, 800

Alfred Claggett, 50, 100, 1500, 30, 315

William Cox, 100, 45, 2000, 50, 350

Richard Edelin, 300, 80, 12000, 150, 1100

Wm. Hanblim, 250, 140, 5000, 100, 800

Joseph Sanders, 200, 250, 8000, 150, 800

E. Smoot, 150, 150, 5000, 75, 800

R. Barnes, 250, 100, 6000, 80, 350

G. Muskett, 150, 50,7000, 250, 1500

Wm. Stone, 400, 300, 15000, 250, 1500

J. Stone, 800, 30, 1500, 100, 400

B. Barnes, 150, 50, 5000, 50, 600

Alex Ferrell, 15, 10, 500, 12, 150

Wm. Hindle, 450, 250, 20000, 500, 3000

Wm. Compton, 600, 100, 30000, 500, 1500

T. Burges, 450, 315, 15000, 400, 1500

Wm. H. Bruce, 260, 100, 12000, 100, 2000

J. F. Bruce, 750, -, 15000, 1200, 3000

C. T. Bateman, 125, 25, 5000, 25, 300

C. Long, 150, 25, 5500, 75, 1000

E. Stonestreet, 400, 50, 10000, 150, 1000

James Howard, 300, 100, 2000, 50, 100

A. L. Welsh, 250, 50, 3000, 30, 300

G. O. Oliver, 200, -, 3500, 150, 500

James Higdon, 200, 200, 10000, 200, 720

J. C. St. Clair, 160, 70, 6000, 100, 1000

C. A. Pye, 60, 165, 10000, 200, 600

A. M. Robertson, 300, 500, 15000, 500, 2500

J. Duckett, 900, 400, 30000, 500, 3000

G. Taylor, 240, 800, 7000, 300, 1000

P. Muskett, 240, 80, 3500, 125, 1125

M. Henson, 150, 150, 5000, 200, 800

E. Perry, 40, 350, 1000, 300, 3000

R. Murdock, 44, 44, 2000, 10, 150

E. Loyd, 190, 190, 2000, 50, 500

O. Dean, 200, 300, 8500, 150, 1000

R. Lloyd, 400, 300, 28000, 300, 2500

J. W. Mitchell, 400, 280, 2000, 5000, 2100

B. Compton, 800, 700, 20000, 1500, 4500

C. Brawner, 65, 65, 25000, 200, 500

J. Brawner, 100, 100, 10000, 200, 600

T. Welch, 75, 125, 1000, 51, 500

J. C. Gray, 250, 600, 4500, 300, 1200

Wm. Owens, 170, 170, 5000, 200, 750

C. Wilit, 200, 100, 2000, 75, 450

J. Roby, 60, 50, 700, 5, 150

J. W. Miller, 250, 350, 10000, 200, 800

R. Baxter, 100, 10, 72000, 200, 450

J. Bison, 300, 100, 5000, 200, 1500

T. Morris, 25, 75, 500, 10, 125

J. T. Price, 250, 280, 3000, 100, 1000

Wm. H. Gray, 150, 450, 6000, 300, 1000

J. F. Gray, 300, 100, 5000, 150, 1200

J. Price, 150, 50, 3000, 500, 800

T. Price, 280, 220, 9000, 500, 1400

Wm. Price, 280, 20, 8000, 400, 1800

J. B. Coby, 175, 180, 5000, 250, 1000

T. P. Gray, 450, 300, 5000, 300, 1000

R. Price, 300, 260, 10000, 400, 1450

H. H. Posey, 150, 50, 1000, 50, 350

J. A. Gray, 200, 500, 7000, 500, 1200

C. Coby, 150, 100, 500, 500, 1200

T. A. Smith, 90, 450, 4000, 190, 800

T. Grove, 50, 150, 2000, 50, 200

Wm. Smith, 100, 200, 2500, 75, 500
T. Dent, 75, 100, 4000, 50, 200
T. Price, 250, 200, 5000, 200, 1200
E. Welch, 100, 125, 1500, 75, 500
Wm. H. Bramen, 200, 100, 4000,
300, 500
N. Naly, 500, 530, 12000, 200, 1200
E. Boswell, 150, 150, 1500, 50, 800
R. Ratcliff, 75, 75, 2000, 50, 200
J. Milstead, 65, 65, 10000, 50, 150
C. Dent, 50, 100, 5000, 75, 150
J. Murdock, 150, 30, 2000, 50, 300

E. W. Wheeler, 250, 250, 2000, 75,
500
O. Bradshaw, 30, 20, 500, 25, 100
J. J. Kenrick, 60, 140, 2000, 200, 150
Samuel Henson, 200, 160, 5500,
300, 1200
James Garrett, 50, 150, 1000, 50,
200
E. Bowie, 100, 60, 1000, 50, 250
J. B. Franklin, 150, 150, 5000, 150,
400

Dorchester County Maryland
1860 Agricultural Census

The University of North Carolina Library under a grant from the National Science Foundation microfilmed agricultural Census records. Records were filmed at the University of North Carolina from original records at the Maryland State Library.

Columns 1, 2, 3, 4, 5, and 13 represent the following information on the census:
1. Name of Owner, Agent or Manager of Farm
2. Acres of Improved Land
3. Acres of Unimproved Land
4. Cash Value of the Farm
5. Value of Farming Implements and Machinery
13. Value of Livestock

Pages in this county are out of sequence on the microfilm.

Wm. M. Price, 150, 100, 3000, 100, 400
Mary Cochran, 150, 100, 5000, 34, 260
Margaret Nicols, 50, 150, 2000, 34, 150
Samuel Neel, 100, 50, 2000, 300, 240
Richard Thomas, 90, 70, 1500, 40, 100
T. Thomas, 100, 34, 1500, -, 1200
J. P. Langford, 45, 23, 800, 12, 150
James Demby, 42, 40, 100, 25, 115
Jos. W. Wheatley, 60, 140 2500, 10, 215
Saml. Vane, 6, 2, 200, 19, 100
Eliza Webster, 6, 7, 150, -, 60
Wm. Murphey, 75, 125, 2000 30, 305
Geo. Robinson, 40, 30, 2000, 20, 185
Jas. Holder, 34, 20, 500, 15, 24
Joseph Sefu(Sefus), 50, 50, 100, -, -
Joseph Sefur, -, -, -, -, -
Jas. Murry, 60, 40, 2000, 25, 280
Daniel Murphy, 40, 260, 5000, 50, 300
Thos. Craft, -, -, -, -, -
Ann Murphy, 34, 60, 165, -, 30

S. Howith (Horvith), 150, 130, 2000, 120, 200
J. Kinakin, -, -, -, -, -
Isaac Taylor, 200, 200, 5500, -, 300
Thomas Deen, 200, 50, 2500, 250, 250
J. Grinnum, 15, 85, -, 1000, 40 150
Wm. Dean, 4, 95, 1000, 200, 75
John Jones, 150, 25, 2000, 50, 300
S. Rhoades, 50, 50, 500, 50, 100
Robt. Haynes, -, -, -, -, -
William Holders (Holder), 45, 15, 800, -, 200
Wm. Spear, 70, 38, 500, -, 50
Wm. F. Spear, -, -, -, -, 20
Jos. Spear, -, -, -, 4, 25
Cyrus Bell, 100, 80, 2500, -, -
Cyrus Bell, 60, 40, 1000, 60, 500
Benj. Bell, 100, 100, 1500, -, 175
James Holder, 30, 40, 400, 1, 40
Jerry Allen, 5, 75, 300, -, -
E. Brinsfield, 150, 150, 2000, 10, 90
Ann Owens, 190, 50, 2500, 50, 300
Wm. H. Brinsfield, 60, 40, 2000, 50, 225
Richard Rhoades, 1, -, 300, -, 90
Wm. Picerim, 1, -, 500, -, -
Thos. Murphy, 1, -, 500, -, -
Danl. Murphy, 1, -, 100, -, -

James Vain, 3, 1, 600, -, -
Nathan Lecompt, 3, -, 600, -, 10
John Walker, 1, -, 200, -, 5
Ezekiel Walker, -, -, -, -, 110
James Hill, 2, -, 100, -, -
Byrod Tilman, 2, -, 300, -, -
Elizabeth Turpin, 50, 25, 3000, 25,
50
Jacob Wilson, 2000, 1600, 30000,
800, 2000
Solomon Russell, 100, 50, 1300, 20,
300
E. Brinsfield, 80, 25, 1500, 10, 60
John Lankford, 90, 50, 2000, 20, 180
Saml. Fisher, 1, -, 1500, -, 40
John Russell, 80, 80, 1500, 25, 150
Cath. Mesick, 40, 100, 1500, 10, 10
Saml. Calaway, 75, 100, 1200, 25,
200
Charles Marine, 30, 70, 600, 25, 150
Wm. Knowles, 50, 50, 600, 25, 60
John Williams, 175, 225, 5000, 50,
350
Ezekiel Coburn, -, -, -, -, -
John Marine, 50, 10, 700, 25, 125
George Murphy, 60, 200, 1500, 30,
250
Tilmon Langford, 51, -, 200, -, 6
Benj. McWilliams, 50, 10, 800, 30,
100
James Marine, 30, 40, 900, 30, 100
Covington Griffith, 30, 5, 300, 10,
150
Wm. Wheatley, 125, 50, 2000, 50,
350
Charles Calaway, 150, 100, 2000,
60, 350
George Wheatley, 40, 40, 1500, 30,
200
Ezekiel Wheatley, 100, 50, 1800, 60,
300
James Paine, 200, 100, 3000, 40, 400
Joseph Wheatley, 150, 50, 2000, 40,
300
Josiah Russell, 35, 1, 550, 10, 100

Josiah Truit, 100, 100, 1000, 100,
300
Charlott Cannon, 2, 8, 300, -, 15
Ed Wheatley, 40, 40, 1500, 25, 170
Silas Wheatly, 95, 40, 2500, 25, 170
Job Russell, 120, 75, 1000, 44, 200
J. Vaughan, 20, 9, 300, -, -
E. Marine, 40, 14, 2000, 30, 150
Catharine Welbey, 100, 40, 2000, 40,
30
Thomas Bradley, -, -, -, -, 5
J. Weatherly, -, -, -, -, -
Asa Hemmons, 7, -, 300, -, 40
William Bell, 2, -, 300, -, -
Constine Smith, 2, -, 400, -, -
T. Russell, 5, -, 1000, -, -
Thos. Wallis, 9, -, 2000, -, 150
William Dunn, 150, 100, 2000, 25,
300
William Butler, 1, -, -, -, 75
Francis Lowe, 1, -, 200, -, 2
Wm. Russell, 2, -, 500, -, 25
Clement Walsten, 1, -, 250, -, 6
Ann Collins, 1, -, 300, -, 150
John Tull, 1, -, 150, -, -
Joseph Bell, 1, -, 800, -, 150
James Whettz (Whelty), 150, 200,
5000, 70, 350
Charles Smoot, 80, 30, 2000, 60, 350
Thos. Truit, 40, 60, 1500, 40, 250
Pery Hacket, 125, 95, 2000, 40, 200
Colin Vincent, 150, 150, 8000, 100,
600
Sarah Russell, 70, 15, 1000, 40, 150
Jackson Vincent, 150, 75, 2800, 50,
125
Jon Calaway, 100, 75, 2800, 50, 275
Solomon Darby, 30, 15, 500, 50, 400
Samuel Tull, 60, 15, 1000, 5, 140
Mose (Mase) Mesick, 70, 10, 1000,
100, 500
James Lines, 100, 150, 2000, 70, 200
Jon Calaway, 50, 60, 1000, 50, 175
Newton White, 60, 25, 1000, 30, 140
George Lowe, 30, 170, 3500, 20, 120
E. Coleman, 40, 25, 1000, 40, 175

John Adams, 70, 5, 1500, 50, 300
Turfrin Langford, 20, 75, 500, 75, 170
Henry Tull, 30, 25, 500, 50, 50
L. Lankford, 150, 70, 250, 300, 400
John Williams, -, -, -, -, 130
Ed Colburn, 70, 10, 1000, 150, 150
Levin Phillips, 40, 20, 500, 40, 125
Paws Phillips, 175, 100, 700, 150, 700
Thos. H. Smoot, 150, 50, 2000, 600, 250
Jos. R. Wheatly, 200, 50, 2500, 100, 275
Henry Robinson, 1, -, 200, -, 50
E. S. Hobbs, 100, 100, 1300, 6, 150
John Graham, 16, -, 200, 6, 170
Asbury Dean, 60, 40, 1200, -, 60
Arthur Wheatly, 80, 50, 1200, 50, 300
William Childs, 1, -, 1600, -, 25
Ibly Jones, 75, 125, 1500, 10, 100
Plemus Russell, 200, 145, 1500, 100, 400
William Whight, 75, 70, 1000, 50, 175
J. Clements, 80, 60, 1000, 35, 140
Thos. Oterbridge, 70, 100, 1200, 50, 160
E. Millikin, 80, 10, 1500, 200, 100
S. A. Williams, 100, 60, 2000, 150, 200
Catharine Mesick, 2, -, 300, -, -
C. G. Wheatley, 150, 30, 150, 30, 350
Edward Wright, 250, 120, 6000, 100, 600
J. N. Wright, 3000, 200, 20000, 400, 1000
Wm. J. Harris, 17, -, 1000, 40, 140
Edward Adams, 120, 70, 2000, 100, 400
John Insley, 75, 75, 1500, 50, 250
George Jones, 40, 15, 700, 25, 115
Alfred Wheatley, 100, 80, 1500, 40, 250

Ed. Tull, 80, 120, 1200, 40, 125
Jas. Moore, 100, 150, 4000, 60, 300
Emaline Hacket, 90, 60, 1500, 50, 450
Wm. Robinson, 90, 50, 1500, 50, 350
Tilman Hacket, 120, 200, 3000, 50, 400
Minos Adams, 500, 900, 7000, 150, 400
Wm. Hayser(Harper), 70, 70, 1500, 50, 200
Wm. A. Wheatley, 100, 150, 3000, 50, 250
Wm. R. Smoot, 80, 20, 700, 25, 75
Isaac Williams, 100, 100, 2000, 30, 250
Isaac Davis, 160, 100, 3000, 780, 700
Wm. Bradly, 75, 40, 1500, 40, 200
Ezekiel Graham, 75, 150, 2000, 20, 60
John Tull, 1, 3, 600, -, -
Samuel Lesure (Jeseure), 17, -, -, -, 50
Mathew Newton, 100, 40, 1000, 20, 100
James Willis, 50, 20, 800, 15, 75
Phil Phillips, -, -, -, -, -
John T. Elliott, 1, -, 1200, -, 200
John Newton, 1, -, 700, -, 50
Joseph Newton, 12, -, 125, 2, 18
John W. Tull, 100, 125, 2000, -, 175
Matthew Marine, 75, 40, 1000, 40, 200
Jacob Nicholls, 75, 50, 1000, 20, 100
Thomas Bradly, 100, 75, 2000, 40, 225
Isaac Davis, 150, 350, 1500, -, -
John R. Wright, 50,80, 1200, 15, 125
William Newton, 90, 150, 3000, 30, 175
Risdon Smith, 100, 200, 2000, 150, 600
Marthe Paine, 2, -, 200, -, 45
Patrick Stack, 40, 25, 800, 30, 175

Melly Taylor, -, -, -, -, 15
Daniel Harper, 10, 30, 700, 20, 150
Stans Bramble, 1, -, 100, -, 30
Isaac Millikin, 50, 60, 2000, 50, 150
Isaac Millikin, 20, 80, 2000, -, -
Thomas Wright, 50, 5, 900, 20, 150
Pery Hubbard, 20, 15, 200, -, 140
James Pitts, 25, 50, 700, -, 125
James Mooney, 20, 125, 1000, -, -
Solomon Millikin, 70, 50, 1000, 50, 150
Ann Millikin, 60, 60, 1000, 50, 125
Wm. W. Hacket, 150, 150, 3000, 70, 400
Henry Bramble, 50, 45, 1000, 30, 150
Allison Wright, 11, -, 600, 10, 175
Davey Andrews, 2, -, 200, -, 4
L. Lermond, 80, 70, 2000, 40, 250
George Smith, 4, -, 3000, 150, -
Benj. Hackert, 100, 90, 800, 30, 600
Henry Miller, 65, 40, 1000, 25, 175
Zadaciah Willey, 100, 40, 900, -, 50
Mary Marble, -, 90, 1000, -, -
Isaac Bray, 2, -, 1300, 40, -
Levin Henry, 100, 50, 3000, 60, -
\Jeremiah Collins, 125, 30, 2000, 40, 150
Nath. Medford, 65, 40, 6000, 30, 250
Elisha Harper, 40, 8, 2000, 40, 100
Turfren Neel, 60, 45, 4000, 100, 150
William Conway, 150, 100, 4000, 100, 150
Ann Smith, -, -, -, -, -
Benj. Conway, 60, 50, 2300, 100, 250
William Wright, 1, -, 100, -, 40
James Hurlock, 120, 50, 2200, 100, 300
William Newbray, 200, 112, 400, 100, 500
Daniel Dean, 200, 100, 3000, 50, 200
Ezekiel Williams, 200, 100, 4000, 100, 450

Nimos Wright, 112, 25, 3000, 40, 300
John Noles, -, -, -, -, 25
Enoch Lowe, 80, 25, 1500, 50, 150
Isaac Lowe, 150, 90, 2000, 50, 2000
William Paine, -, -, -, -, -
William Todd, 15, 20, 1000, 30, 100
William Webb, 25, -, 500, -, 125
James Caudle (Carroll), 200, 52, 3000, 150, 500
William Charles, 50, 50, 3000, 50, 300
Chas. Carroll, 160, 80, 4000, 150, 400
William Medford, 130, 100, 3000, 50, 200
Woolford Mesick, 40, 40, 1400, 40, 150
Thos. Trice, 50, 25, 1200, 60, 170
S ___ Andrews, 150, 50, 4000, 100, 250
Kenly Wright, 150, 50, 3000, 60, 150
Thos. J. Twilly, 40, 50, 450, 50, 150
John Todd, 200, 113, 4000, 100, 500
S. Dawson, 100, 75, 2000, 50, 150
Job Willoughby, 150, 100, 3000, 40, 350
Algerum Cockran, 130, 25, 1500, 50, 350
Thos. Cochran, 80, 50, 1500, 50, 300
William Goslin, 60, 40, 1000, 40, 150
John T. Hacket, 90, 140, 3000, 75, 500
Henry D. Wright, 50, 119, 5000, 100, 500
Willis Hurlock, 20, 75, 600, 20, 50
Edward Wollen, 50, 50, 4000, 50, 150
Jas. Harper, 100, 30, 1000, 50, 350
William Hurst, 150, 50, 3000, 50, 350
Wm. Braughben, 725, 40, 6000, 25, 250

William Higgins, 100, 25, 1500, 5, 200

John Ward, 40, 10, 450, 10, 100

Jas. M. Thompson, 220, 80, 3000, 150, 500

John Read, 150, 20, 1200, 50, 200

William Short, 4, 10, 300, 150, -

Mitchell Thompson, 270, 70, 6000, 150, 600

Alex Clifton, 150, 100, 4000, 50, 500

William Thomas, 50, 150, 1000, 40, 200

Clarin Lard, 150, 100, 4000, 75, 300

Jas. R. Dunaho, 125, 50, 3000, 150, 350

Maciah Vicus, 70, 50, 1500, 40, 140

Thomas Lard, 100, 50, 250, 50, 200

Samuel Wright, 100, 50, 2000, 75, 350

Harriett Dean, 170, 30, 3500, 75, 500

Jno. A. Bramble, 180, 80, 4000, 75, 600

Edawrd Vicus, 100, 70, 1700, 100, 400

Leah Maris, 150, 90, 1500, 25, 250

Elisha Holliday, 30, -, 300, 20, 45

Wesly Sampson, 30, 25, 700, 20, 175

Draper Camper, 40, 20, 800, 25, 175

Joab Coleman, 120, 80, 1500, 25, 125

Charles Jackson, 100, 50, 1000, 25, 125

Jas. Webster, 175, 110, 3000, 75, 400

Jas. Sherman, 160, 100, 5000, 200, 500

Jefferson Hubbard, 30, 30, 900, 25, 100

John Woller (Wollen), 100, 149, 2500, 75, 350

Wm. M. Ross, 150, 75, 3000, 150, 600

Wm. Johnson, 80, 100, 2000, 20, 500

Sophia Powers, 100, 45, 2000, 30, 175

Maria Paine, 120, 140, 4300, 70, 600

Wm. Turner, 160, 160, 3000, 30, 200

John Bell, 70, 40, 1000, 60, 200

Joseph Nesbit, 15, 35, 3000, 20, 150

S. Braughn, 200, 38, 6000, 250, 600

John Whingtison, 200, 101, 7000, 200, 700

John Grove, 20, -, 1000, -, 125

Levin Hudson, 100, 125, 5000, 100, 500

Thos. Gambrill, 150, 150, 5000, 100, 600

Trustin Davis, 150, 150, 2500, 50, 250

Abl. Land, 150, 149, 3500, 40, 150

Thos. Smith, 75, 20, 700, 30, 100

Lanzy Wallace, 120, 100, 3000, 40, 200

Jeremiah Lewis, 60, 50, 1000, 30, 150

E. Christopher, 50, 25, 700, 30, 100

E. Wheatley, 75, 25, 1000, 30, 225

Richard Wheatley, 40, 24, 1200, 10, 350

Josiah Jenkins, 20, 20, 500, 2, 20

John Moore, 35, 5, 1000, 40, 125

John D. Stephens, 150, 100, 5000, 60, 300

W. J. Sampson, 25, 25, 500, 5, 300

George Dunn, 100, 75, 2500, 20, 275

George Hicks, 180, 86, 6000, 150, 100

Wm. Thompson, 7, 3, 300, 20, 60

Jas. Carroll, 150, -, 500, 50, 150

Jas. Pirt, 100, 50, 2000, 100, 250

James Davis, 70, 280, 600, 100, 350

Ebin Paws, 70, 10, 3500, 80, 190

William Vincent, 175, 75, 8000, 300, 900

Isaac Wright, 125, 100, 2200, 50, 450

Ennols Watus (Water), 80, 100, 2000, 30, 225

Clement Hubbard, 250, 150, 5000, 100, 600

Thomas Sellers, 150, 25, 1000, 40, 200

Joseph Twilly, 150, 80, 3000, 40, 300

Jame Samus, 80, -, 1500, 30, 175

William Page, 150, 60, 3000, 60, 200

Turpin Wright, 350, 400, 20000, 150, 300

Harriet Stephens, 25, 35, 500, 25, 125

Lewis Ross, 300, 206, 10000, 800, 2800

Lewis Ross Sr., 150, 150, 5000, 40, 250

Josiah Hooper, 80, 80, 1000, 50, 250

John Stewart, 100, 25, 600, 100, 150

Levin Stewart, 200, 50, 8000, -, -

Isaac Hooper, 150, 75, 5000, 50, 300

Thos. Sherman, 260, 200, 12000, 300, 600

John Patterson, 70, 80, 2000, 30, 200

Richard Dixon, 73, 30, 4000, 100, 520

Wm. Harrison, 200, 124, 6100, 800, 619

John Webb, 150, 100, 5000, 150, 50

Wm. Holland, 100, 150, 6000, 75, 550

Jno. Marshall, 150, 75, 6000, 75, 450

Geo. Staplefoot, 125, 15, 350, 70, 400

Thos. Saxton, 170, 50, 7000, 100, 1000

Jas. Stephens, 150, 100, 6000, 50, 1100

Edward Bramble, 40, 32, 3500, 25, 300

John Harper, 100, 50, 3000, 100, 300

Wm. Straughburg, 80, 15, 3000, 25, 225

Thos. Hicks, 300, 100, 6000, 75, 600

Wm. Christopher, 500, 1, 4100, 30, 175

Wm. Millaby, 100, 100, 1000, 50, 275

Jeffry Stanley, 18, -, 1800, 7, 135

John Baker, 255, 100, 8000, 150, 1070

Draper Jackson, 90, 10, 2500, 30, 20

Saml. Webster, 300, 153, 11000, 100, 750

Tilmon Andrews, 150, 200, 2000, 40, 300

Henry Wright, 260, 60, 10000, 250, 1200

John Webster, 250, 290, 10000, 150, 1100

Lewis Stewart, 125, 125, 8000, 60, 500

Ambrose Wilson, 250, 100, 10000, 150, 1000

Clement Young (Ganz), 100, 40, 1000, 35, 250

John Fletcher, 150, 175, 6000, 50, 450

Elijah Stephens, 225, 130, 6000, 40, 600

Thos. Helsbu, 120, 80, 3000, 60, 375

Jas. Hutchinson, 70, 20, 2500, 95, 250

Jacob Johnson, 120, 80, 4000, 50, 500

E. Harden, 90, 95, 2000, 40, 250

Ennols Mew__, 30, 40, 500, 25, 120

Peter Dokins, 30, 50, 500, 25, 200

Enoch Wilson, 30, 50, 600, 25, 150

Enoch Bailey, 30, 90, 2500, 50, 200

Saml. McBride, 190, 130, 6000, 125, 500

Magor Wheatley, 100, 100, 2500, 50, 300

Richard Lee, 110, 100, 2500, 40, 300

Wm. Henry, 80, 30, 2500, 50, 250

Z. Fountain, 140, 100, 6000, 200, 500

Robt. Rawleigh, 150, 60, 5000, 100, 500

Wesly Parker, 120, 100, 4000, 30, 400

Josiah Elsby, 200, 72, 6000, 100, 450

Thomas Lard, 90, 50, 3000, 20, 150
Moses Jones, 20, 30, 400, 20, 150
Tilmon Hurley, 105, 14, 1500, 50, 600
Ephraim Molick, 70, 140, 1500, 40, 278
Elben Starky, 125, 150, 2000, 40, 200
Charles Denard, 100, 140, 2000, 45, 175
Arnold Horseman, 82, 75, 1000, 20, 200
Jesse Jones, 180, 100, 6000, 100, 550
Robt. Marshall, 120, 70, 4000, 50, 500
Wm. Frazier, 150, 200, 5000, 100, 600
Milly Dennis, 125, 100, 1500, 50, 200
Peter Quash, 150, 100, 2500, 45, 350
J. W. Wrightson, 300, 3000, 60000, 100, 500
Stansbury Mesick, 130, 130, 2500, 50, 450
B. Woolford, 70, 100, 800, 25, 250
Harry Johnson, 65, 70, 1000, 15, 140
Nich. Largfit, 50, 50, 1000, 20, 200
Nich Travers, 90, 210, 1800, 30, 200
Isaac Neal, 120, 149, 1800, 100, 300
Peter Carroll, 40, 90, 2000, 15, 175
Martin L. Wall, 82, 40, 1000, 40, 300
Salomon Melly, 50, 50, 600, 30, 300
Darius Horseman, 25, 60, 800, 20, 160
Mathias Milly (Willy), 150, 150, 2800, 25, 400
Thos. Hurly, 100, 160, 3000, 20, 250
Thos. Vickers, 100, 60, 3000, 40, 250
Arthus Hause, 120, 120, 2500, 50, 300
Robt. Rideout, 70, 50, 1000, 40, 300
Ennols Arch, 90, 50, 1500, 40, 650
Joshia Crockett, 40, -, 500, 25, 100

David Willy, 30, 10, 500, 30, 300
Samuel Wall, 180, 180, 1700, 25, 150
Dennis Hurley, 30, 40, 600, 20, 125
James Pinket, 50, 50, 1000, 30, 150
Derrit C. Handly, 84, 25, 4000, 100, 500
Handly Phillips, 200, 100, 5000, -, -
George Piercy, 150, 57, 4000, 75, 500
Charles Dutton, 70, 50, 1500, 25, 275
George Piercy, 100, 50, 3000, -, -
James Hacket, 100, 100, 2000, 150, 400
Samuel Bell, 45, 50, 800, 40, -
Thomas Hodson, 300, 500 10000, 200, 1200
Shadrack Murphy, 44, 50, 800, 20, 225
Thos. Goold, 100, 10, 1500, 30, 175
Levin Paul, 25, 25, 500, 25, 150
George Carroll, 15, 25, 800, 25, 200
James Goold, 100, 50, 1500, 25, 300
Henry Loard, 15, 25, 500, 20, 90
Washington Load, 15, 5, 450, 20, 50
Robert Sampson, 15, 20, 400, 20, 200
John McAlister, 25, 37, 800, 20 150
John Cochran, 375, 300, 12000, 200, 1000
Thos. J. Lecompte, 200, 140, 5000, 30, 450
William McBride, 130, 50, 5000, 90, 550
James McBride, 120, 50, 3000, 50, 400
Noah Rawleigh, 200, 200, 8000, 100, 1000
Jefferson Hubbard, 40, 50, 900, 10, 110
Stephen Lecompte, 150, 140, 3100, 150, 650
Charles Dear, 100, 100, 4000, 100, 400

Charles Dear, 100, 100, 4000, 100, 400

James Wheatley, 60, 100, 900, 40, 300

Alex. Sherman, 150, 150, 6000, 100, 700

Alex Sherman, 75, 125, 2000, 200, 100

Mathew Howeth, 75, 50, 2500, 75, 450

James Loard, 125, 75, 4000, 150, 700

William Holder, 15, 200, 5000, 50, 300

William Smith, 100, 250, 1800, 10, 140

Lorenzo Hurst, 100, 50, 1000, 20, 150

Jarvus Donn, 100, 50, 1000, 60, 375

Solomon Lecompte, 175, 250, 8000, 250, 1200

William Hrun, 395, 350, 14000, 150, 900

John Demby, 90, -, 1000, 25, 350

Hooper Jolly, 70, -, 1000, 25, 200

Mary Baltimore, 40, -, 500, -, 75

Saml. Thompson, 400, 300, 9500, 125, 650

Josiah Webb, 300, 150, 3000, 300, 300

William Reed, 200, 220, 60, 200, 1000

Ann Pinket, 15, 40, 400, 25, 400

Elizabeth Pinket, 15, 40, 400, 20, 300

Ebin Short, 35, 100, 1500, 50, 450

Thomas McBride, 150, 40, 2500, 100, 400

John Thompson, 250, 260, 7000, 200, 800

James Cochran, 90, 110, 1500, 50, 350

Thomas Ball, 200, 300, 5000, 50, 500

Ezekiel Fleming, 250, 50, 6000, 200, 800

Thomas Kerr, 180, 50, 3000, 125, 900

Algernon Piercy, 450, 150, 15000, 400, 2000

James Patterson, 130, 104, 5000, 60, 700

Charles Patterson, 115, 20, 4000, 60, 800

James Bradley, 105, 50, 4000, 10, 100

Isaac Banks, 115, 30, 2000, 20, 375

Francis Lecompte, 150, 200, 7000, 50, 275

William Hooper, 165, 20, 5000, 80, 350

Alex Hurley, 200, 45, 7000, 200, 1000

George Lecompte, 135, 75, 3000, 40, 450

John Murphy, 180, 80, 5000, 70, 500

John Brawhon, 125, 40, 3000, 35, 345

Edward Hurst, 180, 20, 5000, 30, 450

William Jump, 200, 20, 6000, 50, 50

Hooper C. Hicks, 200, 100, 12500, 350, 1200

Jacob Insley, 30, -, 1000, 50, 300

John Grey, 15, 35, 2500, 30, 125

Levin Langford, 23, 30, 1500, 30, 125

James K. Lewis, 22, -, 3500, 40, 400

Withers Smith, 80, 40, 4000, 100, 500

Elizabeth Craft, 175, 50, 6000, 100, 800

Edward Smithers, 200, 33, 8000, 150, 600

Levin Jolly, 75, 25, 1100, 25, 200

Isaac Law, 125, 100, 300, 100, 1500

James Wrightson, 200, 100, 6000, 100, 800

Ira Cochran, 175, 75, 7000, 100, 650

Marcellus Lacom, 200, 30, 2500, 500, 1200

Zachariah Fooks, 100, 60, 2000, 40, 500

Isaac Camper, 125, 50, 200, 15, 350

William Jolley, 100, 125, 2000, 50, 300

John Parker, 95, 50, 900, 40, 500

Richard Hollen, 150 50, 2500, 50, 600

Levin Rawleigh, 100, 125, 4000, 30, 600

Thomas Hollen, 200, 100, 7000, 150, 800

George Saunders, 240, 19, 8000, 75, 1500

James Waddell, 225, 416, 12000, 300, 2000

Robert Parker, 40, 30, 1000, 40, 500

William Handly, 240, 70, 1200, 75, 800

Zarabable McAllister, 300, 300, 8000, 50, 750

Levin Perry, 70, 50, 1000, 40, 100

Isaac Marlock, 25, 20, 300, 20, 200

James H. Kelly, 260, 470, 10000, 50, 900

Kendall Lewis, 200, 70, 5000, 50, 700

Francis Harry, 200, 100, 4000, 100, 300

William H. Piersey, 400, 350, 1000, 75, 1500

John Henry, 50, 60, 100, 20, 250

George J, Meakins, 150, 75, 3000, 100, 800

James Gore Jr., 225, 100, 5000, 100, 450

James Gore Sr., 225, 135, 9000, 100, 600

Levin Nash, 29, 10, 350, 25, 350

Littleton Woolford, 140, 150, 350, 50, 450

William Kees, 150, 100, 4000, 100, 550

Thomas Twilley, 200, 75, 4000, 70, 600

James Higgins, 120, 380, 10000, 100, 550

John Carroll, 72, 30, 1000, 50, 150

Lewis Ross, 24, 47, 6700, 100, 160

Thomas A. Spence, 17, -, 6000, 50, 400

Francis J. Henry, 47, -, 5950, 100, 1000

Robert Bell, 4, -, 3500, 30, 200

James W. Henry, 120, 27, 6000, 200, 650

Benjamin Jenifer, 16, -, 1000, 50, 250

William A. Silover, 9, -, 2000, 50, 500

Samuel Hopkins, 90, 30, 2000, 70, 250

Jasper Cornish, 31, 3, 1200, 30, 50

Josias Simmons, 7, -, 2000, 60, 2000

William Banks. 20, -, 1500, 30, 150

Thomas J. Dial, 80, 145, 4000, 150, 800

Samuel Byron, 100, 26, 7000, 60, 450

W. C. Littleton, 10, -, 1800, 40, 250

Harnett Frazier, 25, 5, 1000, 40, 250

John W. Dean, 50, 40, 2500, 50, 350

John H. Davis, 75, 145, 4000, 60, 350

John G. Bell, 100, 21, 10000, 70, 350

Thomas Coleman, 30, 40, 700, 30, 100

Mahala Geoghan, 60, 150, 3200, 20, 200

James Geoghan, 30, 30, 1100, 40, 225

Mary Cates (Eater), 60, 20, 2500, 50, 400

William Travers, 100, 50, 4500, 100, 600

Sam. Beannock, 6, 80, 1600, 20, 100

Solomon Kervin, 100, 400, 10000, 300, 800

Travers Tally, 60, 70, 2000, 100, 400

Henry Allers, 75, 70, 200, 75, 350

Levi Travers, 150, 150, 15000, 300, 700

Marselus Willey, 50, 200, 2500, 40, 250

Edward Simms, 60, 350, 2500, 50, 300

Levi Travers, 70, 60, 2500, 50, 400

Mury Keene, 40, 60, 1000, 40, 250

Robert Keene, 50, 75, 2000, 40, 350

William Keene, 75, 75, 5000,150, 700

Thomas Spicer, 40, 20, 2500, 50, 250

Levi Travers, 75, 50, 2500, 40, 350

Bassett Woolford, 200, 82, 8000, 300, 800

Whitefield Woolford, 100, 150, 4000, 25, 300

Otho Litscum, 100, 150, 8000, 100, 350

Stephen Woolford, 130, 40, 8000, 400, 500

Harriet Oscarn (Ascam), 30, 16, 150, 50, 200

Greenbury Thomas, 60, 60, 1800, 60, 200

Sarah Jones, 10, 5, 1200, 20, 150

John Jones, 50, 50, 1800, 50, 350

William Jones, 75, 75, 6000, 200, 400

James Craig, 70, 75, 300, 50, 200

Mary Leonard, 60, 70, 200, 40, 225

Ezekiel Jones, 70, 30, 6000, 100, 450

Saml. Harrington, 100, 5, 5000, 300, 300

Peter Harrington, 110, 90, 8000, 300, 600

Thos. Woolfood, 150, 250, 8000, 250, 350

Alex. Woolford, 80, 100, 2500, 15, 30

Benj. Byers, 180, 50, 3000, 300, 700

Francis Linthicum, 40, 20, 1000, 50, 150

William Claridge, 34, -, 500, 10, 250

William Tall, 145, 145, 3000, 70, 500

Levin Philips, 35, -, 1500, 20, 200

James Hubbard, 60, 60, 500, 27, 10

Levin Simmons, 10, 15, 800, 25, 150

Jos. H. Auld, 75, 80, 3300, 400, 200

Uriah Maguire, 80, 10, 3500, 450, 600

Joseph Brook, 50, 110, 3000, 30, 300

Dewit Kinnamord (Kinnamond), 40, 90, 1000, 25, 250

Hary Stalon, 12, 10, 500, 20, 65

Moses Navoy, 20, 50, 700, 20, -

Levi Bowen, 80, 126, 3000, 25, 100

John Wheler, 90, 120, 7000, 100, 500

Saml. Harris, 40, 50, 1500, 30, 200

Vincent Green, 50, 70, 2000, 40, 200

William Aspen, 50, 75, 1000, 50, 400

Jackson Robinson, 100, 60, 500, 100, 400

James Burton (Benton), 60, 90, 1200, 50, 200

Joseph Henson, 60, 90, 700, 50, 225

Stewart Vicers (Vicus), 80, 60, 4000, 50, 450

John Burton, 29, 68, 800, 20, 200

Thomas Keene, 200, 500, 5000, 100, 800

Henry Shenton, 50, 70, 1000, 50, 300

Charles Cornish, 50, 400, 600, 50, 400

Robt. Binton, 20, 40, 1000, 50, 200

John B. Bradshaw, 27, 10, 500, 40, 150

Daniel Meekins, 40, 60, 1000, 40, 200

Joshua Meaking, 50, 200, 2000, 30, 175

Wesley Keene, 50, 70, 300, 50, 350

Elizabeth Morris, 40, 70, 600, 50, 235

Aaron Wallace, 100, 150, 100, 50, 200

Abraham Bishop, 60, 50, 1000, 40, 200

Jerry Malone, 20, 15, 700, 30, 225

Joseph Marine, 40, 60, 200, 30, 25

Spencer Caton, 15, 5, 600, 50, 275

Andrew Insley, 40, 40, 1000, 50, 300

Geo. Foxwell, 50, 150, 2000, 50, 400

Elijah Barkley, 50, 150, 2000, 50, 150

Levin Owins, 30, 40,700, 20, 150

Marcellas Wroten, 40, 100, 1400, 30, 500

Jackson Slocum, 20, 100, 800, 60, 400

Jeramiah Bramble, 30, 130, 80, 50, 200

Solomon Phillips, 30, 100, 1000,80, 250

Geo. Wroten, 50, 20, 1200, 50, 500

Geo. Booze, 60, 155, 700, 40, 250

John Edger, 20, 100, 2600, 75, 700

Robert Gooter, 60, 110, 1200, 50, 450

Silas Travers, 100, 200, 2700, 80, 600

Moses Shenton, 60, 200, 2700, 80, 600

Jacob Somden, 60, 200, 2000, 65, 500

Thomas. Tyler, 100, 75, 1500, 50, 700

William Gooten, 50, 100, 300, 30, 200

Timothy Bardon, 100, 500, 20000, 200, 1000

Robert Travers, 50, 70, 800, 40, 125

John Woodland, 40, 40, 1000, 100, 300

Mary Haywood, 50, 100, 1000, 60, 300

John Haywood, 40, 100, 800, 70, 350

Joseph Boly (Buly), 30, 100, 500, 50, 200

Robt. Cristopher, 45, 200, 1000, 40, 60

Thomas Stewart, 40, 100, 800, 50, 450

John Slocum, 45, 160, 1000, 25, 150

Obdiah Forrester, 75, 150, 1500, 25, 300

Benjamin Johnson, 35, 75, 1000, 50, 300

George Wallace, 75, 500, 4000, 100, 600

George Pierson, 50, 100, 2000, 100, 500

Robt. Andrews, 60, 200, 1600, 50, 450

George Miston, 70, 200, 700, 200, 1500

Sovereign Mason, 30, 60, 1500, 100, 400

Thomas Pritchett, 30, 60, 2000, 100, 400

Samuel Hanshicon, 55, 200, 2000, 50, 500

Henry Shenton, 50, 100, 1000, 30, 350

John Andrews, 75, 250, 2700, 100,500

Peter Banett, 50, 100, 1400, 50, 400

Denard Johnson, 50, 50, 800, 25, 200

Orash Johnson, 70, 80, 2000, 60, 300

Matthew Meakins, 50, 25, 700, 25, 300

John R. Keene, 100, 50, 3000, 40, 300

Eliza Makins, 90, 75, 4000, 50, 300

Robt. Marshall, 50, 80, 1400, 25, 225

Curtis Williams, 60, 200, 2500, 40, 225

Samuel Keene, 100, 75, 4000, 40, 250

Levin Donnock, 75, 60, 4000, 75, 500

Wash Smith, 90, 40, 350, 60, 250

Elijah Tall, 42, 11, 2000, 80, 200

Saml. Tall, 66, 15, 200, 60, 200

Thomas Keene, 50, 10, 3000, 25, 250

Danl. Thompson, 60, 20, 1500, 30, 200

Saml. Keene, 75, 75, 4000, 30, 300

Levi Leonard, 60, 40, 1500, 25, 200

Eliza Travers, 65, 75, 4000, 75, 700

Thos. B. Travers, 60, 50, 5000, 100, 800

Richard Swift, 30, 130, 3000, -, -

Margt. Stapleford, 90, 45, 400, 60, 375

Jerimiah Pallison, 150, 600, 20000, 400, 1375

Joseph North, 15, 25, 1500, 30, 800

T. Stapleford, 60, 20, 3000, 30, 400

Isaac Davis, 75, 100, 4000, 40, 250

Mary Davis, 30, 25, 1000, 25, 200

Robert Barnes, 50, 50, 2000, 30, 250

Thos. Travers, 75, 70, 6000, 200, 500

Mary Travers, 100, 25, 5000, 150, 450

James Palmer, 50, 40, 200, 40, 250

Jas. Thompson, 100, 114, 3500, 200, 250

William Cates, 100, 100, 10000, 400, 700

Levi Pagan, 100, 100, 2500, 80, 600

J. Edminson, 50, 25, 2000, 75, 300

Thos. Lambden, 70, 110, 3000, 80, 300

Sam. Travers, 185, 130, 17000, 400, 1200

Bazilla Lane, 8, -, 400, 20, 50

Jos. L. Bryan, 150, 70, 8000, 300, 1200

Wm. W. Patterson, 150, 70, 7000, 250, 600

Matthew Frarett, 25, 21, 1100, 50, 150

John Travers, 50, 30, 7000, 40, 300

Emory Cook, 40, 23, 1900, 50, 300

John Rigins, 25, 10, 900, 30, 250

Robert F. Tubman, 175, 225, 10000, 100, 900

John T. Stewart, 300, 40, 15000, 1000, 1200

William Grason, 275, 137, 15000, 100, 1500

Rider Henry, 250, 130, 15500, 800, 2000

Wm. F. Goldsborough, 500, 300, 32000, 100, 4450

James Hadaway, 200, 170, 14800, 600, 1500

Joseph Spedden, 100, 30, 7000, 200, 950

Thomas Hubbard, 175, 65, 9000, 100, 700

Wesley North, 70, 27, 5000, 200, 700

Richard Phillips, 130, 62, 6000, 50, 425

William Lecompte, 150, 42, 8000, 100, 900

John Radcliffe, 175, 150, 10000, 150, 100

John Creighton, 200, 30, 9000, 100, 850

John Byers, 125, 75, 10000, 200, 1000

John Cook, 200, 100, 10500, 500, 500

Washington Trego, 75, 50, 4000, 30, 200

John R. Patterson, 233, 217, 18000, 200, 1050

Joseph Arnett, 73, 100, 4000, 15, 200

Daniel M. Henry, 95, 71, 15000, 500, 900

Samuel Patterson, 200, 180, 20000, 400, 200

James Brannock, 30, 30, 600, 20, 80

Elliott Phillips, 120, 130, 4500, 100, 700

William M. Robinson, 150, 70, 1000, 400, 900

Thomas Dail, 39, 25, 700, 30, 450

Richard Nevit, 150, 79, 7000,75, 650

Jean Vicers, 90, 30, 3000, 30, 275

John Brannock, 20, 12, 800, 30, 160

Solomon Hinson, 40, 130, 2000, 40, 180

William Hopkins, 70, 70, 4000, 200, 600

Levin Jones, 70, 225, 6000, -, -

Levin Jones, 170, 100, 1000, 150, 1200

Eliza Jones, 125, 125, 4000, 50, 750

Levin Skinner, 100, 100, 600, 100, 750

Zachariah Skinner, 125, 7, 10000, 150, 700

Otterbridge Neel, 90, 121, 500, 100, 600

George Orens (Owens), 95, 40, 500, 100, 750

John Kirby, 40, 10, 1000, 50, 200

James Mills, 90, 90, 6000, 50, 350

Thomas Hopkins, 67, 16, 300, 50, 600

William Phillips, 120, 92, 4300, 100, 500

Charles Bnerwood, 100, 93, 5000, 300, 800

Joseph Robinson, 125, 50, 7000, 300, 900

Willis Branock, 75, 19, 2500, 100,700

Zachariah Skinner, 75, 50, 6000, 100, 350

Levin Skinner, 108, -, 4500, 75, 350

Edward Johnson, 125, 212, 8000, 200, 1000

William Dail, 170, 256, 15000, 300, 1100

Moss Ophen, 60, -, 450, 40, 400

John W Dail, 125, 244, 12000, 200, 800

Joseph McAllister, 75, -, 600, 25, 80

Charles Vicers, 140, 110, 7000, 100, 750

Thomas Flint, 125, 100, 7000, 100, 750

James Jones, 10, 60, 6000, 100, 550

Ullison Monroe, 80, -, 5000, 100, 600

John Phillips, 150, 150, 4500, 80, 450

William Langford, 150, 75, 8000, 50, 400

Tilghman Cheesman, 100, 50, 300, 50, 400

Thomas Wright, 125, 45, 2500, 50, 450

Jeremiah Wright, 150, 75, 10000, 250, 1500

James Hurley, 150, 150, 8000, 40, 750

John Jolly, 225, 150, 6000, 50, 600

Levin Mears, 100, 75, 2000, 100, 500

Elisha Jolly, 150, 100, 3000, 22, 200

Fralter Mowbray, 150, 150, 4500, 20, 250

Richard Smith, 120, 250, 8000, 70, 400

David Cochran, 250, 250, 500, 300, 2000

Jared Shorter, 150, 150, 4000, 50, 650

Charles Austin, 175, 125, 4000, 100, 1000

William Davis, 325, 200, 500, 150, 750

Jeremiah Holland, 325, 200, 5000, 150, 750

Greenbury Holt, 200, 200, 5000, 175, 1400

George Meredith, 150, 100, 4000, 45, 800

Pritchet Meredith, 350, 280, 6000, 100, 1200

John Stanley, 200, 250, 4500, 50, 450

John Jackson, 100, 150, 2000, 40, 45

William Shorter, 100, 200, 4000, 100, 200

Joseph Bespitch, 150, 400, 800, 100, 850

Edwin Dashields, 200, 200, 5000, 100, 500

Joseph Bespitch, 150, 400, 800, 100, 850

Edwin Dashields, 200, 200, 5000, 100, 500

Thomas Meredith, 300, 200, 500, 100, 1500

Banamon Mills, 235, 236, 3000, 100, 650

John Mears, 100, 100, 4000, 50, 600

Polish Mills, 212, 100, 8000, 200, 100

Noah Strawberry, 100, 100, 2500, 50, 500

George Warren, 100, 75, 1000, 50, 450

John Cowarll, 200, 150, 10000, 800, 1200

John Bradas, 70, 200, 3000, 25, 400

Peter Borrough, 160, 150, 4000, 50, 500

John House, 200, 200, 5000, 50, 850

James Harding, 100, 90, 2000, 40, 450

John Bassett, 80, 147, 1000, 40, 250

William Shorter, 8, -, 200, 20, 300

John Paul, 160, 80, 5000, 40, 300

Jas. H. Eggleston, 300, 200, 8000, 100, 800

Mary Wrightson, 250, 200, 4000, 50, 500

Daniel Creighton, 100, 150, 2000, 50, 350

Edwin House, 150, 100, 3000, 50, 500

James Sampson, 150, 250, 5000, 50, 650

James Murphy, 200, 20 2500, 50, 900

William W. House, 150, 100, 3000, 100, 800

Shadrack Cromwell, 100, 98, 2000, 50, 600

William W. House, 70, 150, 1500, -, -

William Roszell, 90, 30, 2000, 75, 500

Alfred Mowbray, 220, 80, 6000, 150, 700

Wm. C. Huffington, 275, 125, 20000, 400, 2000

Levin Jolly, 20, 40, 1200, 30, 300

Noble Wright, 100, 200, 3000, 40, 450

Job Bacquith, 100, 43, 300, 40, 600

Joseph Stewart, 125, 400, 5000, 150, 500

Joshua Stanley, 100, 60, 2000, 30, 300

Casse Baltimore, 70, 500, 3000, 40, 400

William Billips, 125, 235, 5000, 100, 1000

James Marshall, 120, 30, 1500, 50, 700

Jacob Williams, 50, 80, 2300, 100, 750

John Meredith, 30, 50, 1000, 25, 300

William Mowbray, 250, 200, 800, 50, 800

George Austin, 400, 250, 20000, 500, 2000

Josiah House, 150, 200, 5000, 100, 800

John McGarth, 45, 83, 500, 125, 300

Richard Binder, 40, 70, 700, 25, 150

William C. Hurley, 100, 100, 3000, 100, 400

Abraham Lee, 100, 100, 4000, 70, 300

Benjamin Fentris, 70, 90, 2000, 50, 300

Robert Vincent, 30, 40, 1000, 300, 250

Daniel Hubbard, 90, 60, 4000, 50, 500

Thomas Stewart, 100, 100, 4000, 50, 700

Stephen Jackson, 90, 75, 2500, 40, 175

Zedakier Williams, 50, 150, 1500, 30, 150

William F. Hicks, 130, 20, 3000, 50, 800

Joseph Rook, 100, 100, 2000, 50, 200

Henry Hooper, 130, 50, 6000, 100, 1000

John P. Hooper, 113, 65, 6000, 250, 900

Samuel Mowbray, 150, 125, 400, 50, 600

Samuel Smith, 150, 150, 5000, 6, 800

James Mowbray, 6, 7, 1000, 15, 150

James Stanley, 125, 250, 1500, 50, 600

Levin C. Davis, 140, 41, 3000, 125, 1000

John Sheke (Lheke), 80, 46, 1500, 20, 175

Josiah Carroll, 150, 150, 6000, 100, 800

John Williams, 90, 100, 2500, 50, 500

Francis H. Vincent, 200, 200, 12000, 200, 800

William Cabin, 80, 125, 2500, 50, 300

Jas. R. Stephens, 175, 225, 12000, 75, 500

Harrison Mears, 85, 33, 2000, 50, 500

William North, 100, 70, 2000, 150, 850

William Langford, 50, -, 900, 40, 350

Charles P. Straughan, 116, 84, 7000, 100, 800

James L. Colston, 70, 30, 7000, 200, 900

Henry Nicholson, 163, 103, 12000, 125, 850

John F. Kurtz, 125, 175, 10000, 300, 1000

Levin S. Dail, 90, 50, 8000, 200, 700

Francis P. Phillips, 256, 136, 16000, 500, 1500

Eugean Tubman, 125, 194, 5000, 50, 500

James Mills, 150, 200, 9000, 75, 1000

George Rue (Rul), 100, 50, 5000, 50, 500

William Reese, 13, -, 1500, -, 70

Edward Hopkins, 70, 60, 3000, 50, 400

William Millse, 180, 100, 10000, 100, 800

Charles B. Jones, 40, 10, 400, 200, 400

Thomas Hobbs, 95, 20, 6000, 100, 450

Worth Goldsborough, 140, 45, 10000, 150, 1000

John Spence, 143, 102, 14000, 400, 1000

William Rook, 100, 100, 4000, 50, 450

Stephen Hurley, 90, 100, 350, 50, 800

Francis Phillips, 130, 20, 8000, 100, 700

Thomas McAllister, 180, 50, 2500, 75, 550

Caleb Shepherd, 150, 193, 1000, 400, 1500

Wingate Wallis, 80, 53, 2000, 40, 400

William Stephens, 80, 61, 5000, 100, 450

John Slalor, 15, 5, 800, 30, 250

William W. Bryan, 117, 24, 20000, 500, 1350

James Wallis, 150, 40, 7000, -, -

James Wallis, 240, 175, 8000, 400, 2000

James Woolford, 125, 120, 7000, 600, 1200

Lewis Ross, 24, 47, 6700, 100, 160

Thomas Wheatley, 20, 50, 1200, 50, 75

William Bennett, 30, 70, 1000, 50, 350

James North, 60, 24, 2500, 30, 400
John Hubbard, 50, 40, 1500, 50, 80
Thomas Bernnett, 20, 50, 1500, 40, 300
James North, 80, 50, 5000, 40, 350
Jas. H. Radcliff, 140, 40, 5000, 300, 500
George Applegarth, 30, 20, 2000, 100, 500
John H. Thomas, 50, 50, 2000, 30, 250
William North, 30, 70, 1600, 30, 225
Levin Clark, 25, 60, 1000, 30, 240
Levi W. Clark, 20, 40, 600, 20, 230
Josiah Thomas, 30, 47, 1500, 30, 250
William T. Figs, 10, 20, 500, 20, 450
Richard Phillips, 25, 90, 3000, 50, 475
Robert Hubbard, 40, 104, 3000, 50, 450
Samuel Corner, 60, 60, 5000, 50, 450
John Marshall, 60, 30, 1500, 40, 300
Thomas Cantwell, 30, 40, 1500, 50, 275
William Stephens, 20, 10, 500, 30, 250
Rebecca Wheler, 60, 20, 2500, 40, 400
Arthur Travers, 60, 26, 1800, 25, 310
William Travers, 7, 165, 6000, 50,700
Nehemiah Becquith, 100, 25, 6000, 50, 550
Charles Hubbard, 20, 20, 600, 20, 400
John F. North, 50, 40, 1500, 44, 350
Henry Hubbard, 60, 50, 2500, 50, 400
William Wheler, 30, 50, 2000, 10, 150
George Applegarth, 35, 95, 3000, 75, 350
Levin Lenard, 100, 33, 2000, 40, 350
Richard Lenard, 15, 10, 400, 20, 70

Thomas Frego (Trego), 20, 33, 1200, 40, 175
James Reddit, 74, 31, 300, 300, 600
Robert W. Williams, 75, 75, 3000, 150, 600
Sarah Wrightson, 160, 75, 6000, 100, 700
Josiah Land, 125, 60, 4000, 50, 600
John Robinson, 150, 100, 6000, 150, 600
William F. Yeates, 200, 200, 10000, 100, 800
Charles Phillips, 190, 100, 6000, 200, 900
John Baker, 40, 50, 1000, 40, 375
D. Lintscum, 75, 50, 2500, 50, 500
Summer Cornish, 65, 60, 1800, 40, 150
Hiram Moore, 45, 20, 800, 50, 300
Josiah Marine, 75, 150, 150, 60, 300
Zach. Lintscum, 150, 600, 6000, 200, 700
Henry Henson, 30, 20, 550, 30, 150
James Keene, 40, 49, 1000, 50, 300
Vance Camper, 125, 150, 2250, 60, 30
Gabriel Huet, 25, 20, 50, 20, 125
John Woolford, 110, 80, 10000, 400, 800
Ezra Jolly, 110, 150, 800, 100, 400
Martin Wright, 60, 125, 3600, 30, 250
Lewis Thomas, 100, 120, 4000, 75, 450
Jeremiah North, 60, 50, 200, 30, 375
John H. Norath, 50, 60, 200, 25, 300
John Dorsey, 100, 120, 8000, 500, 835
Precilla Bowdle (Boudle, Bondle), 90, 65, 7000, 700, 450
Thomas Willis, 150, 150, 900, 75, 750
Skinser Richardson, 60, 40, 2000, 30, 300
John W. Bell, 75, 75, 2000, 40, 225

Thomas Mael, 100, 54, 7000, 150, 550

John R. Martin, 150, 50, 10000, 100, 1000

Wm. Richardson, 37, 20, 1000, 25, 25

Stanley Richardson, 18, 6, 800, 15, 200

George Vickers, 70, 100, 400, 15, 300

Zach Brannock, 30, 30, 200,40, 200

Levin Richardson, 120, 180, 3000, 70, 575

Charles Keene, 50, 25, 1000,70, 270

Skinner Richardson, 35, 40, 200, 40, 25

Jesse Richardson, 13, 4, 1500, 30, 30

Richard Lintscum, 200, 100, 10000, 100, 700

Zach. Lintscum, 10, 15, 300, 50, 125

James Parker, 50, 20, 1000, 40, 250

Josiah Lintscum, 100, 133, 15000, 600, 1060

Thomas J. James, 133, 170, 9000, 200, 1200

John Neal, 100, 100, 350, 100, 350

W. W. Crawford, 80, 60, 500, 300, 1000

Thomas J. James, 35, 35, 1200, 25, 300

Hughey Neal, 100, 100, 5000, 100, 6000

Whitefield Woolford, 75, 25, 500, 400, 600

William Lang, 27, 19, 700, 25, 40

Levin W. Fitchen, 60, 60, 1200, 150, 60

Levin L. Lintscum, 75, 12, 1000, 50, 500

John Jones, 70, 90, 11000, 50, 500

Joseph Woolford, 80, 300, 2500, 100, 350

John D. Parker, 250, 100, 5000, 400, 1000

Henry Calender, 25, 50, 1000, 30, 200

Solomon Woolford, 60, 60, 1000, 50, 500

Jerry Matthews, 70, 500, 5000, 50, 520

Harris Jackson 85, 140, 2500, 48, 600

John R. Meekins, 70, 100, 2000, 50, 175

William Harding, 90, 125, 2000, 60, 500

Neal Moore, 75, 75, 2000, 50, 600

Samuel Todd, 75, 300, 5000, 50, 400

William W. Jones, 100, 450,7000, 100, 900

George Richardson, 40, 40, 1500, 50, 256

Levin Bespitch, 10, 28, 700, 25, 250

John Clark, 12, 100, 600, 40, 225

Thomas D. Willis, 125, 120, 5000, 100, 450

William Mace, 125, 100, 2500, 125, 750

Dennis Cooper, 100, 100, 2000, 50, 350

Ibin Nichols, 80, 40, 3000, 60, 250

John Mace, 200, 100, 6000, 100, 500

Peter Ennols, 100, 50, 200, 50, 200

Edward Moore, 100, 100, 2500, 40, 250

William Lewis, 75, 75, 800, 30, 350

Lazrius Powel, 250, 475, 10000, 300, 1200

Levin Donnock, 250, 250, 5000, 300, 1000

Samuel Donnock, 125, 125, 300, 200, 100

Thomas Bradley, 150, 300, 4000, 75, 325

John Land, 140, 620, 3000, 75, 625

James Reed, 100, 200, 5000, 75, 800

Littleton Huse, 120, 200, 4000, 50, 400

James Clark, 120, 250, 3000, 50, 500

Lloyd Harris, 100, 150, 2500, 75, 425

James Cromwell, 50, 200, 1600, 50, 200

Benjamin Reed, 50, 200, 1600, 50, 200

Robert Spear, 100, 150, 2000, 150, 450

James Huse, 100, 100, 1500, 50, 400

George Hart, 150, 100, 3000, 100, 600

Alexander Langrell, 40, 60, 1600, 50, 200

William Fallan, 20, 50, 1000, 20, 250

William Dayton, 30, 100, 3000, 40, 200

Tyler Bradshaw, 25, 100, 2000, 20, 250

Henry Fallan, 30, 60, 800, 24, 400

Joseph Langrell, 30, 40, 1200, 30, 150

Elias Elliott, 50, 20, 2000, 20, 200

William Warfield, 150, 300, 7000, 100, 500

Jno. Staplefoot, 45, 200, 1800, 200, 600

Chas. Lake, 45, 200, 4000, 75, 500

William Stokes, 45, 200, 800, 60, 240

Henry Matney, 90, 200, 2000, 70, 500

Wrightson Willey, 45, 150, 1200, 75, 600

Daniel Worton, 40, 100, 1000, 50, 500

Isaac Moore, 24, 35, 1000, 60, 340

Marcellus Roberson, 35, 25, 800, 55, 200

Wm. Andrews, 40, 110, 1000, 50, 340

Martin Foxwell, 45, 160, 2000, 55, 500

Fanny Mister, 50, 100, 2500, 60, 500

Catherine Insley, 50, 100, 2000, 60, 700

William Andrews, 50, 100, 2500, 60, 700

Albert Johnson, 150, 950, 6000, 200, 1000

Goodman Gooter, 25, 150, 1000, 200, 450

John Alpegrath, 100, 200, 2000, 300, 600

Vachell Keene, 150, 350, 5000, 150, 600

Washington Gooter, 9, -, 800, 30, 300

Levin Keene, 70, 600, 1500, 50, 400

William Jones, 80, 200, 3000, 50, 700

Matthew Wroten, 40, 60, 2000, 50, 500

Amelia Tubman, 60, 300, 4000, 60, 300

Zebider Harper, 15, 75, 600, 30, 250

David Tyler, 60, 30, 3000, 100, 800

Moses Banks, 50, 100, 1400, 100, 600

Adam Keene, 40, 700, 10000, 50, 200

William Cooper, 75, 600, 7000, 200, 700

Joseph Pinder, 100, 100, 2000, 50, 275

Josiah Horseman, 100, 100, 2000, 75, 200

Charles Johnson, 70, 100, 1500, 50, 450

Levin Stanley, 75, 80, 800, 25, 400

James Hill, 56, 40, 600, 25, 100

Alfred Chase, 60, 45, 600, 20, 175

John Lewis, 120, 265, 400, 50, 250

William Hurley, 40, 56, 1800, 50, 350

Sarah Hurley, 25, 150, 1000, -, 75

Harmon Hurley, 35, 40, 800, 50, 175

John Nevel, 200, 200, 6000, 40, 500

Hudson Hurley, 30, 50, 3000, 50, 500

Elizabeth Donohue, 60, 80, 3000, 50, 350

Alfred Lewis, 400, 1750, 8000, 50, 600

Joel G. Hurley, 45, 70, 1000, 40, 150

Josiah Hurley, 40, 160, 1000, 30, 150

Joshua Townsend, 18, 10, 300, 40, 150

William Laton, 150, 100, 4000, 71, 300

Alfred Hurley, 50, 50, 500, 40, 125

Job Horseman, 70, 100, 1000, 40, 190

George Hurley, 23, 50, 800, 40, 140

John Beard, 90, 100, 1900, 50, 175

Henry Hurley, 50, 25, 1200, 40, 175

J. Wainwright, 150, 200, 6000, 200, 600

William Donohoe, 150, 125, 4500, 10, 600

Uriah Hurley, 60, 25, 700, 25, 175

James Jones, 112, 20, 1000, 30, 125

Cain Hurley, 45, 150, 2500, 40, 200

Eaphany Hurley, 75, 1000, 1500, 50, 250

James Mccready, 7, 100, 1000, 25, 125

M. Travers, 75, 125, 1800, 30, 250

Major Marshal, 15, 2, 500, 20, -

Robert Dilaha, 140, 100, 4000, 75, 400

Levin Lewis (Lervis), 40, 20, 2000, 75, 400

Jeremiah Bramble, 90, 45, 100, 40, 250

Henry Dockins, 35, 40, 800, 25, 150

John Wing, 45, 50, 800, 25, 175

James Nichols, 100, 4, 1400, 20, 250

Francis Kenly, 150, 100, 4000, 50, 250

Zachariah Paul, 80, 50, 800, 30, 200

Samuel Lawson, 100, 100, 1000, 50, 278

William Parsons, 50, 100, 1500, 40, 200

William Husly (Hurly), 70, 100, 1000, 50, 200

Geo. Morelock, 100, 100, 1500, 50, 350

William Marine, 50, 40, 600, 25, 120

Jas. Montgomery, 200, 150, 6000, 75, 500

Noah Pinder, 100, 75, 1500, 50, 250

Eabin Stanley, 100, 90, 1500, 50, 300

Gracy Applegarth, 100, 100, 1800, 50, 300

Charle Earle (Easle), 100, 75, 100, 50, 200

John Twilly, 100, 200, 3100, 100, 400

Joseph Stafford, 150, 150, 3000, 50, 200

Joseph Stanley, 150, 75, 2800, 40, 200

John Tubman, 60, 360, 2500, 30, 300

John Dunnock, 40, 100, 2500, 30, 350

George Meekins, 100,100, 400, 100, 450

William Mills, 24, 20, 1000, 30, 250

Wm. W. Meakins, 45, 45, 1000, 30, 250

Tary Ennols, 80, 100, 2000, 30, -

Robert Lewis, 16, 32, 800, 15, 100

Jeremiah Craiton, 13, 30, 800, 10, 100

Robert Meekins, 50, 100, 2600, 75, 200

Marselas Aaron, 30, 20, 1200, 40, 200

Samuel Canon, 30, 20, 700, 20, 150

George Phillips, 30, 60, 750, 10, 250

Solomon Taylor, 35, 59, 900, 30, 150

Samuel Hooper, 30, 25, 1000, 30, 100

Benjamin Travers, 100, 400, 4000, 25, 150

Henry Meekins, 25, 25, 3000, 30, 200

Thomas Travers, 15, 50, 500, 25, 150

Joseph Meekins, 45, 50, 1000, 25, 140

Robert Craiton, 50, 40, 1000, 25, 40
William Barkley, 40, 50, 1000, 25, 150
Robert Craiton, 16, 14, 1400, 20, 100
Macey Adams, 20, 20, 50, 20, 175
Thomas Creighton, 40, 40, 1000, 25, 250
John Aaron, 30, 100, 1500, 30, 175
James Maclaine, 40, 80, 1500, 25, 175
John Brohawn, 40, 80, 1500, 25, 200
Sarah Slacom, 40, 90, 2500, 30, 250
Elias Todd, 50, 25, 1000, 40, 150
Joshua Noble, 170, 75, 4000, 150, 500
Daniel Neal, -, -, -, -, 25
Joseph M. Kenney (McKenney), 70, 30, 2000, 75, 30
John Williams, 11, -, 2000, 5, 200
Samuel Elliott, -, -, 400, -, 125
David Elliott, 5, -, 1000, -, 225
Wm. Brown, 70, 50, 1000, 50, 125
Edward Goslin, 30, -, 3000, 50, 150
Robert Fish, -, -, -, -, -
Paul Conway, 25, 200, 5000, 55, 200
Martin Smith, 150, 80, 5000, 100, 300
William Noble, -, -, -, -, -
Sylvester Smith, 70, 130, 5000, 200, 300
Solomon Richards, 10, 10, 300, 20, 200
Wm. Mowbray, 50, 6, 700, 30, 150
Hooper Hubbard, 50, 20, 1000, 70, 250
Ennols Collins, 25, 15, 800, 40, 50
Arietta Smith, 40, 45, 1000, 30, 80
Margaret Raines, 150, 300, 5000, 50, 300
John Browdle, 100, 90, 2000, 60, 450
Nicholas Wright, 200, 100, 700, 100, 600
Edward Parvin, 150, 75, 2000, 100, 350

Joseph Thomas, 150, 80, 2000, 60, 250
Alex Trice, 150, 75 300, 75, 350
James Sullivan, 2, 10, 300, -, 150
Michael Charles, 100, 50, 2000, 50, 250
Wm. Parker, 100, 20, 2000, 50, 250
Willis Noble, 100, 15, 200, 40, 150
James Lowe, 50, 50, 1000, 40, 150
Josiah Cannon, 100, 75, 3000, 50, 275
James Lowe, 100, 90, 2000, 50, 250
William Charles, 75, 125, 1500, 50, 200
Algernon Thomas, 200, 60, 6000, 300, 800
Samuel Collins, 100, 30, 2000, 50, 200
Jas. Thomas, 150, 75, 5000, 60, 600
Henry Camper, 25, 10, 6000, 10, 135
Wm. Thomas, 100, 90, 2000, 40, 175
Willis Carroll, 140, 200, 4500, 40, 275
Thomas Elsley (Elsby), 150, 150, 4000, 125, 350
John Cochran, 100, 50, 3000, 50, 150
J B. Cochran, 100, 100, 2500, 100, 400
Cely Bush, 130, 28, 2500, 60, 175
Poland Collins, 150, 25, 2200, 50, 200
Keroy Wright, 80, 20, 1500, 40, 175
Samuel Paine, 100, 65, 2000, 70, 250
Jas. H. Williams, 90, 40, 3000, 150, 450
William Hurlock, 15, 10, 150, 40, 150
Elzey Hill, 15, 6, 600, 50, 140
Cyrus Neal, 100, 40, 100, 60, 125
Jos. H. Mann, 140, 60, 3000, 50, 200
Jas. Simms, 2, -, 50, -, 30
Michael Colburn, 100, 50, 3000, 50, 475
Conlan Wright, 50, 100, 500, 50, 225
Thomas Gray, 8, 50, 3500, 30, 100

Daniel Cannon, 150, 50, 300,70, 600
Francis H. Jones, 12, 75, 1700, 40,
190
Ann Hubbard, 80, 35, 1000, 50, 175
William Tilmon, 70, 80, 1000, 30, 70
Perry Flowers, 100,77, 3000, 30, 300
John H. Noble, 80, 60, 2500, 150,
400
Joseph Paine, 40, 60, 100, 30, 150
Jas. Mowbray, 80, 50, 2000, 60, 275
Willis Andrews, 60, 50, 900, 40, 250
John P. Wright, 70, 50, 1500, 50,
200
Josiah Charles, 75, 75, 800, 40, 150
Geo. Kennedy, 75, 45, 700, 15, 200
William Lain, 40, 7, 2000, 50, 100
Elijah Colburn, 125, 75, 500, 10, 250
Jeremiah Nicols, 40, 15, 1000, 50,
150
Samuel Trice, 75, 25, 3000, 70, 200
Jams Stark, 150, 100, 700, 30, 350
Caleb Stark, 70, 20, 2000, 50, 150
Caleb Bowdle, 80, 20, 1200, 40, 250
James Johnson, 80, 50, 500, 150, 120
Jonah Kelly, 20, 135, 1200, 40, 500
Silas Nicols, 40, 100, 1000, 50, 750
Silas Lane, 60, 75, 1000, 50, 100
Danl. Whitely, 15, 20, 400, 20, 6
Cain Lane, 50, 100, 2000, 50, 175
James Nichols, 120, 80, 4000, 150,
400
E. T. Banner, 75, 100, 900, 25, 275
Ourice (Orena) Mills, 25, 20, 500,
40, 125
Samuel Thomas, 30, 7, 1200, 40, 200
Thomas Chesmer, 40, 25, 600, 30,
125
Jos. Fletcher, 25, 50, 500, 40, 125
Wm. Becquith, 120, 30, 3000, 50,
500
Isaac Henry, 100, 96, 3500, 50, 300
Charles Brown, 75, 120, 2000, 50,
300
Richard Stephens, 200, 150, 8000,
100, 350

Wm. Littleton, 75, 250, 3000, 40,
100
James Cochran, 40, 15, 1200, 25,
125
Kendall M. Jacobs, 275, 135, 6000,
200, 800
Daniel Hubbard, 40, 4, 1000, 30, 125
Levi Cannon, 25, 20, 500, 30, 125
H. Willaby, 1000, 1000, 2000, 25,
150
Daniel Wright, 75, 25, 1100 50, 250
William Jestor, 120, 40, 1500, 40,
250
James H. Jester, 40, 50, 80, 30, -
James F. Wright, 100, 55, 2500, 50,
350
John Arnett, 125, 50, 2000, 40, 150
John Jones, 30, 35, 600, 20, 100
John M. Hurlock, 150, 100, 4000,
75, 300
John Dean, 200, 50, 2500, 65, 375
Nathan Williams, 125, 20, 3000,75,
475
John Hutchison, 200, 100, 3000, 40,
200
Thomas Cochran, 80, 20, 100, 50,
175
Price Cochran, 70, 60, 1200, 40, 200
Elisha Phillips, 100, 75, 1500, 40,
200
John Cochran, 90, 60, 1000, 30, -
Nathan Vicers (Vicus), 230, 7, 4000,
100, 500
Susan McCollister, 20, 14, 400, 25, -
Jabez Wright, 100, 50, 1250, 50, 300
Horice Dobson, 30, 20, 400, 25, 125
Thos. J. Watkins, 300, 250, 4400,
100, 100
Thos. Willaby, 70, 40, 100, 4, 200
John Goslin, 100, 100, 2500, 40, 200
Henrietta Prous, 100, 25, 1500, 50,
200
Charles Pinket, 60, 25, 800, 25, 150
Clement Bradley, 100, 50, 2500, 51,
175

# Frederick County Maryland
## Part I
### 1860 Agricultural Census

The University of North Carolina Library under a grant from the National Science Foundation microfilmed agricultural Census records. Records were filmed at the University of North Carolina from original records at the Maryland State Library.

Columns 1, 2, 3, 4, 5, and 13 represent the following information on the census:
1. Name of Owner, Agent or Manager of Farm
2. Acres of Improved Land
3. Acres of Unimproved Land
4. Cash Value of the Farm
5. Value of Farming Implements and Machinery
13. Value of Livestock

Pages for this county are out of sequence.

Thos. Clagett, 308, 25, 30000, 400, 1100

Manasses Grove, 60, -, 5000, 100, 350

Sam Grinder, 13, -, 2000, 100, 325

Thos. R. Jarboe, 205, 12, 16000, 300, 1400

Wm. P. Webster, 175, -, 13000, 400, 1000

J. L. Davis, 152, -, 12000, 380, 600

David T. Jones, 210, -, 14000, 500, 1800

Danl. Baker, 100, 14, 6400, 300, 300

Wm. Richardson, 256, -, 21000, 300, 1040

Danl. Weaver, 160, 25, 14000, 200, 500

Jas. T. Day, 257, 143, 20000, 350, 1500

J. Davis Richardson, 215, -, 21500, 200, 500

B. D. Duvall, 138, 20, 9000, 150, 1010

Danl. Michaels, 248, 30, 24000, 700, 1900

Edwd. McGill, 240, 80, 23000, 70, 1800

Jno. A. Staley, 185, 15, 18000, 600, 1100

August Nicodermis, 200, -, 2000, 700, 1500

Jno. Jarboe, 140, 5, 12000, 400, 1125

Jno. H. Detrick, 153, 25, 14000, 200, 925

Jacob M. Bushey(Buckey?), 180, 8, 10000, 500, 1340

Jacob M. Buckey, 270, 12, 20000, 1000, 7000

W. S. Miller, 200, 22, 18000, 600, 1800

Abner Cassell, 116, 7, 11000, 300, 750

Jas. Finney, 80, 10, 6600, 100, 440

Adam Shaeffer, 75, -, 4000, 200, 500

John Shaeffer, 50, -, 2000, 100, 370

Thos. S. Simmons, 175, -, 16000, 500, 1350

Wm. Funk, 245, 15, 21000, 300, 1400

Richd. Simmons, 148, 10, 12000, 200, 900

Geo. & Peter Thomas, 203, 50, 20000, 500, 1000

Jos. N. Chiswell, 345, 55, 20000, 1325, 3400

Jno. W. Dutrow, 157, -, 13000, 400, 1200

Jas. L. Belt, 190, 20, 8000, 400, 500

Jacb Crist, 80, 20, 1500, 600, 800

Robt. Butcher, 13, -, 400, 25, 120

J. L. Crist, 110, 6, 2700, 100, 550

Dick Offutt, 15, -, 200, 10, 100

Basil Delashmutt, 198, -, 12000, 500, 1100

Wm. Eagle, 300, 155, 10000, 200, 1300

Benj. Moffett, 196, 25, 16200, 700, 1600

Wm. Alnut, 30, -, 24000, 700, 2610

Jno. R. Belt, 70, 10, 1600, 100, 350

Jas. Spencer, 9, 2, 300, 80, 125

Chas. W. McAbee, 80, 20, 3000, 200, 350

Danl. T. Jones, 262, 50, 12500, 600, 2660

Richd. Thomas, 250, 40, 12000, 800, 2400

Peter N. Leapley, 100, 30, 3000, 100, 500

Mary A. Piking, 100, 8, 1500, 50, 450

L. C. Beall, 100, 19, 1500, 340, 680

Jos. H. Beall, 218, -, 12000, 150, 50

Cobb Beall, 155, -, 9000, 100, 550

Benj. Stewart, 95, 38, 3200, 100, 350

Edwd. Nichols, 200, 50, 500, 300, 1700

Richd. Jones, 200, 50, 8000, 200, 600

Nathan Talbott, 250, 81, 12000, 75, 600

A. T. Snouffer, 325, 81, 15000, 500, 2400

Richd. Dutrow, 230, 56, 15000, 500, 1300

Stephen Thomas, 190, 40, 10000, 30, 1300

Henry Drusching, 300, 101, -, 2750, 200, 400

Jos. H. Besant, 86, -, 1000, 80, 320

Wm. H. Smith, 18, -, 1160, 20, 300

B. F. Gatten, 17, -, 900, 50, 220

Jos. Tawley(Fawley), 166, 22, 10000, 300, 900

Jno. Hester, 17, -, 400, 50, 250

Jos. Fawley, 22, 3, 600, 20, 200

Chas. E. Thomas, 338, 12, 17500, 800, 2200

Otho Thomas, 225, 25, 9000, 500, 1900

Haw Ranneberger, 150, 30, 6000, 50, 450

Fred Tongstorm, 56, -, 1200, 150, 600

W. Douglass, 34, -, 15000, 50, 160

Atram Brown, 14, 3, 850, 260, 350

Peter Roelky, 200, 100, 9000, 250, 1400

Guiseppe Michael, 20, 5, 800, 20, 80

Spencer Minor, 120, 14, 5000, 100, 600

Geo. Snouffer, 116, -, 1900, 300, 900

Geo. Snouffer, 160, 10, 3400, 400, 1120

Geo. Snouffer, 260, 84, 17200, 200, 1000

Geo. A. Brady, 412, 40, 36000, 1000, 2800

David Spacht, 59, 40, 4000, 50, 500

Jacob Spacht, 150, 44, 14000, 500, 1000

John Snouffer, 264, 86, 20000, 600, 1250

Jno. A. Trundle, 200, -, 15000, 500, 2150

Jno. A. Trundle, 100, -, 7000, -, 400

Saml. Jarboe, 491, 209, 42000, 600, 1535

Jno. B. Snouffer, 450, 50, 35000, 1000, 2500

Thomas. Harwood, 135, -, 8000, 200, 1800

Grafton Plummer, 135, -, 8000, 100, 600

Edwd. T. Hebb, 283, 17, 15000, 500, 2200

Jas. H. Elgir, 100, -, 7000, 200, 350

Danl. Rhodes, 207, -, 20000, 500, 1400

Geo. W. Copland, 27, -, 2000, 50, 130

Jno. C. Osborn, 228, -, 7500, 400, 1300

Jno. C. Osborn, 135, -, 2500, 100, 500

Lloyd T. Duvall, 100, 100, 8000, 350, 1000

Daniel Duvall, 180, 96, 1100, 150, 825

Jacob Wirts, 97, 27, 6000, 400, 1015

Otho W. Trundle, 200, 122, 11000, 500, 1000

Wilson Trundle, 65, 25, 2700, 100, 600

Clarence Hompstin, 50, -, 1250, 100, 450

Thos. V. Thomas, 200, 400, 9000, 400, 1400

Jacob Spacht Jr., 220, 80, 12000, 300, 800

Jas. Walling, 86, 44, 5200, 150, 450

Jas. Delashmutt, 100, 35, 4000, 200, 500

Perry Jackson, 30, -, 650, 150, 275

Ezra Michael, 180, 38, 8500, 400, 950

Cath. Johnson, 120, 40, 7500, 300, 500

Jos. Carey, 50, 5, 2500, 100, 250

Green Fouts, 30, -, 2800, 50, 400

Jno. W. Cook, 240, 70, 10000, 400, 1100

Arthur Johnson, 180, 44, 10000, 300, 650

Peter Meyers, 150, 20, 4000, 200, 950

Geo. Meyers, 150, 50, 4000, 250, 750

Leond. Waskey, 45, 10, 1000, 100, 500

Lewis Thomas, 135, 35, 8500, 300, 900

Hamilton Gisebertt, 100, 18, 400, 200, 600

Saml. Dutrow, 290, 15, 16000, 500, 2130

Jos. Thomas, 96, -, 7000, 300, 700

Thos. Smith, 270, 30, 15000, 400, 1200

Jno. B. Thomas, 315, 7, 22540, 800, 2400

John H. Unsell, 302, -, 24000, 500, 1200

Geo. W. Padgett, 215, 100, 21500, 700, 1700

Geo. Thomas, 182, 3, 16000, 7800, 2350

Jas. A. Cook, 98, 66, 8000, 300, 800

Jno. Thomas, 230, 70, 20000, 200, 900

Jonathan Thomas, 131, 19, 8000, 300, 1050

Frances Blessing, 105, 2, 5000, 200, 600

Saml. Vohip (Vohis), 120, 30, 6000, 300, 970

B. Cunningham, 260, 200, 1900, 800, 2500

Saml. Linder, 150, 53, 8000, 200, 900

Wm. H. Blessing, 130, 40, 6000, 300, 700

Jno. Wrenn, 200, 100, 9000, 300, 1100

Jno. Degrange, 170, 100, 10000, 300, 1100

Danl. Thelman, 100, 9, 4000, 200, 1000

David Zimmerman, 70, 10, 3000, 100, 500

Henry Rohr, 102, 8, 4000, 150, 450

Abm. & Peter Beard, 110, 10, 4500, 750, 600

David Degrange, 60, -, 2400, 50, 400

Thos. McPherson, 16, -, 900, 10, 300

Henry Zimmerman, 164, 40, 12000, 600, 850

Edwd. Zimmerman, 130, 20, 120000, 750, 900

Horace Zimmerman, 100, 18, 6500, 300, 800

Geo. Willard, 39, -, 2500, 100, 350

Wm. Zimmerman, 60, 2, 4200, 150, 700

Jas. H. Hargate, 70, 4, 3500, 100, 600

Eli Shaeffer, 22, -, 2000, 50, 200

Jno. Whip, 28, -, 2800, 20, 200

Saml. Whip, 120, 28, 6000, 200, 800

Saml. Thomas, 1230, 50, 17000, 600, 1200

Nick Cromwell, 366, 60, 34400, 500, 1800

Wm. Thomas, 241, 20, 21000, 400, 2000

Jno. A. Hedges, 268, 50, 28000, 400, 300

Jno. Smith, 8, 2, 7000, 40, 750

Jno. A. Hedges, 268, 50, 28000, 400, 2300

Jno. Smith, 8, 2, 700, 40, 750

Joshua Zimmerman, 90, 15, 5400, 125, 900

Jos. Zimmerman, 55, 20, 2000, 80, 450

John Zimmerman, 56, 12, 2500, 100, 400

Jno. W. Hargate, 216, 125, 15000, 500, 1400

B. A. Cunningham, 130, -, 8000, 100, 500

Benjamin. J. Snouffer, 203, 20, 17000, 700, 2000

Geo. Blessing, 80, 140, 7000, 200, 700

William Green, 40, -, 800, 100, 400

David Grossnickle, 100, 100, 4000, 75, 500

Geo. Grossnickle, 60, 20, 2500, 150, 500

David Delader, 20, -, 400, 50, 250

Daniel Hoover, 20, 30, 1000, 40, 250

S. Keesetring, 12, 88, 1500, 150, 530

Peter Grossnickle, 100, -, 3000, 200, 890

Daniel Arnehem, 100, 120, 6600, 200, 600

J. Grossnickle, 45, 30, 3000, 200, 500

T. G. Bleckenstaff, 100, -, 2000, 150, 400

E. Grossnickle, 90, 34, 5000, 150, 100

J. Stottlemeyer, 63, 48, 5000, 300, 800

J. Grossnickle, 89, -, 4000, 200, 400

J. Shanafeltz, 18, 14, 1000, 150, 221

E. Bleckenstaff, 70, 100, 4000, 300, 742

Fredk. Biser, 109, -, 4500, 200, 560

D. Stottlemeyer, 100, 400, 4000, 100, 770

Jacob Ludy, 100, 155, 880, 100, 440

E. Leatherman, 40, 27, 200, 10, 181

J. Stottlemeyer, 25, 111, 1200, 35, 300

Solomon Forrest, 200, -, 4000, 200, 700

Joseph Stottlemeyer, 100, -, 2000, 50, 400

Saml. Bremdoby(Sremdoby), 100, -, 3000, 300, 600

S. Manganus, 70, 700, 4000, 200, 600

John Beeker, 58, 40, 1040, 150, 400

John Donpple, 40, 40, 1300, 140, 300

John Hissong, 90, -, 2600, 200, 400

Jonathan Fry, 70, 50, 2000, 200, 500

Peter Sharp, 57, -, 600, 50, 150

Dr. Wurnerfelzts, 180, -, 3400, 200, 600

J. Hooker, 100, 41, 5000, 200, 800

H. H. Mullen, 33, -, 1500, 80, 300

J. Wolfe, 63, 40, 5000, 100, 300

L. W. Hays, 30, -, 900, 55, 240

Saml. Hoover, 75, 45, 3000, 200, 500

J. Harsheman, 225, -, 9500, 300, 800

C. Herscheman, 150, -, 4500, 200, 700

E. Hersheman, 44, -, 1700, 50, 400

Ezra Smith, 30, -, 1000, 100, 300

John Donpple, 60, 8, 1660, 75, 269

Joel Garnerd, 70, -, 2000, 800, 400

L. Earsterdy, 60, -, 180, 150, 200, 500

Solomon Hoftman, 100, -, 3000, 150, 300

J. Palmer, 40, -, 1500, 75, 150

M. Simmer, 19, -, 600, 50, 300

George Summers, 95, -, 3000, 150, 500

E. Rontzahm, 180, -, 1000, 150, 969

Jacob Rontzahm, 220, -, 8800, 150, 1119

S. Simmer, 130, -, 850, 150, 1060

Adam Leatherman, 68, -, 4000, 150, 700

J. Leatherman, 48, -, 3000, 150, 700

J. L. Warenfeltz, 125, -, 7800, 50, 260

J. Warenfeltz, 141, -, 10000, 100, 400

Enous Marker, 45, -, 2000, 50, 300

J. T. Warenfetz(Warenfety), 95, -, 4500, 100, 400

Rebecca Easterdy, 100, -, 4000, 100, 400

Rebecca Easterdy, 120, -, 6000, 150, 400

G. M. Easterdy, 20, -, 1700, 50, 320

Samuel Palmer, 18, -, 1000, 15, 50

George Ansboro, 100, -, 3000, 100, 600

Peter Marker, 100, -, 3500, 100, 500

Sarah Marker, 50, -, 3000, 50, 400

Daniel Garver, 50, -, 300, 50, 300

John Shroger (Shroyer), 60, -, 1600, 50, 440

John Devilbin, 110, 82, 5000, 100, 1000

William Butler, 116, 10, 2500, 20, 500

Sarah Gunder, 110, -, 2500, 50, 500

William H. Grimes, 245, 15, 5200, -, 700

Adam Black, 75, -, 1500, 30, 100

Philip Matthias, 63, -, 1250, 15, 100

Doct. G. Zimmerman, 39, -, 2000, 25, 200

Henry Whitmore, 60, 50, 5000, 250, 900

Elizabeth Myers, 120, 20, 6500, 150, 800

Henry Schryock, 120, 22, 2000, 100, 700

Ephraim Myers, 190, 25, 6000, 200, 100

Jacob McDonald, 58, 5, 1100, 50, 61

Jacob McDonald, 100, 10, 1700, 200, 800

Jacob McDonald, 59, 5, 1000, -, -

William Bell, 150, 30, 5500, 200, 850

Henry Bowersox, 22, -, 2000, 50, 200

Sarah Cramer, 126, 40, 4000, 100, 350

Samuel Kinsly(Kinely), 100, 14, 4000, 200, 600

John Stottlemeyer, 70, 13, 2200, 100, 700

Samuel Hoffman, 100, 10, 3400, 75, 600

Daniel Barkman (Bachman), 21, -, 550, 50, 150

John Flanagan, 60, 20, 800, 50, 600

Solomon Ohler, 34, -, 500, 50, 200

Michael Hackler, 150, 20, 3500, 50, 500

Henry Morningston, 40, -, 1000, 10, 75

Jacob Wachter, 40, -, 1000, 50, 200

John Saunders, 70, -, 3000, 150, 800

William Todd, 75, 74, 8000, 300, 800

Peter Degrange, 25, -, 1000, 50, 300

John Willson, 120, 100, 3000, 150, 800

Samuel Cline, 46, 12, 2000, 250, 300

Henry Shield, 75, 10, 2000, 150, 450

George Wachter (Machter), 36, 50, 900, 50, 600

Eph. Deleplane, 80, -, 2000, 50, 600

Joshua Deleplane, 100, -, 2200, 35, 400

William Keefer, 62, -, 1500, 50, 350

Frederick Hunhey(Hunkey), 127, -, 2000, 100, 800

John Putman, 118, 4, 3500, 150, 800

Aaron Baltzell, 200, -, 5000, 100, 900

Ephraim Ridge, 130, -, 2500, 150, 470

John Curtis, 65, -, 1800, 50, 300

Greenbury Ridge, 60, 10, 1500, 50, 400

James Null, 60, -, 1000, 20, 100

Eph. Graskan (Graskar), 21, -, 60, 10, 150

George A. Graham, 17, 20, 4000, 25, 300

David Cassell, 336, 50, 9000, 300, 1150

Baltzer Fox, 350, 10, 12000, 300, 1000

Samuel Keefer, 300, 150, 9000, 250, 1000

Washing Miller, 155, 20, 8000, 150, 800

Eph. Develbin, 39, 21, 2700, 25, 200

Benjamin C. Flowers, 66, -, 3000, 25, 200

George H. Powl, 54, 6, 1400, 30, 300

George Meixell, 55, 30, 2400, 75, 500

Daniel Gaugh, 175, 30, 5000, 150, 1200

Samuel Favorite, 164, 20, 400, 100, 900

James Null, 125, 15, 4000, 175, 1200

Eli Cramer, 135, 50, 500, 150, 1000

Jonathan Gaugh, 125, 2, 4500, 200, 900

Eve Hill, 250, 50, 12000, 125, 1050

Eve Hill, 70, 10, 2000, 100

George Meixell, 80, 30, 2450, 150, 600

Henry Eaton, 100, 32, 4000, 50, 1550

Jacob Bishop, 34, -, 1000, 30, 100

George W. Faishour, 65, 20, 1400, 100, 600

Henry A. Freshour, 65, 40, 1400, 75, 400

William Long, 68, -, 1500, 125, 500

Michael Isengle, 175, -, 5000, 100, 650

Jacob Ramsburg, 58, -, 20, 2500, 200, 500

Alexander Ramsburg, 150, 57, 5000, 200, 600

Hezekiah Shaw, 120, 10, 3000, 50, 550

George W. Moot, 60, -, 1400, 50, 400

John Gusey, 157, -, 6000, 250, 900

Grafton Craver, 35, -, 1500, 50, 60

Jacob Clem, 111, 22, 2500, 100, 600

Daniel Snook, 210, 60, 10000, 300, 1400

Cornelius Barrick, 20, -, 850, 40, 200

William A. Shaffer, 70, 57, 2000, 100, 700

Christian Bowersox, 75, -, 1800, 50, 200

William Hill, 234, 54, 10000, 300, 1000

Joshua Briggs(Biggs), 270, 60, 8000, 400, 800

Joshua Biggs, 31, -, 1000, 50, 390

Frederick Crouse, 150, 50, 12000, 200, 1000

Daniel Ramsburg, 160, 100, 2500, 50,700

Ezra Michael, 120, 46, 4000, 150, 903

Samuel Rice, 23, -, 800, 50, 150

Adam Albert, 65, 20, 1000, 25, 200
Joseph King, 50, -, 500, 20, 600
Daniel Leatheman, 150, 15, 3000, 100, 700
Jacob Crouse, 167, 5, 11500, 200, 900
Jacob Harp, 250, -, 12000, 100, 600
Zebulon Kettymale, 80, 89, 2600, 200, 900
Cornelius Ridge, 146, 21, 4500, 100, 700
John Speak (Spiak), 130, -, 1500, 50, 600
William Hall, 171, -, 3500, 200, 700
George Hankey, 123, 12, 3000, 75, 950
George Layman, 155, 15, 4500, 140, 750
John W. Staub, 110, 10, 3500, 100, 650
Henry C. Waxman, 180, 100, 10000, 500, 1000
George Miller, 137, 26, 3000, 150, 800
George Miller Jr., 136, 10, 2500, 50, -
Henry Ramsburg, 140, 10, 4500, 175, 100
Henry Willhide, 100, 10, 2200, 100, 400
Henry Willhide Jr., 50, -, 1000, 60, 500
William Jackson, 300, 50, 12000, 180, 1000
William Roberts, 100, 40, 2000, 50, 600
Samuel Eichelburger, 100, 50, 2000, 100, 550
John Harity, 70, 5, 1500, 75, 900
Henry Layman, 153, 17, 3500, 100, 700
George Groshan, 100, -, 2500, 100, 745
David Kolb, 26, -, 1410, 20, 45
Johnathan Roughet, 300, 30, 1000, 450, 1000

David Koontz, 170, 20, 4000, 300, 800
Edwin Springer, 180, 40, 8000, 350, 550
William Gilbert, 140, -, 3000, 270, 600
Sylvester Valentine, 135, 18, 3000, 175, 700
Abraham Anders, 100, 15, 2000, 200, 650
Abraham Troxel, 227, 25, 3000, 250, 300
Elias Kriso, 100, 21, 4000, 300, 1000
Jacob Hinea (Hines), 137, 6, 2000, 100, 800
John Hoffman, 125, 15, 4200, 200, 700
Michael Eichelburger, 248, 40, 11000, 500, 1220
Noah Miller, 130, 30, 3200, 200, 500
Samuel J. Beteler, 124, -, 2500, 100, 700
Elis Otto, 215, 40, 10700, 450, 1200
Joseph J. Crane, 143, 10, 2800, 300, 900
John Wachter, 115, 15, 2500, 150, 600
George W. Appold, 60, -, 1500, 100, 275
Isaac Saylor, 202, 20, 8000, 400, 1000
Peter Harkey, 200, 15, 10000, 500, 1100
Peter Hankey, 104, -, 2500, 100, 500
Dixon Eichelburger, 278, 100, 12000, 400, 1400
Elizabeth Hoffman, 100, -, 1200, 50, 600
Elias F. Valentine, 100, 73, 3000, 150, 1000
Jacob Sate, 127, 5, 4000, 250, 800
Warner T. Grimes, 120, 32, 5000, 100, 600
Charles Stephens, 39, -, 800, 125, 400

James Valentine, 128, 20, 4500, 120, 400

Thomas Elder, 60, -, 1200, 50, 400

John Fisher, 132, 15, 1600, 100, 1000

Solomon Ridenour, 130, -, 1500, 50, 350

Ephraim Crouse, 65, -, 1250, 75, 300

Frederick Troxel, 26, -, 1000, 50, 100

Samuel Beteler, 180, 60, 8000, 200, 900

Thomas Pusey, 27, -, 1000, 50, 250

George W. Barrick, 225, 50, 15000, 500, 1500

Mary Speeker, 105, 15, 1000, -, 50

Henry Hinge, 127, 10, 3000, 100, 700

Jacob Ecker, 107, 10, 3500, 100, 500

Amon T. Norris, 86, 25, 2500, 75, 800

George Valentine, 112, 10, 3000, 75, 420

William Krise, 127, 10, 5000, 150, 500

John W. Albaugh, 150, 5, 3500, 120, 490

John L. Little, 50, -, 4000, 50 300

Michael Roberts, 5, 2000, 100, 700

M. C. Adelspufer, 93, -, 500, 50, 460

Jos. Hospelhorn, 160, 59, 3500, 148, 517

Henry Gilwick, 37, -, 1200, 50, 280

Patrick Sargi, 80, 20, 2000, 30, 242

Samuel Daprow, 46, -, 2500, 200, 832

Matthias Martin, 115, 8, 1440, 50, 527

Jos. Swearry, 73, 24, 2800, 35, 116

Henry Hann, 94, 12, 3500, 100, 408

Michael Rider, 90, 6, 3500, 150, 603

Jacob Robertson, 200, 60, 7000, 400, 435

Wm. Gilson, 172, 45, 6000, 175, 800

Peter Graybill, 200, 50, 6500, 100, 640

Jos. Culbertson, 70, 90, 1500, 50, 470

Christian Carrell, 370, 130, 30000, 200, 760

Adam Bowers, 75, 12, 3000, 340, 670

Alfred B. Black, 120, 10, 2000, 15, 460

Geo.Shaeffer, 165, 50, 4500, 200, 727

Wm. P. Gardner, 100, 39, 2800, 160, 622

Jno. Hackersmith, 180, 40, 3500, 100, 550

Lydia Fraser, 125, 15, 1500,75, 692

Jacob Lynn, 100, 40, 3000, 190, 395

Jno. J. Kelley, 225, 60, 8000, 83, 558

Isaac Fisher, 150, 25, 4375, 275, 908

Jno. Oler, 67, 23, 840, 75, 491

Wm. Walker, 125, 27, 4560, 200, 515

Peter Tell, 70, 31, 5000, 200, 442

Saml. Valentine, 110, 19, 2500, 75, 608

Andrew J. Oler, 114, 20, 2546, 50, 357

Geo. Null, 86, 8, 1800, 50, 352

Reuben Fleming, 130, 82, 6000, 175, 310

Jacob Bollinger, 90, 30, 5000, 150, 410

Absolem Smith, 60, 19, 2000, 250, 470

Samuel Ott, 130, 15, 3500, 118, 740

Jacob Shoemaker, 224, 56, 5600, 300, 449

George Moore, -, -, -, 352, 542

Isaac Oler, 93, -, 2000, 15, 459

Lewis P. Shriver, 130, 60, 5000, 350, 832

Hamilton Martin, 150, 20, 4000, 100, 722

Saml. G. Oler, 316, 100, 7600, 150, 1330

Chas. Warick, 165, 35, 4000, 288, 672

Sol Slabang, 58, 75, 3000, 150, 293
Saml. Eckenrode, 155, 40, 7000, 56, 385
Jacob Gillelean, -, -, -, -, 666
William Gillilean, 206, 37, 7135, 228, 598
Jno. Trostle, 90, 6, 2000, 70, 405
Jac. Baumgarnod, 90, 30, 3500, 175, 483
J. Stansbury, 100, 13, 2800, 106, 660
Jesse Clopt (Close), 148, 48, 3000, 405, 722
Jos. Terrell, 149, 30, 6000, 196, 895
Jesse Hoover, 142, 18, 2800, 126, 491
Saml. Flegil, 50, 23, 2000, 44, 260
Elijah Close, 120, 20, 420, 50, 550
Wm. Wealty, 170, 20, 2000, 11, 154
David Morrison, 238, 18, 5300, 216, 928
Edmond E. Biggs, 75, 9, 2000, 135, 545
Isaac Barton, 107, 4, 2896, 115, 649
Nichl. Stansbury, 160, 40, 2800, 75, 566
Jacob Munshorer, 55, 30, 4000, 105, 842
Nichl. Seabalt, 150, 35, 2800, 96, 375
David Witmore, 125, 36, 3220 60, 460
Josiah Doterer, 132, 30, 2000, 167, 570
Chas. Zacherius, 275, 125, 9000, 276, 1050
John Walker, 165, -, 2000, 30, 119
John Dorsey, 330, 121, 8000, 282, 1186
Chs. Hobbs, 140, 5, 3000, 75, 568
Jno. F. Elder, 170, 65, 4000, 500, 10
Dennis McCanin, 42, 12, 1500, 20, 219
Daniel Esrook, 145, 5, 2800, 60, 489
H. G. Ovelman, 240, 35, 6000, 107, 805
David Serfs (Seifs), 113, 18, 238, 201, 486

Abiah Martin, 111, 14, 2000, 215, 668
Wm. Sipes, 150, 26, 3500, 115, 410
Joseph Martin, 135, 27, 3500, 194, 515
Mary A. Biggs, 103, 15, 2690, 75, 415
Jno. Martin, 107, 7, 2000, -, 95
Jas. Griffin, 110, 26, 2144, 40, 182
Wm. Malter, 243, 10, 8000, 196, 919
Jno. Hobbs, 100, 5, 2500, 50, 352
Wm. S. Black, 160, 30, 3000, 100, 620
Hezekiah Doterer, 78, 17, 2000, 40, 571
Jacob Overhtlyer, 115, -, 3000, 68, 502
Wm. R. Beall, 45, 111, 1200, 30, 402
Henry Little, 50, 100, 900, 30, 194
Christian Lanly, 75, 300, 3800, 80, 448
Cole R. Annan, 85, 225, 7000, 190, 543
David Gamble, 75, 85, 3000, 250, 490
Michael Hooke, 200, 25, 12000, 73, 280
William Mason, 200, 10, 5250, 130, 818
Richard Gibson, 240, 20, 5500, 357, 9400
John Fass, 131, 16, 3000, 100, 250
James Ohler, 132, 35, 3450, 150, 552
Ephraim Saffington (Buffington), 150, 25, 3200, 50, 490
Felise B. Taney, 176, 50, 2188, 165, 564
John Walter, 47, 150, 2000, 70, 474
Sally Nicles, 175, 50, 10000, 250, 1020
Lewis Elder, 60, 4, 4000, 80, 297
Chas. A. Manning, 290, 2, 7000, 500, 1196
Dr. Jas. A. Short, 150, 50, 4000, 100, 410

George Worthing, 30, 20, 1600, 109, 533

Joseph Dyers (Syers, Byers), 200, 23, 5000, 200, 625

Solomon Prise, 290, 4, 5750, 80,879

Samuel Masserwill, 65, 10, 5000, 75, 700

George P. Beam, 136, -, 5000, 200, 266

Joshua Matter, 100, 84, 4840, 200, 655

Dr. A. Annan, 90, 70, 6400, 300, 1025

Lewis A. Matter (Malter), 100, -, 15000, 300, 720

Samuel Matler, 100, 70, 3000, -, -

James D. Hickey, 48, -, 200, 50, 250

Mt. St. Mary's College, 250, 500, 7000, 375, 2750

Henry Deilman, 47, 17, 2500, 60, 100

St. Joseph Seminary, 279, -, 8928, 570, 3520

Hezekiah Bailey, 40, -, 400, 125, 230

Alfred Brengle, 26, 45, 2800, 150, 900

William Reich, 200, -, 18000, 600, 1000

Caspar Cline, 86, 10, 6000, 150, 600

Jno. A. Steined, 10, -, 3000, 30, 450

A. D. O'Leary, 30, -, 10000, 100, 400

Dr. Wm. Tyler, 107, 40, 21800, 200, 1000

Richd. Potts, 210, -, 15500, 1000, 1050

Jno. Loats, 178, -, 35600, 1000, 3000

Andrew Bozel (Boyle), 20, -, 3000, 50, 2000

Wm. Dean, 115, 15, 7400, 200, 1100

Jas. Carlin, 11, -, 2200, 70, 250

Saml. Wolf, 72, -, 10000, 400, 900

Saml. Maight, 18, -, 3000, 150, 250

Jno. W. Birely, 110, -, 7700, 50, 200

Jno. Sifford, 20, -, 4500, 150,600

J. Sifford, 120, 50, 1000, 500, 300

John Sifford, 160, 35, 10000, 500, 500

Jno. Wilcoxen, 60, 20, 6600, 200, 500

Wm. H. Ramsburg, 165, 35, 14700, 300, 2000

Jno. Zimmerman, 31, -, 5000, 150, 200

G. H. Rizer, 139, -, 12000, 270, 1000

Jno. L. Mires, 183, 30, 19500, 600, 1000

Geo. Font, 150, -, 14000, 400, 740

David Boyd, 110, -, 15000, 500, 700

Dr. W. D. Garrison, 20, -, 5000, 75, 175

Jacob Lewis, 248, -, 34700, 600, 2600

Philip Killian, 120, -, 12000, 500, 1200

Jno. Thomas, 37, -, 3700, 75, 320

N. D. Hand, 27, -, 2700, 50, 150

Lemard C. Molineaux, 197, -, 19700, 500, 1400

Martin Tonson, 190, 30, 20000, 400, 1025

Cath. Brunner, 73, 60, 9300, 50, 400

J. Oliver Myers, 225, 30, 18800, 500, 1730

David Kemp, 220, 120, 28000, 500, 1290

Chas. Myers, 104, -, 10000, 300, 120

G. W. Cronise, 105, -, 10000, 100, 450

Theo. Shults, 76, 56, 8200, 200, 560

Joshua Dill, 35, -, 7000, 50, 550

John Font, 11, -, 2200, 20, 300

Valentine Albaugh, 50, 12, 8400, 200, 500

Jos. Lightner, 49 ½, -, 5000, 100, 1100

Wm. Kolk, 27, -, 5000, 50, 560

Jas. Steiner, 46, -, 6000, 125, 350

Jacob Steiner, 72, -, 8000, 150, 500

Ezra Staley, 41 ¾, -, 5000, 150, 350

Isaac Brunner, 95, -, 7600, 250, 875

Issac Howard, 335, -, 35500, 200, 1425

Simon Crosnise Sr., 67, -, 6700, 100, 510

Simon Cronise Jr., 20, -, 2000, 20, 300

Jonathan Crosnise, 61, -, 3500, 125, 250

Jno. Crosnise, 25, 5, 1500, 75, 150

David Best, 375, 50, 25500, 600, 1265

Valentine Adams, 364, 34, 30000, 350, 2000

Wesley Griffin, 150, -, 9000, 150, 1100

Wm. H. H. Adams, 140, -, 14000, 170, 600

Nathan Davis, 298, -, 26820, 500, 1500

Danl. Downs, 20, -, 2000, 50, 350

Mary Font, 100 50, 13000, 200, 800

Chas. W. Burns, 123, -, 12300, 300, 530

Geo. Gallion, 100, 5, 10000, 128, 500

Geo. Gallion, 111, -, 11103, 100, 632

Danl. Scholl, 20, -, 2000, 50, 36

Jno. A. Smith, 125, -, 10000, 500, 1045

Edwd. Howard, 316, 50, 20000, 300, 1832

Jno. Stone, 170, -, 14450, 300, 1335

Jno. Cassel, 250, 50, 10000, 400, 1150

Lewis G. Kemp, 148, 8, 14800, 400, 1550

Jno. Phligew (Phligen), 22, -, 3000, 300, 280

Isaiah Medy, 710, 26, 13600, 200, 1180

Phillip Collier, 257, 60, 15850, 200, 1195

Saml. Zimmerman, 110, -, 5500, 130, 800

Geo. Zimmerman, 110, 6, 5500, 170, 700

Elias Zimmerman, 150, -, 7500, 500, 1200

Cath. Stockman, 100, 40, 7000, 200, 625

Cath. Cassell, 120, 80, 8000, 200, 350

Abner Hargath, 24, -, 900, 50, 250

Jesse King, 6, -, 600, 50, 200

Lydia King, 98, -, 800, 70, 180

Saml. Crosnise, 155, 26, 10000, 425, 800

Fred K. Friday, 45, 27, 2300, 50, 150

Jno. Friday, 24, 1, 1250, 30, 125

Wm. Fallman, 10, -, 500, 20, 140

Henry Smith, 12, 9, 700, 15, 136

Wm. Kemp, 35, -, 450, 10, 130

Jacob Motler, 11, -, 350, 10, 110

Jno. A. Degrange, 60, 30, 2500, 100, 250

Jno. Leakins, 14, -, 420, 90, 130

Peter Holter, 345, -, 700, 50, 250

Danl. Smith, 150, 15, 8250, 300, 768

Jno. Smith, 1125, 20, 8700, 200, 610

Mahlon Roderic, 172, 34, 8500, 200, 1063

Christian Greenwall, 34, 4, 1120, 50, 140

Jacob Michols, 100, -, 7500, 200, 500

Alf Staley, 100, -, 7000, 150, 500

Chas. E. Howard, 75, 12, 4350, 300, 1230

Edwd Howard, 50, -, 3500, 300, -

Joshua Rhodes, 34, -, 2500, 75, 240

Jos. Rhodes, 26, -, 1500, 50, 235

David Derr, 148, 6, 7700, 200, 960

Edward Buckey, 180, 57, 11200, 300, 1400

Saml. Stup, 61, -, 3900, 70, 390

David Shaeffer, 148, 10, 8880, 150, 1290

Jos. Zimmerman, 165, 25, 10700, 400, 1210

Jacob Getzendaner, 90, 14, 2590, 200, 350

Wm. Crampton, 85, 19, 2100, 100, 600

Ezra Burkhart, 18, -, 1500, 100, 110

Geo. Burkhart, 26, -, 1500, 100, 100

Jas. Getzendaner, 34, -, 1100, 30, 135

Chas. Eader, 260, 44, 15000, 300, 1290

Lloyd Whip, 50, 69, 2500, 50, 2500

Saml. Getzendaner, 60, -, 1200, 200, 420

Sarah Getzendaner, 12, 9, 500, 40, 200

Mary Fraley, 108, 2, 4032, 200, 610

Sam. Hargate, 250, 60, 16000, 400, 1980

Geo. W. Smith, 183, -, 16000, 350, 1300

Wm. P. Maulsby, 50, -, 10000, 200, 500

David Stup, 84, 30, 5100, 200, 600

Henry Ohler, 23, 6, 630, 20, 160

Jno. Head, 12, 8, 450, 10, 80

Philip Dennis, 17, 5, 800, 40, 130

Lewis Ramsburg, 130, -, 8450, 150, 750

Jas. Heine, 14, 3, 350, 10, 180

Fredk. Kimball, 20,- , 500, 10, 120

Jno. Beck, 19, -, 500, 50, 220

Eli Stup, 18, -, 500, 21, 120

Jacob Stup, 32, -, 1000, 10, 110

Jas. Smith, 17, -, 600, 5, 98

Jacob Staub, 23, -, 1000, 50, 60

Elias Stein, 75, 7, 2500, 150, 650

Wm. Mercer, 12, -, 150, 70, 230

Fredk. Cline, 40, -, 2120, 50, 230

Elias Weddle, 17, -, 1000, 5, 150

Jas. Stevens, 30, 15, 1400, 15, 160

Benj. Keyser, 30, -, 1200, 30, 210

Michl. Hagen, 21, 40, 2000, 100, 250

Danl. Gambrine, 15, -, 1200, 40, 110

Jno. Hagen, 100, 4, 10000, 75, 350

Jno; Hagen, 75, 114, 7560, 200, 390

Chas. Lightner, 180, 20, 12000, 200, 1100

Isaiah Ramsburg, 55, 7, 2000, 100, 600

Henry Smith, 16, 4, 500, 10, 110

Jas. Wagner, 14, -, 500, 10, 88

Elis Owings, 22, -, 2200, 20, 150

Thos. Webster, 35, -, 1200 50, 400

Danl. Nann (Hann), 21, -, 600, 10, 140

Elias Yost(Bost), 18, 2, 900, 15, 128

Eph. Riddlemoser, 22, 18, 1500, 95, 120

Joseph Stup, 175, 22, 17500, 300, 1100

Andw. Smith, 26 ½, -, 2000, 30, 150

Jno. C. Flemming, 70, 6, 6000, 100, 350

Jno. Smith, 160, 50, 12000, 300, 700

David Miller, 158 ½, -, 13000, 250, 1050

Jno. H. Stup, 100, -, 10000, 100, 600

Joseph Crosnise, 60, -, 6000, 150, 800

Wm. C. Preston, 240, 60, 17000, 400, 2300

Chas. H. Burkhart, 120, 20, 5500, 100, 500

A. W. Burkhart, 20, 10, 2000, 50, 150

Elias Delashmut, 125, 12, 18000, 200, 1100

Philip Reich, 520, 30, 5000, 1500, 300

Dennis Stubl, 158, -, 15800, 200, 700

Rezin Hobbs, 14, -, 1400, 10, 100

Thos. Simpson, 17, -, 3000, 50, 750

Cristn. Folk, 20, 80, 500, 25, 250

Jno. Stein, 30, 95, 800, 100, 350

Wm,. Knell, 20, 30, 400, 50, 250

Theo. Hartzock, 13, 10, 250, 10, 140

Jno. Michael, 30, 72, 500, 20, 100

Jno. Derr, 150, 20, 17000, 700, 1300

David Kemp, 206, -, 2100, 1000, 1350

Andw. J. Worman (Norman), 10, -, 1500, 10, 325

Jno. Marker, 230, 15, 24500, 800, 1400

Jacob Walker, 200, -, 20000, 800, 1800

Wm. White, 64, 5, 7000, 200, 600

Jno. Hynes, 110, -, 11000, 250, 350

Geo. Rhoderick, 160, -, 16000, 300, 850

W. D. Bowers, 18, -, 1800, 100, 450

Jno. Engle, 70, -, 2100, 50, 250

Jno. McDeritt, 120, 12, 7000, 150, 700

Emanl. Feeser, 26, -, 2500, 100, 300

Albert Ramsburg, 118, -, 4000, 100, 525

Urias Ramsburg, 120, 20, 4000, 200, 700

Wm. F. Zimmerman, 82, -, 3000, 100, 600

Chas. Broadrup, 70, 70, 6000, 200, 500

Michael Steel, 118, -, 4000, 150, 1200

Jno. Steel, 105, 30, 4000, 150, 800

Leander Steel, 119, 10, 3000, 150, 500

Michael Henck, 102, 20, 3000, 100, 500

Abner Meixell, 92, 3, 2500, 50, 350

Fredk. Stull, 150, 50, 8000, 300, 1000

Danl. Eyler, 100, 12, 5600, 200, 350

Cornelius Staley, 106, 35, 1100, 100, 800

Cornelius Staley, 201, 40, 17300, 80, 200

Benj. Hull (Hall), 110, -, 1200, 200, 630

Sol. Love, 174, -, 17000, 300, 1500

Wm. Hildebrand, 210, -, 2100, 400, 1100

Jno. Reich, 170, 16, 18000, 250, 800

Philip Hutshen, 44, -, 5000, 100, 325

Richd. W. Holland, 128, -, 10000, 200, 650

Dan Getzendaner, 163, -, 10000, 500, 1375

F. N. Getzendaner, 137, -, 11000, 500, 1300

Caspar Merhling, 30, -, 6000, 60, 300

C. Artz, 100, -, 6500, 200, 900

W. Burrick, 20, -, 3000, 50, 194

Wm. Worman (Norman), 320, 80, 32000, 1000, 2000

Lewis Thomas, 190, 50, 17000, 500, 1800

David Thomas, 270, 50, 33000, 500, 2000

Wm. Hough, 197, -, 17000, 200, 600

Wash Smith, 205, -, 20000, 200, 700

Augt. Staley, 188, -, 18000, 300, 900

William Eader, 300, -, 30000, 150, 1000

Geo. Gittinger, 137, 60, 12000, 1500, 1200

Geo. R. Dennis, 300, 80, 38200, 1500, 5000

John Gallion, 266, 6, 30000, 1000, 2500

Danl. Maerhat, 150, 60, 1600, 1000, 1300

Henry Perry, 140, 60, 1400, 800, 1000

Sol. Barrick, 185, 15, 1200, 350, 600

Thos. Buxton, 205, 10, 3000, 200, 1500

David Glaze, 200, 20, 30000, 200, 1600

Wm. Perrill, 100, 20, 8500, 40, 1000

Enos Hedges, 42, 12, 2800, 40, 200

Henry Railings, 40, -, 2400, 50, 200

Peter Brunner (Banner), 230, 20, 15000, 600, 800

Amos Cramer, 224, 20, 30000, 300, 900

Joseph Glaze, 227, 20, 30000, 400, 500

Ezra Warrenfeltz, 120, 10, 8200, 300, 500

Danl. Hedges, 222, -, 17000, 300, 2000

Joshua Craven, 90, 10, 2000, 200, 300

Henry Bruner, 175, 18, 11000, 650, 1200

David Mayne, 90, 30, 2000, 50, 150

Alfred Grove, 158, 14, 8000, 300, 700

Dan Shook, 100, 14, 3000, 150, 600

Edwd. Kemp, 87, -, 2000, 40, 350

Jno. Shurdler, 26, 10, 1300, 50, 275

Henry Brane, 15, -, 1000, 15, 150

Jno. Buck, 90, 40, 3000, 40, 200

Lewis Blank, 21, -, 600, 30, 150

Wm. Mayne, 31, -, 1800, 50, 175

Danl. Smith, 16, 28, 1400, 40, 175

Fredk. Cline, 13, -, 400, 30, 120

David Houck, 52, -, 3000, 75, 600

Sol. Shroyer, 10, 4, 400, 30, 100

Michael Hink, 13, -, 600, 47, 154

Mary Zimmerman, 18, -, 900, 20, 110

Jno. Zimmerman, 34, 40, 800, 35, 150

Wm. Zimmerman, 30, 2, 1000, 35, 160

Michael Zimmerman, 34, 2, 1000, 50, 150

Chas. McDevritt, 39, -, 2000, 50, 200

Nathan Neighbors, 70, 500, 5000, 300, 1200

Elias Grove, 100, 90, 6000, 100, 1300

Jos. Ungelberger, 40, 25, 2000, 150, 300

Adam Steel, 200, 30, 9000, 300, 1500

Jno. Ceysin, 15, -, 350, 20, 100

Perry Smith, 3, 4, 2000, 100, 300

Jacob Zimmerman, 65, 30, 2000, 75, 900

J. Zimmerman, 90, 10, 3000, 800, 500

Chas. Font, 17, -, 800, 50, 100

Geo. Stull, 100, 70, 5000, 100, 800

Jno. Stull, 18, 800, 50, 100

Jno. Shankle, 13, -, 600, 25, 90

Philip Shankle, 9, -, 400, 25, 150

Joshua Firestone, 10, 14, 1000, 40, 140

Jno. Zimmerman, 50, 40, 4750, 150, 200

Jno. Cutsole, 12, -, 800, 25, 150

Jacob Staley, 15, -, 800, 50, 100

Geo. Dutrow, 30, 90, 1200, 25, 175

Philip Wachter, 143, 33, 5000, 200, 1500

Peter Houck, 10, -, 550, 25, 80

Abraham Michael, 25, 8, 800, 12, 250

Wm. Dixon, 275, -, 30000, 200, 1100

Grafton Clagett, 52, 15, 9500, 200, 250

Danl. Schaeffer, 133, -, 12000, 600, 200

Fredk. Meazle, 63, -, 1000, 50, 250

Saml. Shook 20, -, 1500, 100, 150

Mary Bobpit, 30, 100, 1500, 40, 250

Geo. Kinkle, 21, -, 1200, 100, 180

Conrad Smith, 45, -, 7700, 50, 200

Maria King, 16, -, 1000, 10, 180

Saml. McDevritt, 29, 52, 1900, 100, 300

Mich. Whitmore, 60, -, 3000, 50, 250

N. Whitmore Jr., 190, -, 7000, 150, 600

Elis Staley, 75, -, 3000, 100, 300

Jacob Stull, 200, 100, 6000, 100, 700

Nelson Ramsburg, 200, -, 4000, 100, 300

Simon Craven, 75, -, 2000, 50, 250

Saml. Heffner, 107, 27, 5000, 200, 600

William Stiger, 90, 37, 3000, 150, 400

S. P. Heffner, 48, -, 3000, 40, 300

Dennis Ramsey, 125, 20, 6400, 100, 600

Geo. Measel, 21, -, 1500, 75, 180

David Measel, 17, -, 750, 10, 145

Mehry Wachter, 63, -, 2600, 50, 600
Joshua Wachter, 16, -, 800, 20, 75
Jacob Stull, 50, -, 1000, 40, 80
Wm. Nichols, 27, -, 1000, 75, 250
Lewis A. Wachter, 36, 11, 2500, 100, 500
Lewis Wachter, 70, -, 2100, 10, 450
Jacob Holtz, 54, -, 2000, 100, 400
Albert Holtz, 75, -, 2000, 20, 350
Jno. Stup, 51, -, 2500, 100, 400
Sam. Ramsburg, 48, 28, 2000, 120, 300
Jno. Holtz, 100, 44, 4000, 100, 400
Benj. Bertges, 28, 24, 1550, 40, 100
Sarah Cline, 50, -, 1500, 100, 125
Chas. Albright, 9, -, 1000, 10, 120
Danl. Strafir, 22, -, 1500, 100, 140
Wm. Ungleberger, 15, -, 1000, 10, 100
Wm. Fenger (Ferger), 80, -, 3500, 75, 790
David _. Staley, 150, 29, 5000, 100, 500
Jn. Feazy, 32, -, 6000, 15, 100
Geo. Feazy Sr., 15, -, 600, 10, 210
David Mantz, 125, 15, 5000, 200, 800
Geo. Ungleberger Sr., 66, -, 3000, 75, 300
Saml. Keyser, 70, 9, 2000, 27, 375
Jno. Shafer, 121, 56, 6700, 90, 380
Joshua Staley, 50, 10, 1300, 60, 290
Geo. S. Miller, 133, 55, 14800, 600, 1150
Jno. Hildebrand, 50, -, 2000, 75, 287
Ezra Hildebrand, 60, 20, 1500, 150, 800
John Twenty, 20, 5, 800, 75, 200
Philip Burris, 12, -, 500, 10, 150
Adam Staley, 33, -, 700, 30, 100
Joshua Mayne, 21, -, 1000, 40, 130
Jacob Mayne, 54, -, 200, 100, 500
David Mayne, 19, -, 1200, 50, 200
Elias McKenzie, 52, -, 1100, 30, 325
Geo. Henshaw, 32, -, 600, 50, 200

Fredk. Kintz, 137, 12, 5000, 250, 1135
Fredk. Kintz, 20, -, 1400, 50, 300
Jno. New Brand, 18, -, 700, 30, 100
Geo. Boteheimer, 12, -, 700, 25, 120
Jno. Mayer, 13, -, 850, 30, 135
Solomon Mayer, 60, 40, 4000, 125, 750
Henry Abb, 25, 3, 1000, 50, 200
Joseph Hart, 18, 6, 600, 75, 156
Jonathan Mayer, 75, 44, 5000, 57, 489
Conrad Shultz, 82, -, 1200, 80, 200
George Smith, 58, 7, 7000, 200, 400
G. A. Smith, 174, -, 16000, 600, 1000
Chas. Howard, 80, 5, 5486, 150, 400
Josiah Miller, 194, 6, 6000, 300, 300
Jno. W. Miller, 109, 14, 3000, 103, 100
Julia Hildebrand, 100, -, 2000, 100, 250
Joshua Hildebrand, 100, -, 2000, 150, 275
Danl. Sunday, 9, -, 2500, 50, 150
Richd. Harper, 26, -, 2600, 100, 350
Chas. Keller, 55, -, 6000, 250, 500
Joshua Stull, 20, -, 600, 30, 160
Tobias Staley, 43, -, 600, 10, 150
Jas. Briner, 51, -, 4000, 50, 200
Peter Mantz, 6, -, 2000, 40, 250
Geo. Metzger, 37, -, 3700, 60, 300
J. Nichols, 30, -, 5000, 60, 350
Jno. W. Miller, 8, -, 1600, 50, 320
M. H. Haller, 14, -, 1200, 30, 370
Lewis Dill, 26, -, 5000, 50, 290
John Young, 100, -, 10000, 300, 850
Charles Smith, 100, -, 2000, 150, 800
Richd. Waltman, 40, -, 600, -, 190
Val. Hasbough (Harbaugh), 100, 40, 2800, 20, 717
Leonard Harborough, 175, 85, 10000, 300, 1080
Henry Harbaugh, 125, 60, 5500, 500, 584

D. D. Brown, 70, 30, 2500, 75, 430

Samuel Rozer (Royer), 50, 20, 800, 25, 323

Israel Williard, 35, 105, 1200, 150, 32

Elias Harbaugh, 75, 100, 2000, 100, 347

Ephraim Harbaugh, 109, -, 3000, 200, 735

Sanford Harbaugh, 112, -, 3000, 200, 735

Hiram Harbaugh, 250, -, 5000, 150, 544

John Gladhill, 70, -, 1500, 50, 270

Adam Tressler, 140, -, 3500, 250, 618

John Miller, 170, -, 5500, 250, 430

John McClaine, 150, 100, 3300, 250, 570

Daniel Harbaugh, 116, -, 3000, 200, 297

Lewis Wertenbaker, 160, -, 3000, 150, 400

Elias Harbaugh, 62, -, 2500, 125, 470

Jacob Harbaugh, 130, -, 3820, 200, 900

John H. Harbaugh, 20, 120, 2000, 30, 210

John Harbaugh, 204, -, 4080, 150, 698

William Seisman, 43, -, 950, 40, 200

George Harbaugh, 220, -, 3000, 100, 596

Henry Harbaugh, 159, -, 3000, 100, 435

Yost Harbaugh, 150, -, 3000, 150, 545

Daniel Hardangl (Harbaugh), 315, -, 7000, 500, 1161

John Lantz, 20, 27, 1000, 100, 325

James E. Shultz, 238, -, 3500, 150, 380

Jacob Wellias, 200, -, 3000, 100, 573

Daniel Boyer, 105, -, 5000, 200, 825

Commenus Willias, 197, -, 6000, 200,672

Henry Buhrman, 29, 300, 2850, 150, 460

Wm.Buhrman, 50, 50, 2800, 100, 410

John Krist, 90, 50, 3880, 150, 550

Geo. H. Fox, 133, -, 1500, 250, 980

David Wolf, 40, 120, 3000, 75, 573

Geo. P. Fox, 50, 200, 1000, 200, 945

Wm. B. Brown, 80, -, 1600, 150, 970

Henry Brown, 130, -, 2000, 50, 331

Daniel Buhrman, 70, 230, 2000, 200, 477

Daniel Thomas, 66, 13, 1200, 150, 432

Jacob Ridenour, 90, 21, 1800, 95, 230

Barney Lewis, 100, -, 2000, 150, 380

Geo. Buhrman, 61, -, 800, 25, 59

Jacob Buhrman, 20, 125, 1200, 7, 477

Henry Buhrman, 126, -, 1000, 150, 440

Joshua Moser, 70, 150, 1200, 150, 630

Abrm. Toms, 144, -, 2000, 150, 725

William Toms, 240, 50, 6000, 300, 475

Index

Arndt, 8
Arnehem, 194
Arnett, 181, 190
Arney, 69
Arnold, 6, 18, 29-30, 80, 116, 126-127
Art, 47
Arter, 129
Arther, 120
Artz, 203
Ascam, 179
Ash, 15-16, 54, 138
Ashbey, 17
Ashby, 17
Asher, 77
Aslough, 128
Aspen, 179
Athey, 13
Atkinson, 146-147
Atwell, 37-38, 40
Augustine, 16
Auld, 179
Ault, 20
Austin, 71, 182-183
Avilt, 116
Avitt, 116
Avrey, 48
Baan, 136
Babylon, 118-122
Bachman, 114, 195
Bachtell, 130
Backloss, 139
Backman, 124
Backs, 63
Bacon, 7, 62, 71
Bacquith, 183
Bael, 124
Baer, 80
Bail, 115-116
Baile, 115
Bailey, 9, 44, 58-59, 113, 159-160, 175, 200
Baily, 50, 56, 162
Baker, 19, 21, 44, 56, 58, 76, 81, 105, 112, 125, 145, 175, 185, 191
Bakney, 29

Balderson, 26
Balderston, 150-151
Baldwin, 26, 29, 32, 35, 118
Bales, 36
Ball, 23, 46, 177
Ballenburger, 66
Balser, 8
Baltimore, 177, 183
Baltzell, 196
Balwin, 74-75
Banblitz, 52, 57
Band, 134
Banett, 180
Banger, 67
Banit, 106
Bankan, 5
Bankard, 129
Banker, 9, 35
Bankers, 125
Bankert, 119, 121, 124, 129-130
Banks, 121, 177-178, 187
Banner, 190, 203
Bannock, 186
Banst, 119
Barber, 26, 125-126, 159
Bard, 13, 80
Bardon, 180
Bare, 120
Barger, 79
Barhart, 129
Baringer, 67, 74
Barker, 35, 50, 157
Barkholder, 9
Barkley, 167, 180, 188
Barkman, 195
Barley, 113
Barman, 5
Barnard, 5, 19
Barnes, 11, 31, 41, 61, 104-105, 107, 111, 115, 123, 125-127, 150-151, 168, 181,
Barnet, 107
Barns, 47, 145, 152
Barnum, 65, 74
Barr, 104, 139, 148
Barrall, 139

Barret, 156
Barrett, 148, 155
Barrick, 196, 198, 203
Bart, 138
Bartles, 164
Barton, 77, 199
Baseman, 108-109
Basford, 26
Basie, 29
Bassett, 183
Bassford, 24
Baston, 25
Bateman, 159, 161, 168
Bathwell, 139
Battee, 42
Batter, 166
Baublitz, 57-58
Baugardner, 124
Baughman, 6
Bauser, 20
Baum, 129
Bauman, 43-44
Baumgardner, 130
Baumgarner, 131
Baumgarnod, 199
Baumgartner, 116, 118
Baust, 119
Bawls, 73
Bawyer, 12
Baxter, 75, 168
Baylor, 129
Bayne, 67, 69
Baynes, 43
Beacham, 46
Beachley, 2
Beachy, 7-9
Beaks, 121
Beale, 131, 167
Beall, 5, 7, 21-22, 192, 199
Beallman, 33
Beallmear, 33
Beallmore, 33
Beam, 44, 109, 163, 200
Beamer, 43
Bean, 18, 46, 161, 164
Beannock, 178

Beard, 27-28, 72, 77, 126, 151, 188, 193
Beardan, 123
Beasman, 109
Beaston, 138, 148
Beatty, 64
Beaver, 122-123
Beavers, 123
Beck, 202
Beckley, 51
Beckman, 17-19
Becks, 133
Beckum, 116
Becquith, 185, 190
Becraft, 109
Beddle, 144, 147
Bedford, 77
Bedwell, 138
Beecham, 139
Beeker, 194
Beggs, 124
Bell, 23, 34, 42-43, 59-60, 170-171, 174, 176, 178, 185, 195
Belt, 33, 46, 51, 54, 65, 109, 111, 192
Beltz, 115
Belzard, 127
Bender, 12
Benedict, 122
Beng, 73
Benge, 142
Benison, 135
Benjamin, 146
Benner, 77, 80, 119
Bennet, 34
Bennett, 12, 16, 43, 49, 77, 105-106, 108-109, 141, 184
Bennington, 50
Bennison, 44
Bensinger, 42-43
Benson, 20, 30, 33-37, 49, 55
Benton, 179
Bentz, 109
Berchears, 105
Berchly, 20
Bergatts, 126

Bondle, 185
Bonham, 156
Bonthlon, 126
Boodough, 147
Boone, 29-30, 162
Boony, 42
Boop, 122
Boose, 114
Booth, 74-75, 138, 141
Booze, 180
Bopp, 121
Borchus, 7
Bordler, 32
Boresly, 66
Borig, 54
Boring, 115
Borkum, 122
Borley, 113
Borman, 43
Borrough, 183
Borroughes, 162
Borsus, 47
Boseman, 43-44
Bosey, 160
Bosley, 19, 62-63, 66, 69-70, 110
Bosly, 62, 65, 70
Boss, 121
Bosson, 54
Bost, 202
Boston, 120
Boswell, 164, 169
Boteheimer, 205
Boteman, 161
Botwell, 67
Bouchell, 138
Boucher, 146
Boudle, 185
Bouene, 123
Bougus, 134
Boulden, 137, 139
Boulder, 137
Bouldon, 148
Boulen, 139
Bourer, 8
Bourke, 29
Bourland, 143

Bourns, 44
Bourroughs, 159
Bowage, 115
Bowdle, 185, 190
Bowen, 26, 49, 68, 140, 179
Bower, 117, 161
Bowerman, 164
Bowers, 44, 113, 115, 117, 121
Bowersox, 129-130, 195-196
Bowes, 198, 203
Bowie, 169
Bowling, 162, 166
Bowlski, 138
Bowman, 2, 20-21, 114-115, 130-131, 162
Bown, 166
Boyd, 31, 137, 151, 153, 200
Boyer, 113, 124, 127, 206
Boyer, 2-3, 20, 32, 37, 46
Boyle, 24,122, 155, 200
Boyles, 143
Bozel, 200
Br__, 5
Brace, 54, 67
Brackenridge, 52
Bradas, 183
Bradbery, 70
Bradenbaugh, 107
Bradford, 69
Bradley, 171, 177, 186, 190
Bradly, 120, 172
Bradshaw, 169, 179, 187
Brady, 10, 20, 47, 120, 150, 192
Bramaman, 20
Bramble, 173-175, 180, 188
Bramen, 169
Brand, 80
Brandard, 73
Brandenbaugh, 107
Brane, 204
Braner, 79
Branfield, 154
Brannan, 163
Branning, 14
Brannock, 181
Branock, 182

Branst, 17
Brant, 21
Brashears, 24-25, 28
Brasheur, 106
Brashiers, 9
Brasly, 1
Brasman, 42
Braughben, 173
Braughn, 174
Brawhon, 177
Brawner, 164, 168
Bray, 18, 173
Brayfield, 163
Brde, 78
Bread, 45
Breckinridge, 69
Bremble, 18
Bremdoby, 194
Brengle, 200
Brennan, 124
Brenter, 76
Brewer, 25, 32
Brian, 36, 138
Brice, 29
Brickley, 146, 150-151
Brickman, 42, 67
Brierly, 48
Briggs, 196
Brimenan, 9
Briner, 205
Brinker, 14
Brinkly, 43
Brinkman, 73, 115
Brinsfield, 170-171
Brinton, 150
Briscoe, 67, 149
Brittenburger, 64
Brittenhouse, 30
Brittian, 156
Broadbent, 78
Broadrup, 203
Broadwater, 7-8, 11
Brocan, 145
Brock, 147
Brod, 6
Broders, 129

Brodie, 68
Brodwater, 6-7
Brogden, 24
Brohawn, 189
Bromet, 122
Brook, 4, 39, 65, 179
Brooke, 22
Brooks, 36, 45, 47, 65, 76
Brosk, 4
Brothers, 123, 127
Broughter, 152
Browdle, 189
Brown, 1, 7, 9, 23-26, 28-32, 47-48,
50-51, 65, 67-69, 72, 78, 106-108,
110, 112-113, 121-122, 124, 127-
128, 130,131, 140-141, 145-146,
149, 151-157, 166, 189-190, 192,
206
Brownaman, 7
Browning, 4, 16-17
Brownwell, 152
Bruce, 10, 13, 65, 78, 168
Bruel, 35
Bruner, 204
Brunner, 200, 203
Bruster, 51
Bryan, 28, 32, 138-139, 166, 181,
184
Bublites, 57
Bucey, 22
Buchanan, 21, 69
Bucher, 113
Buck, 79, 204
Buckey, 191, 201
Buckingham, 105, 108, 115, 123,
125-127
Buckland, 157
Buckle, 110
Buckler, 109
Buckley, 128, 148, 151
Bucklie, 47
Buckly, 20, 139
Buckman, 67
Budd, 160
Buel, 22
Buell, 29

Buey, 12, 14-17
Buffington, 48, 132, 142, 199
Buhrman, 206
Buley, 136
Bull, 53, 57, 59, 61-62
Bullen, 144
Bullinger, 62
Buly, 180
Buman, 5
Bunn, 110
Bunseman, 46
Bunst, 119
Burch, 160, 163-164, 166
Burdock, 46
Burgan, 74
Burgeon, 130
Burges, 168
Burk, 60, 141
Burke, 41, 46, 50, 57, 117
Burkett, 24
Burkhart, 202
Burks, 121
Burlin, 152
Burmeda, 61
Burnell, 26, 110
Burnett, 49
Burnibe, 144
Burnic, 120
Burnitz, 68
Burns, 61, 76-77, 127, 146, 157, 201
Burnstine, 5
Burrick, 203
Burris, 205
Burrus, 61
Burtin, 62
Burton, 30, 76-77, 179
Busack, 42
Busby, 125
Buse, 137
Busey, 25, 60
Bush, 28, 106, 127, 189
Bushey, 51, 191
Buskirk, 4-5, 138
Bustee, 68
Butcher, 192
Buthart, 115

Butler, 8, 34, 48, 128, 147, 162, 164,
166, 171, 195
Butskey, 74
Butt, 59
Buxton, 203
Bydey, 76
Byerly, 51, 131
Byers, 124, 179, 181, 200
Byron, 178
Cabin, 184
Cadden, 77
Cadel, 27
Cadwaleder, 77
Cage, 164
Calahan, 6
Calaway, 171
Caldwell, 80
Calender, 186
Calk, 42
Callan, 8, 14, 19
Callaway, 108
Callow, 119
Cally, 138
Calton, 160
Caltriton, 111
Calwell, 133
Cambell, 153
Camble, 42
Cameron, 135, 145-146, 151, 155-
157
Camerson, 136
Camp, 20
Campbell, 43, 108, 140, 143, 150,
153
Camper, 174, 177, 185, 189
Canby, 41
Cann, 51, 142
Cannon, 142, 171, 189-190
Canon, 188
Canter, 162-164
Cantwell, 140-141, 185
Capler, 125
Caples, 36, 125, 127-128
Carback, 78
Carey, 193
Carhart, 156-157

213

Cooper, 43, 52, 56-59, 146, 151, 186-187
Cope, 19
Coothers, 155
Coottriton, 115
Copland, 193
Cople__, 54
Coppersmith, 122, 132
Corbin, 69, 110
Corcoran, 62
Corier, 153
Cormer, 32
Cornelius, 60
Corner, 48, 185
Cornish, 178-179, 185
Corns, 75
Coroman, 33
Corse, 73
Corsey, 80
Cortes, 13
Cosgrove, 153
Cosney, 80
Cotes, 139
Cother, 151
Cotrighter, 124
Cottitor, 119
Couden, 154
Coulson, 149, 1510152
Coulter, 147
Coumine, 114
Councilman, 47
Counselman, 46, 79
Courmine, 113
Couts, 114
Cover, 105, 130
Covington, 162, 164
Cowan, 46, 144
Cowarll, 183
Cowley, 78
Cox, 49, 52, 54, 59, 66, 110, 160, 165-166, 168
Cpets, 124
Crabbs, 19
Crabtree, 12-13
Craddock, 48
Craft, 59, 170, 177

Craggs, 37
Craig, 48, 70, 145, 152-154, 179
Crain, 160, 162
Craiton, 188
Cramer, 195-196, 203
Cramford, 15
Crampton, 8, 202
Crandall, 37-38
Crandle, 26
Crane, 31, 69, 197
Cranston, 28
Crapster, 119
Craven, 204
Craver, 196
Crawford, 49, 104, 119, 137, 146, 186
Crawmer, 59
Creame, 114
Cregg, 138
Creighton, 181, 183, 189
Cresap, 10, 12
Cresnel, 125
Criager, 52
Criemind, 136
Criner, 7
Crismond, 160
Crisp, 36
Crist, 192
Cristopher, 180
Criswell, 104, 139, 144
Crockett, 134, 176
Crogan, 36
Cromhart, 52
Cromwell, 35-36, 68, 183, 186, 194
Cromwelt, 68
Crone, 5
Crook, 37, 45
Crooks, 45-46, 125
Crookshank, 134-135
Crosby, 41-42
Crose, 5
Crosgrove, 146
Crosklas, 134
Crosly, 12
Crosnise, 201-202
Cross, 57-58

Dornin, 22
Dorsett, 24
Dorsey, 27, 34-35, 37, 42, 44-46, 56,
58, 105-108, 123, 185, 199
Dosey, 106
Doterer, 199
Dotey, 106
Dougherty, 9, 67
Douglass, 192
Dove, 39
Dowell, 39
Downey, 106
Downing, 145, 164
Downs, 34, 36, 59, 166, 201
Doyle, 123
Drake, 22, 122, 138
Dran, 15
Drdem, 30
Drelson, 28
Drenon, 144
Drin, 17
Drogie, 19
Drumon, 144
Drury, 24, 39-40
Drusching, 192
Duce, 51, 129
Duckett, 168
Duckey, 51
Duckman, 52
Duckworth, 8, 13
Dudderer, 105
Dudrer, 131
Dudrow, 124
Duff, 75, 143
Duffy, 167
Dugger, 160
Duleplan, 120
Dulzour, 124
Dumer, 34
Dunaho, 174
Dunahoo, 157
Dunall, 26
Dunbar, 31
Duncan, 51
Dungan, 80
Dunn, 41, 110, 171, 174

Dunnock, 188
Durall, 26-29
Durbin, 123
Durett, 23
Durey, 56
Durst, 2, 6-9, 22
Dusey, 71
Dushane, 137
Dutrow, 130, 192-193, 204
Dutton, 159, 176
Duval, 45
Duvall, 33-35, 40, 106, 123, 191,
193
Dye, 5
Dyer, 164
Dyes, 200
Dyhoff, 54
Dyson, 37, 159, 161, 164
Eader, 202-203
Eagle, 192
Eagleston, 73
Earl, 78
Earle, 139, 188
Earsterdy, 195
Easle, 188
Easterdy, 195
Eastman, 22
Easton, 125
Eater, 178
Eaton, 56, 196
Ebaugh, 109, 111
Echenberger, 8
Eck, 130
Eckard, 120
Eckart, 63
Eckels, 10
Eckenrode, 199
Ecker, 198
Eckert, 121
Eckinrode, 117
Edelen, 45, 160
Edelin, 161-163, 165, 168
Eden, 107
Edgar, 4
Edge, 47
Edger, 180

Faultestine, 114
Favorite, 196
Fawley, 192
Fazee, 1
Feagle, 120
Feams, 4
Feast, 81
Feazy, 205
Feeser, 203
Fegley, 5
Feiser, 117, 130
Felts, 80
Femanus, 43
Fendall, 69
Fenger, 205
Fentris, 183
Fenwick, 38
Ferby, 79
Ferger, 205
Fergerson, 76
Ferguson, 142
Ferlon, 11
Fermake, 120
Ferrell, 9, 160, 162, 168
Feson, 159
Fete, 45
Fickman, 21
Ficktig, 21
Fields, 146
Figs, 185
Fike, 18
Filter, 137
Finch, 110
Finley, 142
Finn, 80
Finney, 191
Firestone, 204
Firte, 6
Fish, 189
Fisher, 20, 22, 47, 57, 62, 68, 72, 111, 122, 135, 171, 198
Fiste, 6
Fister, 148
Fitch, 79
Fitchen, 186
Fite, 109, 124

Fitter, 137
Fitz, 119
Fitzgonder, 17
Fitzgouder, 17
Fitzpatrick, 61
Fitzsimmons, 62, 72
Fitzsimons, 141
Fitzwater, 142
Flagel, 1, 4
Flanagan, 195
Flank, 126
Flatter, 128
Fleagle, 119
Fleckenger, 20
Flegil, 199
Flegle, 118, 129
Flemen, 108
Fleming, 105, 177, 198
Flemming, 202
Fletcher, 16, 74, 175, 190
Flickinger, 118, 129
Flint, 47, 182
Flinthand, 133
Flounders, 140-141, 144
Flowers, 74, 190, 196
Floyd, 162
Flurry, 76
Foard, 27, 134, 143-144, 148
Fogeslong, 120
Fogle, 46, 105, 116, 119
Folck, 9, 22
Folk, 6, 53, 202
Fongue, 28
Font, 200-201, 204
Fooks, 177
Forbes, 80
Ford, 50, 74-75, 77, 134, 136, 145, 147, 167
Forde, 147
Forder, 137
Foreman, 31, 42, 58, 114, 135
Forester, 115, 123
Formatt, 120
Formett, 120
Forrest, 194
Forrester, 180

Fursting, 42
Furting, 42
Fuser, 31, 130-131
Fuzel, 125
Fuzzle, 105, 122
Gailor, 37
Gainer, 32
Gairy, 35
Gaiter, 107
Gaither, 27, 35, 37
Gaithers, 29
Gale, 68, 157
Gales, 62
Galisbenner, 11
Gallaher, 144, 149
Gallaway, 71
Galleher, 141, 144-145
Gallias, 108
Gallion, 201, 203
Galloway, 62, 77
Galt, 116, 118
Gamber, 127
Gamble, 42, 76, 157, 199
Gambol, 157
Gambrill, 34, 174
Gambrine, 202
Gamics, 113
Gance, 118
Ganer, 18
Ganes, 18
Ganett, 26, 114
Ganner, 38
Ganz, 113, 175
Garcy, 10
Gardener, 38
Gardiner, 50, 52, 160, 162-163, 166-167
Gardner, 17, 24, 32, 35-36, 39, 46, 108, 126-127, 149, 163-164, 167, 198
Garet, 144
Garey, 32
Gargan, 73
Garison, 134
Garner, 8, 118, 132, 165
Garnerd, 195

Garraghan, 3
Garraphan, 3
Garret, 61
Garrett, 19, 77, 114, 142, 144, 169
Garrison, 200
Garritson, 50
Garry, 3
Gartely, 7
Gartley, 7-8
Garver, 195
Garvigle, 20
Gary, 39
Gatch, 79
Gatchell, 142, 145, 149
Gath, 118
Gatten, 192
Gaugh, 196
Gaulden, 118
Gauser, 13
Gauz, 113
Gay, 153, 167
Gayty, 152
Geary, 50
Geddis, 76
Gee, 108
Gegner, 73-74
Geiger, 115
Geiman, 115
Geist, 66
Gelhause, 10
Gelley, 70
Gellot, 75
Gellott, 72
Gelston, 41
Gemmell, 60
Genff, 9
Gent, 49, 52, 63, 66
Gentrey, 75
Gentrun, 73
Geoghan, 178
George Creek Co., 10
George, 2, 105, 107, 139, 147-148
Gephardt, 52
Gerack, 20
German, 120, 124
Gerot, 65

223

Gottz, 119
Gough, 161
Goul, 148
Gould, 78
Gover, 39, 42, 51
Grabenhunt, 68
Grace, 147
Gracy, 8
Graham, 5, 11, 42, 47, 67, 145, 150, 172, 196
Gramer, 121
Grant, 141, 145-146
Granter, 73
Graskan, 196
Graskar, 196
Graso, 21
Grason, 48, 181
Grass, 8, 14, 127
Graves, 9, 11
Graw, 70
Gray, 25, 27, 30-33, 143, 168, 189
Graybill, 198
Green, 1, 3, 11, 38, 60, 64, 67, 73-74, 110, 128, 133-134, 140, 145, 165, 179, 194
Greenwade, 10-11
Greenwall, 201
Greenwood, 104
Greer, 60
Gregg, 142
Gregory, 73
Grey, 177
Greystone, 140
Grider, 20
Grier, 146
Griffen, 127
Griffey, 127
Griffin, 37, 71-72, 126, 199, 201
Griffith, 3, 19, 34, 46, 48, 52, 55, 63, 66, 71, 138, 156, 171
Griffy, 152
Grilst, 74
Grim, 52
Grimage, 36
Grimes, 19, 25, 105, 128, 195, 197
Grinder, 191

Grinnum, 170
Grissel, 79
Griswold, 68
Groff, 50-51
Grogan, 139
Grogg, 110
Grombeck, 22
Groome, 136
Grose, 131
Groshan, 197
Gross, 14, 110-111
Grosse, 20
Grossnickle, 194
Grove, 108, 114, 126, 168, 174, 191, 204
Groves, 112, 141
Growe, 106
Grubb, 154
Gruffy, 143
Grumbine, 123
Gruser, 66
Gruserd, 156
Grussan, 156
Guergey, 18
Guilen, 67
Guilew, 67
Guin, 55
Guinan, 124
Gunder, 195
Gunn, 105
Gusey, 196
Gust, 66
Gusuch, 60
Guting, 129
Gutzy, 17
Guyton, 74-75
Hackersmith, 1989
Hackert, 173
Hacket, 171-173, 176
Hackler, 195
Hadaway, 181
Hadock, 155
Hagan, 6
Hage, 4
Hagen, 202
Hague, 146

Hagy, 51
Hahn, 42, 118, 130-131
Haines, 108, 112, 115, 120-121, 126-127, 129, 148-149, 151, 157
Haird, 129
Haire, 129
Hairo, 192
Hal, 26, 130, 142, 155
Halbert, 76
Hale, 53, 55, 63, 75
Haley, 76
Hall, 4, 8, 26-27, 32, 38-40, 63, 73, 79-80, 112, 119, 124, 134, 151-152, 155, 162, 197, 203
Haller, 205
Hallerberry, 118
Halson, 133
Halterbrick, 117
Halurik, 119
Hamberson, 5
Hambleton, 36
Hamil, 19
Hamilton, 15, 24, 44-45, 48, 59, 72, 157, 161-162, 165-167
Hamman, 74
Hammell, 18
Hammersmith, 150
Hammond, 26, 30, 33-34, 36-37, 106
Hamner, 5
Hampshire, 56, 58
Hamrick, 12
Hance, 35
Hancock, 1, 31-32, 109, 163-165
Hand, 41, 53-54, 200
Handly, 176, 178
Hane, 138
Haner, 117
Hanes, 115, 152
Hanet, 57
Haney, 46
Hanfed, 120
Hankey, 197
Hankins, 5
Hann, 4, 120, 132, 198, 202
Hannah, 141, 148, 154
Hannon, 166-167

Hanshaw, 155
Hanshicon, 180
Hantank, 9
Happe, 124
Harbaugh, 205-206
Harbin, 163
Harborough, 205
Hardangl, 206
Harden, 108-109, 175
Hardesty, 23, 33, 38
Hardgraves, 141
Hardin, 6, 20
Harding, 141, 183, 186
Hardman, 22
Hardon, 33
Hardsock, 14
Hardy, 165
Hare, 53-54, 56-58, 111, 115
Harem, 8
Hargate, 194, 202
Hargath, 201
Harget, 68
Hariekoo, 6
Harigan, 156
Harison, 28
Harity, 197
Harker, 44
Harkes, 165
Harkey, 197
Harkley, 105
Harkness, 60
Harlan, 140, 154
Harland, 75
Harley, 15
Harlin, 162
Harman, 9, 32, 35, 36, 48, 52, 115, 118, 131
Harmer, 117-118
Harmon, 37, 131-132
Harner, 117
Harnet, 13
Harney, 23
Harneyman, 63
Harnish, 118
Harp, 105, 107, 109, 197
Harper, 165, 172-173, 175, 187, 205

Hemmons, 171
Hempfield, 139
Hempsey, 135
Hemt, 53
Henck, 203
Henderson, 22, 35, 72, 104, 139, 143
Hendric, 59
Hendricher, 137
Hendrickson, 9
Hendrixson, 15
Henersly, 159
Henies, 66
Henkle, 36, 50, 80
Henlde, 22
Henler, 68
Hennedoy, 133
Henry, 58, 79, 105, 125, 136, 173,
175, 178, 181, 190
Henshaw, 31, 205
Henson, 65, 168-169, 179, 185
Heran, 164
Herbert, 42
Herd, 130
Hergenrother, 73
Hering, 27
Herman, 13, 20, 32
Hermet, 57
Hernt, 53
Herrit, 53
Herscheman, 195
Hervitt, 49
Hess, 116-117, 130, 132, 144
Hessey, 133, 135-136
Hesson, 119-120, 130
Hester, 51, 192
Hewett, 106-107
Hewlett, 43
Hick, 58
Hickey, 200
Hicks, 59, 167, 174-175, 177, 183
Hiddong, 11
Hide, 74
Hiel, 46
Hielman, 21
Higdon, 160, 166, 168
Higgins, 14, 26, 28, 174, 178

Higgs, 160
Highthait, 67
Hilbrand, 120
Hildeband, 121
Hildebrand, 203, 205
Hile, 58
Hileman, 20-21
Hilker, 58
Hill, 9, 30, 40, 56, 68, 117, 124, 154,
171, 187, 189, 196
Hillyard, 30
Hilterbrick, 118
Hindle, 168
Hindman, 150
Hinea, 197
Hiner, 42, 119
Hines, 31, 46, 115, 119, 153, 197
Hinge, 198
Hink, 204
Hinkle, 9, 14, 21, 112
Hinkly, 46
Hinson, 182
Hipsley, 63, 76
Hires, 16
Hirs, 69
Hiseley, 115
Hiser, 42, 117
Hiss, 74
Hissong, 194
Hitch, 77, 81
Hitchcox, 148
Hitchen, 121
Hiteshue, 116, 132
Hoale, 131
Hobbrock, 128
Hobbs, 172, 184, 199, 202
Hocke, 51
Hodges, 27, 32, 35
Hodinott, 67
Hodson, 176
Hoe, 69
Hoen, 67
Hofacker, 58, 112
Hoferkamp, 73
Hoff, 4, 58, 108, 128
Hoffacker, 57, 113

Humbertford, 162
Humer, 78
Humphrey, 33
Humphries, 52
Hunault, 67
Hunhey, 196
Hunt, 53, 57, 71, 75, 114, 148, 160, 166
Hunter, 30, 61, 164
Huntley, 5
Huntt, 167
Hurbert, 160
Hurdl, 113
Hurles, 33
Hurley, 176-177, 182-184, 187-188
Hurlock, 173, 189-190
Hurly, 176, 188
Hurst, 173, 177
Huse, 43, 186
Husfelt, 135
Husley, 188
Huss, 147
Hutchens, 71
Hutchinson, 133, 173, 190
Hutshen, 203
Hutson, 41
Hutton, 66
Hyde, 29, 37
Hydes, 120
Hyland, 147-148
Hynes, 203
Idle, 137
Iglehart, 23, 28-29
Igleman, 80
Imwold, 68
Ingham, 57, 66
Inglhart, 45
Insley, 172, 177, 180, 187
Ireland, 109
Irons, 14
Irwin, 143
Isaacs, 32, 157
Isengle, 196
Isennoff, 75
Issanple, 52
Jabb, 31

Jackson, 32, 35, 55, 62, 73, 77, 140, 142, 145, 149-150, 153-154, 163, 174-175, 182-183, 186, 193, 197
Jacob, 31, 109
Jacobs, 11, 13, 31, 33-34, 39, 56, 125-126, 190
James, 55, 69, 148, 186
Jamison, 3, 15, 116, 160-161, 163
Jane, 53
Janney, 146, 155, 157
Jarboe, 191-192
Jarvis, 134
Jean, 45
Jeffers, 37
Jeffries, 4, 139, 147
Jenifer, 76, 178
Jenkens, 76, 80
Jenkins, 2, 19, 26, 32, 35, 72, 107, 165, 174
Jenkson, 107
Jenness, 153
Jermess, 153
Jervis, 136
Jeseure, 172
Jesor, 190
Jessop, 64, 75
Jester, 138
Jesuits, 134
Jilo, 17
Jobe, 156
Joens, 30, 39
Johes, 132
Johns, 52, 65, 134
Johnson, 5, 9, 16, 20, 22, 30-31, 41, 47, 52, 64, 67, 75, 77, 79, 109, 132, 139-141, 144, 146, 148, 157, 160-161, 174-176, 180, 182, 187, 190, 193
Johnston, 59, 71-72
Joice, 30
Jois, 72
Jolley, 178
Jolly, 177, 182-183, 185
Jones, 10, 21, 23-24, 27, 29, 32-33, 36-37, 40, 42-43, 50, 57, 59, 62-64, 68-69, 72, 114, 116-117, 128-130,

Low, 17, 50
Lowback, 70
Lowdenlager, 109
Lowe, 17-18, 171, 173, 189
Lower, 17
Lowery, 123
Lowise, 135
Lowman, 105
Lowry, 45
Loyd, 136, 168
Lucabaugh, 113
Lucas, 29, 36, 52, 68, 162
Luchabaugh, 113
Lucis, 136
Ludwic, 120
Ludy, 194
Lurfel, 123
Lusby, 27, 80, 109, 133-136
Luster, 119
Lyman, 134
Lynch, 31, 42, 46, 123-124, 130, 141-142, 146
Lynn, 131, 198
Lynthicum, 35
Lyon, 47, 152, 163, 167
Lyons, 160
Lytte, 156
Lyttle, 70
Mace, 79, 186
Macgraff, 20
Machian, 17
Machter, 196
Mackber, 33
Mackee, 143
Mackel, 142
Mackery, 11
Mackey, 143
Mackie, 142
Macklin, 67
Maclaine, 189
Maddox, 159-160, 165, 167
Made, 71
Maden, 51-52
Madkins, 179
Mael, 185
Maerhat, 203

Magee, 128
Mager, 128
Magill, 136
Maglat, 66
Magory, 67
Magruder, 27
Maguire, 179
Mahan, 52, 142, 144
Maheny, 147
Mahoney, 141, 148, 155
Maight, 200
Mail, 19
Maine, 130
Mainrd, 130
Major, 144
Majors, 34
Makall, 24
Makey, 57
Makins, 180
Malchorn, 123
Malcome, 70
Maley, 166
Mally, 166
Malone, 11-12, 180
Malte, 199
Malter, 200
Manaham, 123
Manald, 35
Mandiford, 59
Mandray, 113
Manganus, 194
Manglelin, 64
Mangum, 25
Manley, 141
Manloff, 135
Manloff, 136
Manly, 52
Mann, 189
Mannikling, 74
Manning, 122, 199
Manold, 35
Manor, 47
Manroe, 109
Mansfield, 44
Mantz, 205
Marbe, 173

Meitzell, 6
Meixell, 196, 203
Mekins, 179
Melcher, 62, 73
Melly, 176
Mendenall, 143
Mendenhall, 141
Menshaw, 30, 33
Meradith, 135
Mercer, 25, 109, 138, 202
Merchant, 37
Meredith, 60, 121, 182-183
Merhling, 203
Meridith, 145
Mering, 131
Merrett, 137-138
Merrikin, 29
Merrill, 7, 19
Merritt, 35-36
Merry, 148
Merryman, 46, 53, 59, 63-64, 66, 68, 70-72
Merser, 33
Mery, 68
Mesick, 171-173, 176
Messick, 171
Messinger, 130
Messminch, 130
Metzger, 205
Mew__, 175
Mewshaw, 33
Meyers, 193
Mezzick, 29
Michael, 13, 192-193, 196, 202, 204
Michaels, 13, 191
Michel, 13
Middleton, 12, 15, 33, 160, 163
Midline, 43
Miksell, 129-130
Milbourn, 146
Milburn, 136
Miles, 48, 67, 70-72, 160, 163
Milidero, 53
Millaby, 175
Millaman, 10
Millar, 124

Millenday, 78
Miller, 3, 5-8, 10-113, 15, 18, 20, 28-29, 32, 34, 42-44, 48, 55, 58-58, 62, 67, 71, 75, 78, 109, 113, 115, 124-126, 128, 139-141, 144, 168, 173, 191, 196-197, 202, 205-206
Milligan, 47
Millikin, 172-173
Mills, 160, 182-184, 188, 190
Millse, 184
Milly, 176
Milstead, 169
Milton, 125
Minchy, 110
Minety, 60
Minglin, 143
Minkroff, 2
Minnor, 142
Minor, 192
Minter, 115
Mirbriddle, 124
Mires, 200
Misering, 131
Misers, 121
Mister, 187
Miston, 180
Mitchell, 20, 24, 61, 39, 168
Mitchner, 150-151
Mitten, 122
Mizel, 125
Mjoers, 34
Moat, 144
Moffett, 139, 141, 192
Moffit, 67, 146
Mole, 19
Molick, 176
Molineaux, 200
Molter, 130-131
Money, 136
Monmonier, 41, 45
Monod, 161
Monroe, 165, 167, 182
Montgomery, 16, 144, 152, 162, 164-165, 188
Montling, 73

Moon, 18-19, 147, 153, 155-156, 163
Mooney, 78, 173
Moore, 4, 17-19, 25, 42, 47, 67, 71-72, 76, 79, 121, 135, 140, 143-144, 149-151, 163, 167, 172, 174, 185-187, 198
Moot, 196
Mootrick, 118
Morais, 165
Moran, 51, 162
Mordock, 119
Morduck, 123
More, 18, 106, 151
Morelock, 119, 123, 188
Morgan, 12, 33, 64, 73, 75, 108, 135-136, 138, 159
Mories, 16
Moring, 131
Morison, 143
Morningston, 195
Moroe, 164
Morrell, 71
Morris, 34, 48, 57-59, 147, 166, 168, 179
Morrison, 13, 43, 72, 134, 199
Morrow, 13, 50, 77
Mort, 133
Morton, 133-134
Moser, 206
Moses, 2
Moss, 29
Motler, 130, 201
Motter, 130
Moukley, 120
Moul, 123
Mousall, 147
Mouton, 134
Mowbray, 182-184, 189-190
Mowell, 71
Mryer, 2
Mt. St. Mary's College, 200
Muce, 7
Mudd, 164, 166-167
Mudock, 120
Mufaker, 20

Mulinson, 106
Mullen, 13, 149, 194
Muller, 46, 79
Mullican, 30
Mullikin, 33, 42
Mullin, 142
Mullineau, 33
Mullrony, 5
Mumagh, 128
Mummy, 73
Munday, 56
Munshorer, 199
Munt, 149
Murdock, 28, 123, 131, 165, 168-169
Muring, 119
Murphey, 170
Murphy, 73, 126, 138, 143, 163, 170-171, 176-177, 183
Murray, 25-26108-112, 127, 162, 165
Murrey, 73
Murrin, 117
Murry, 34, 126, 157, 170
Musgrove, 63
Musselman, 51, 65
Muskett, 168
Musser, 20
Myerley, 123-124
Myerly, 119
Myers, 10, 15, 33, 48, 51, 53, 68, 106, 112, 119, 121-122, 124, 129-131, 134, 200
Myes, 120, 132, 195
Myler, 8
Nachime, 111
Naly, 169
Nanen Factory Farm, 64
Nann, 202
Nash, 59, 178
Naudain, 140
Naughton, 21
Nausbaum, 104
Nausbum, 121
Navoy, 179
Naylor, 33, 162
Neagle, 69

Olhon, 134
Oligrath, 56
Oliphant, 10
Olir, 48
Oliver, 19, 34, 46, 144, 160, 166, 168
Omalley, 5
Once, 6
Onderdink, 47
Onion, 76
Opel, 20
Ophen, 182
Opher, 179
Oram, 48
Ore, 5
Orendorff, 25, 123-125
Orion, 76
Orme, 25
Orndorff, 19
Orr, 151
Orrick, 65
Orter, 129
Osborn, 31, 51, 54, 193
Osborne, 31, 62, 165
Oscarn, 179
Osler, 58, 109
Oterbridge, 172
Ott, 8, 198
Ottey, 140
Otto, 7, 197
Otts, 4
Ourings, 106
Oursler, 128
Ovelman, 199
Overhtlyer, 199
Owen, 165
Owens, 24-27, 32, 35, 38-39, 139, 144-145, 156, 168, 170, 182
Owings, 44-46, 48-50, 107, 122, 202
Owins, 180, 182
Paca, 154
Padgett, 164-166, 193
Padgran, 62
Pagan, 181
Page, 166, 175
Paine, 48, 110, 171-174, 189-190

Paist, 58
Palmer, 55-57, 60, 77, 146, 181, 195
Panebaker, 112
Pankey, 56
Panther, 53
Papp, 5
Paradee, 138
Paranett, 28
Parish, 61
Park, 61, 144
Parker, 5, 28, 33, 69, 72, 75, 78, 109, 140, 163, 175, 178, 186, 189
Parks, 62-64, 66, 78
Parlett, 78-79
Parrish, 37, 108
Parson, 71
Parsons, 48, 71, 188
Parver, 66
Parvin, 189
Paschall, 157
Pass, 142
Passmore, 149
Paterson, 76
Pathwell, 139
Patterson, 11, 27, 42, 52-53, 70, 73, 80, 107, 109, 153-154, 175, 177, 181
Patton, 153
Paugh, 19
Paught, 105
Paul, 146-147, 176, 183, 188
Pauly, 46
Paws, 174
Peace, 70, 107, 138
Peake, 24
Pearce, 61, 65, 71-72, 74, 78, 106, 134, 136-138
Pearch, 147
Pearrey, 148
Pearson, 48, 139
Peck, 8-9
Peddiarid, 62
Pendleton, 18
Penn, 105, 109
Pennington, 106, 108, 123, 126, 134-136, 145, 147
Pennock, 145

Redgrave, 134
Redgraves, 29
Redish, 36
Redmiles, 32
Reed, 43, 67, 116, 136, 139-140, 146, 177, 186
Reeder, 146, 161
Reese, 50, 78, 112, 124-125, 139, 184
Reeves, 117
Register, 74
Reich, 200, 202-203
Reidd, 164
Reidisul, 116
Reifsnyder, 132
Reindollar, 117-119, 130
Reinecker, 130
Reineker, 43
Reisler, 156, 157
Reister, 51
Rekert, 134
Remare, 62
Remington, 74
Rems, 160
Rence, 53
Rencher, 161
Reniman, 110
Repp, 115
Rest, 9
Reuter, 68
Reybold, 135
Reyer, 70
Reynolds, 10, 69, 74, 143, 147-150, 153-154
Rhinehardt, 43
Rhoades, 30, 170
Rhoads, 138
Rhoderick, 203
Rhodes, 7, 115, 138, 193, 201
Rhoth, 18
Rhule, 55-56, 58
Ribald, 135-136, 147
Ricard, 147
Ricards, 146
Rice, 9, 14, 21, 62, 136, 196
Ricek, 22

Rich, 20, 26
Richard, 19, 68, 109-110, 128
Richards, 50, 149-150, 154, 162-163, 189
Richardson, 25, 29, 37, 66, 71, 107, 115, 136, 147, 185-186, 191
Richel, 4
Richmond, 41
Richner, 3, 42
Richnor, 134
Richstien, 43
Ricketts, 140
Rickner, 7
Ridd, 164
Riddey, 7
Riddle, 18, 76, 145-146
Riddlemoser, 202
Ridell, 47
Ridenour, 198, 206
Rideout, 29, 176
Rider, 37, 63-64, 198
Ridge, 196-197
Ridgley, 49, 107
Ridgway, 10
Ridinger, 116
Ridout, 29
Riel, 12
Riffith, 137
Riggs, 81
Righam, 44
Right, 135
Rigins, 181
Rigri, 10
Rilay, 147
Rile, 156
Riley, 3, 16, 18, 36, 41, 61
Rill, 109-110, 156
Rinehart, 110, 114, 119, 123, 127
Rineman, 111
Ring, 80
Rinkle, 4
Rinner, 118
Riphon, 6
Ripler, 42
Ripley, 46
Riply, 42

Riring, 67
Risener, 42
Riser, 70
Rish, 53
Ristain, 69
Ristean, 75
Ristum, 69
Rittenhaus, 148
Rittenhouse, 80
Ritter, 45, 50, 109, 112
River, 6
Riving, 67
Rizer, 200
Roach, 148
Roades, 129
Roath, 18
Roberson, 7, 42, 187
Roberts, 16, 109, 122, 127, 138, 144, 197-198
Robertson, 123, 159-161, 168, 198
Robeson, 6
Robets, 122
Robey, 164-166
Robinett, 11-12, 14-15
Robinson, 13, 24, 30-32, 44, 67, 135-136, 148, 162, 170, 172, 179, 181-182, 185
Roby, 9, 161-162, 164-168
Rochester, 133
Rockhold, 30
Rocklitz, 30
Rodeham, 19
Roderic, 201
Roelky, 192
Roff, 43, 63
Roger, 111, 118
Rogers, 26, 45, 47-48, 52, 52, 106, 112, 143, 156
Rogs, 112
Rogs, 123
Rohr, 193
Rohrbaugh, 114
Roley, 14
Rolf, 18
Rolley, 111
Rollins, 77

Roman, 10
Romer, 154
Roney, 143
Rontzahm, 195
Rook, 183-184
Rooney, 4
Roons, 130-132
Roop, 115, 122, 124, 131
Rorback, 130
Rose, 147
Roseburg, 125
Rosengarth, 79
Roser, 59
Rosnic, 59
Ross, 3, 5, 174-175, 178, 184
Rosson, 54
Roster, 166
Roszell, 183
Roughet, 197
Routh, 18
Routson, 131
Rowe, 61, 70, 117
Rowland, 152, 155
Roy, 28, 32
Royer, 124, 206
Royston, 53, 55, 64, 72-73
Rozer, 206
Ruby, 11-12, 110-111
Rudemark, 21
Rudolph, 129
Rue, 14, 184
Rufsnyder, 131
Ruggle, 124
Ruhrback, 130
Ruidollar, 117
Rul, 184
Ruling, 56
Ruly, 136
Rumbio, 129
Rupert, 46
Ruphel, 76
Ruppert, 46
Rush, 7, 54
Rushel, 76
Rusk, 80
Rusler, 156-157

Russell, 46, 146-147, 152-153, 157, 171-172
Russler, 145
Rust, 41, 58, 128
Rutherford, 67
Ruthledge, 60
Rutledge, 60
Rutten, 43
Rutter, 47, 147, 152-153
Ryland, 3
Ryley, 3-4, 21
Ryon, 14, 153
Saffington, 199
Saffle, 30, 44
Saffles, 34
Sailor, 37
Saithers, 29
Salsbury, 77
Samma, 142
Sampson, 56, 59-60, 174, 176, 183
Samuel, 123
Samus, 175
Sanble, 57
Sanders, 73, 165, 168
Sanderson, 48
Sandes, 131
Sanford, 26
Sannden, 17
Sapp, 73, 78
Sappington, 33-35
Sargeant, 44
Sargi, 198
Sassan, 166
Sate, 197
Satie, 59
Satio, 59
Sauble, 57, 115, 122
Saunders, 178, 195
Savage Company, 21
Savage, 3, 16-17, 19, 26
Savey, 166
Sawyer, 124
Saxton, 175
Sayler, 131
Saylor, 197
Scak, 41

Scarborough, 140-141, 156-157
Scarf, 73
Schaeffer, 124-125, 204
Schafer, 56
Schaffer, 48
Scheib, 80
Schirer, 9
Schlyder, 131
Schmidt, 8
Scholl, 201
Schooley, 19
Schrib, 80
Schriver, 80
Schrock, 2
Schroeder, 68
Schroyer, 3
Schryock, 195
Schumacher, 69
Schwartze, 42
Schwergart, 123-124
Scott, 28, 34, 37, 52, 55, 66, 71, 131, 139, 144-144
Scovans, 60
Scribner, 106, 109
Scrivener, 38-39
Seabalt, 199
Seak, 41
Seatten, 145
Secter, 2
Seddon, 77
Sedon, 146
Sedwell, 149
Seeley, 154
Sees, 16
Sefu, 170
Sefur, 170
Sefus, 170
Segmore, 79
Seifles, 28
Seifs, 199
Seigler, 27
Seigman, 58
Seisman, 206
Seitz, 60
Selby, 1, 9, 108
Seling, 79

Sell, 120, 131
Sellers, 111-112, 175
Selley, 9
Sellman, 23-25, 28, 39
Selman, 104-105, 107
Sendall, 67
Senicker, 133
Sentman, 146
Sepad, 121
Seppe, 124
Serfield, 54
Serfs, 199
Sersh, 51
Sevier, 27
Sevirson, 49
Sewarageri, 10
Sewell, 26, 42
Shade, 22, 124
Shaeffer, 124, 191, 194, 198, 201
Shafer, 16, 56, 123, 126, 205
Shaffer, 2, 18, 78, 110-113, 196
Shaffers, 4
Shaler, 34
Shales, 33
Shamburg, 8, 128
Shamlager, 57
Shamloffie, 51
Shanafeltz, 194
Shane, 60
Shaner, 111
Shanett, 132
Shankle, 204
Shanklin, 66
Shanks, 19
Shannon, 114
Shanon, 112
Shanybrook, 79
Share, 117
Sharer, 58
Sharp, 194
Sharpless, 19, 144
Shaster, 11
Shaw, 11-12, 14, 16, 34, 76, 116, 145, 160, 165, 196
Shawer, 111
Shawl, 55

Shealey, 69
Shearer, 155-156
Sheckell, 30
Sheckels, 33
Sheckles, 28
Sheets, 116, 118
Shek, 2
Sheke, 184
Shelby, 148
Shellcross, 148
Shenton, 179-180
Sheperd, 122
Shephard, 22
Shepherd, 121, 184
Shepley, 32, 123
Sheppard, 24, 26, 38-39, 157
Shepperd, 71-72
Sherbert, 30
Shere, 58
Sheridan, 153
Sherman, 174-175, 177
Shetzer, 1
Sheur, 155
Shewery, 115
Shield, 196
Shields, 33
Shilling, 6, 127
Shipgager, 109
Shipley, 36-37, 44, 46, 60, 63, 106-109, 115, 122, 126, 128
Shiply, 43
Shire, 58
Shireliff, 14
Shirly, 48
Shoals, 34
Shock, 56, 71
Shoeburn, 159
Shoemaker, 35, 116-118, 198
Shook, 204
Shorborough, 70
Short, 116, 174, 177, 199
Shorter, 182-183
Shote, 146
Shott, 130
Showers, 115
Showk, 117

248

Strong, 78
Stroud, 155
Stuart, 24
Stubbs, 148
Stubl, 202
Stuck, 16
Stucker, 109
Study, 130
Stull, 128, 203-205
Stultz, 119
Stuly, 121
Stulz, 121
Stump, 12, 47, 55, 154
Stup, 201-202, 205
Styer, 17-18
Suatbury, 113
Suit, 40
Sulivan, 114
Sullivan, 60, 189
Summers, 195
Sumwalt, 41, 44, 49
Sunday, 205
Sundergill, 108
Suter, 3
Sutton, 80, 140, 145
Swalzbaugh, 125
Swan, 41, 160-162, 164-165
Swane, 161
Swann, 161
Swartz, 8
Swatts, 52
Swearry, 198
Sweatman, 133
Swerden, 128
Swift, 181
Swisher, 150
Switzer, 1, 12, 105, 121
Swoke, 43
Syes, 200
Syttle, 70
Taft, 59
Tagg, 69
Taggert, 47, 62
Talbert, 74
Talbott, 23, 33, 62-63, 68, 192
Tall, 179-180

Tally, 178
Tamerry, 152
Tancry, 5
Taney, 199
Tanney, 127
Tarmor, 56
Tarner, 161
Tarno, 50
Tasker, 17-18
Tasto, 114
Tate, 112
Tawley, 192
Taylor, 8, 10, 28, 36, 41-43, 46, 56, 67-68, 128, 137, 143, 149, 150-151, 157, 168, 170, 173, 188
Teackle, 68
Teaner, 106
Tee, 13
Tegley, 5
Tell, 198
Telyard, 39
Templeton, 136
Tenison, 161
Ter___, 140
Terme, 107
Terrell, 70, 199
Terry, 146-147
Thachery, 139, 147
Thayer, 3, 19
The American Co., 11
The Lenaway Co., 11
Thelman, 193
Thielman, 35
Thiertson, 75
Thomas, 1, 3-4, 16, 19, 26, 35-37, 42, 47-48, 63, 66, 145, 147, 149, 152, 159, 162, 164, 170, 179, 185, 189-194, 200, 203, 206
Thompkins, 159
Thompson, 17-18, 43, 55, 57, 57, 74, 108, 117-118, 135, 137, 143, 145-146, 151-154, 161-162, 167, 174, 177, 181
Thoms, 174
Thomson, 77
Thornton, 35

Ward, 4, 39, 44, 60, 62, 65, 69, 127, 135, 139-141, 152-153, 161, 174
Warden, 66
Ware, 44, 160
Wareham, 111
Warehan, 116
Warenfeltz, 195
Warenfety, 195
Warenfetz, 195
Warfield, 35, 37, 42, 75, 104, 107-108, 187
Warick, 198
Waring, 43, 151
Warner, 2, 111, 113-114, 120, 122, 129, 131
Warnick, 8
Warren, 50, 137, 142, 144, 183
Warrenfeltz, 203
Warsche, 131
Warson, 3
Warters, 104-105
Wartes, 104
Warts, 78
Wartsman, 80
Waskey, 193
Wason, 150
Water, 107, 174
Waters, 25, 27-28, 33, 36, 45, 51, 69, 107, 162
Wathen, 160
Watinfelt, 125
Watkins, 23, 25, 27, 66, 190
Watson, 15, 26, 28, 32, 37, 42, 64, 138-139, 145, 154, 161-162
Watt, 135
Watts, 32-33, 35, 49, 51-52, 113, 137-138
Watus, 174
Waugh, 34
Wavage, 3
Waver, 2
Waxman, 197
Way, 152, 155
Wayde, 144
Ways, 108
Wayson, 39

We__am, 10
Weaber, 122, 125
Weagley, 46
Weakson, 137
Wealty, 199
Weatherly, 171
Weaver, 9, 13, 114, 130, 141, 191
Webb, 32, 34, 39, 107, 120, 173, 175, 177
Webster, 31-32, 62, 64, 164, 170, 175, 191, 202
Weddle, 202
Weedon, 29
Weeks, 124
Weems, 37-38, 161
Wehan, 124
Weib, 135
Weighright, 131
Weims, 25
Weir, 153
Weiskettle, 22
Weitz, 3
Welbey, 171
Welch, 10, 16, 26-27, 31, 38-39, 78, 168-169
Welcher, 73
Weld, 6
Weldman, 144
Welds, 135
Weller, 3
Wellias 206
Wells, 28, 33, 37-38, 40
Welsh, 27, 33, 146, 168
Weltz, 117
Wemer, 46
Wemp, 74
Wentler, 9
Wents, 114
Wentz, 24, 56, 112, 124
Wert, 134
Wertenbaker, 206
Wertleman, 9
Wesley, 35-36
Wesly, 78
West, 2, 145, 155
Wethered, 107

www.ingramcontent.com/pod-product-compliance
Lightning Source LLC
Chambersburg PA
CBHW080416270326
41929CB00018B/3045